560 Ig lc

Lecture Notes in Earth Sciences

Edited by Somdev Bhattacharji, Gerald M. Friedman,
Horst J. Neugebauer and Adolf Seilacher

8

Global Bio-Events

A Critical Approach

Proceedings of the
First International Meeting of the IGCP Project 216:
"Global Biological Events in Earth History"

Edited by Otto H. Walliser

Springer-Verlag
Berlin Heidelberg New York London Paris Tokyo

Editor

Prof. Dr. Otto H. Walliser
University of Göttingen
Institute and Museum for Geology and Palaeontology
Goldschmidt-Str. 3, D-3400 Göttingen, FRG

A000004778936

ISBN 3-540-17180-0 Springer-Verlag Berlin Heidelberg New York
ISBN 0-387-17180-0 Springer-Verlag New York Berlin Heidelberg

© Springer-Verlag Berlin Heidelberg 1986
Printed in Germany

Printing and binding: Druckhaus Beltz, Hemsbach/Bergstr.
2132/3140-543210

PREFACE

This volume contains the contributions which have been presented at the
5 . ALFRED WEGENER-Conference , held in Göttingen, Federal
Republic of Germany, 21 - 24 May 1986. This conference was the first
international meeting of the IGCP Project 216 : "global
biological events in earth history".

The aim of the conference was, to discuss
(a) the state-of-the-art in respect to the recognition of
    bio-events and to the analysis of their causes
(b) the presentation of new data
(c) the strategies which are needed for further research, carried out in
    the international cooperation programme of Project 216.
It was intended to achieve with these discussions a more critical
approach to the problems of global bio-events.

In addition to the members from Göttingen University, about 100
participants from 24 countries have been registered.

During 3 days , 7 key-notes and 34 further contributions have been
discussed, according to the intention that the discussion should prevail.
At the fourth day, a few Devonian events have been demonstrated at outcrops in the Rheinische Schiefergebirge.

The conference was sponsored by the following institutions:
UNESCO/IUGS through IGCP
IPA (International Palaeontological Association)
Paläontologische Gesellschaft
Deutsche Forschungsgemeinschaft
IGCP National Committee of the Federal Republic of Germany
State of Niedersachsen
University of Göttingen

The conference would not have been possible without the support of
these institutions. But great thanks is also due to all those staff
members and students of my institute, who supported me before and
during the conference. This volume would not have been prepared in such
a short time without the extraordinary effort of Mrs. Martina Noltkämper.

Otto H. Walliser

CONTENTS

---

* Contribution which has been presented as key-note during the Bio-
Event Conference

TERTIARY

INTRODUCTION

THE IGCP PROJECT 216 "GLOBAL BIOLOGICAL EVENTS IN EARTH HISTORY"

WALLISER, Otto H. *)

A contribution to Project
GLOBAL BIO-EVENTS

Abstract: The IGCP Project 216 is concerned with worldwide traceable exceptional changes ("events") within the biosphere. The principle objectives of this project, which needs interdisciplinary cooperation, are the following:
(1) Study of those abiotic (geologic) processes and events which cause global biological events (geological level);
(2) Reconstruction of the overall effect of global geologic events on the biosphere or parts of it (ecological level);
(3) Evaluation of the influence of global events on evolution and evolutionary mechanisms (evolutionary level);
(4) Refining of stratigraphical scales and of correlation methods by combination of biostratigraphy and event-stratigraphy (chronological level).

The establishment of two IGCP projects which both are concerned with global events -- Project 199 "Rare events" and 216 "Global biological events" -- reflects the broad interest which grew within the geoscience community during the last few years. In that time, the relevant publications stressed on spectacular events, such as impacts of extra-terrestric bodies. These have been assumed on account of new geochemical data. But in respect to the biosphere only a few new data -- such as about planctic microfossils at the Cretaceous/Tertiary boundary -- has been available. Therefore, a main goal of the IGCP Project 216 is the elaboration of new data which should be as precise as detailed as ever possible by using all available modern methods. In so far, this project is mainly based on field- and lab-work and less on hypothetical assumptions. The new data then should be used for a renewed discussion about causes, processes and effects. In the meantime the members of the Project 216 should be open for all theoretical possibilities. They should not hesitate to prove or to disprove their own working hypothesis. A few of the latter are discussed in the following paragraphs.

Geological level

Different global abiotic events may have different causes. These causes may have a direct effect on the biosphere or they may trigger processes, which finally then lead to an event. We also should consider that two or even more events and/or processes may overlap or amplify each other.

*) Institut und Museum für Geologie und Paläontologie der Universität, D-3400 Göttingen, F.R.G.

Lecture Notes in Earth Sciences, Vol. 8
Global Bio-Events. Edited by O. Walliser
© Springer-Verlag Berlin Heidelberg 1986

The processes may be very complex with manifold actions, reactions and interactions. Furtheron, the processes and/or events may be episodic or periodic or both kinds may overlap; they may be very short or even geologically long lasting. Last but not least we should be aware that there exist events of extremely different effectiveness.

In spite of this possible multicausality, we observe that in most cases of events sea-level changes or/and black shales play an important role.

In order to evaluate the whole complex chain or network of causes and processes, a close cooperation with other geosciences, such as geochemistry, oceanography, sedimentology, tectonics, geophysics, and even with astrophysics, is necessary.

Ecological level

We have to consider that a global event affects only certain parts of the biosphere, i.e. only one or a few special biotopes or facies within the whole range of biotopes or facies. Those biotopes in which a biological event happened, may then influence neighbouring, primarily unaffected biotopes.

Certainly, an important and often deciding role in respect to major changes within ecosystems, i.e. to bio-events, play the relations within the food-chain. Therefore, in the marine realm, we shall pay special attention to the investigation of phytoplancton.

Evolutionary level

With all hitherto known events we observe a comparable biological sequence: in the time-span of an event there occur extinctions or even mass extinctions, followed -- after a certain interval -- by radiations. This clearly indicates the strong influence which global events have on the evolution. It will be a main task of the project to elaborate the differences between "normal" evolutionary steps and mechanisms on the one hand and those additional ones which have been caused by global events, on the other hand. Already now it is quite obvious that global bio-events, especially extinctions, accelerate the evolution.

Chronological level

Many of the hitherto discovered biological events are connected with changes in lithology, sedimentology and facies, and for example with a sudden occurrence of black shales. Often, these litho-events occur only

in a certain part of a biozone, i.e. this event is shorter as a biozone and can be used for a refinement of the chronological scale. Furthermore, these events can often be traced globally within the relevant
facies.

Another aspect is the fact that in certain times there are worldwide developed the same types of facies, which on the other side differ
from those in other times. These **time-specific facies** can also
be used for correlation and for chronological refinement.

A further possibility for the refinement of the chronological scale
lays in the decipherment of periodic features, such as climatic cycles.
If the latter are short-termed, they may be used for the subdivision of
smallest biostratigraphical units. Often these microcycles are documented
in the lithology as well as in the ecological pattern. In contrast to
that, long-term climatic cycles may well contribute to the occurrence
of bio-events.

If we combine this **event-stratigraphy** with other stratigraphical scales, such as biostratigraphy, chemostratigraphy, lithostratigraphy, volcanostratigraphy, magnetostratigraphy, etc., we receive at
least a multistratigraphical, but finally a **holostratigraphical
scale** . Such a combined scale will provide us with an hitherto un-
reached accuracy in respect to time resolution and to the worldwide
correlation of smallest time units. In addition, the holostratigraphical
scale contains so much and manifold data, that they will play an extraordinary role in recognizing and deciffering global events as well as
the causes and processes which led to these events.

Subprojects

All the above mentioned questions and problems will be investigated by
concentrating to certain time levels, to certain groups of fossils and
to certain problems. From the events those have been chosen, which
assumedly are different of each other in respect to causes as well as
to effets on the biosphere. Some of the key-studies are, for example:
-- Late Precambrian events (as cases of biological events, in which biological innovations play an important role);
-- Ordovician/Silurian boundary event (as a case of biological events
which are triggered by a glaziation);
-- The events within the Devonian and its boundaries. The Devonian has
been chosen for an overall analysis of all events within a Palaeozoic
System, which then can be compared with corresponding investigations in
the much younger Cretaceous System. Of special interest are: the otomari
Event within the Middle Devonian as a case of a black shale event which

caused only minor biotic changes; the Kellwasser Event at the Frasnian/Famennian boundary as a black shale event with enormous consequences to the biosphere, such as the worldwide extinction of biohermal reefs; the Devonian/Carboniferous boundary event as a case of short-time change in sea level;
-- Climatic changes and the appearance, acme and extinction of plant taxa in the late Palaeozoic;
-- The Palaeozoic/Mesozoic (Permian/Triassic) boundary event (as an example for an event which has eventually been caused by long-term processes and changes in climate and sea level);
-- Relation between abiotic events and renewal and evolution of Jurassi ammonoids;
-- Establishing, refining and worldwide correlation of an event-stratigraphy in the Cretaceous (as an example within the Mesozoic, and for comparison with equivalent investigations within the Devonian);
-- The Cretaceous/Tertiary boundary event (as a case in which probably telluric and extratelluric events met together);
-- Relation between quaternary glaziation and the speciation and extinction of mammals.

All these case studies have to be carried out with comparable metho in as many as possible regions of the world. Only with this internation cooperation we shall receive clear indication about the global synchronism of the events and about as many details of the reaction of the biosphere to these events as possible.

Even if the project is mainly based on palaeontological methods, the expected results are also of great importance for the applied geosciences. The refining of stratigraphical scales and methods will facil itate world-wide correlation, partly with time-intervals much smaller than until now possible. Furthermore the project will provide us with facts about long-term processes in the biosphere, which are triggered by a grave disturbance of the ecosystems. These aspects might be also of great value to consider the long-term effects of recent interference into the given ecosystems by existing human population. In so far this investigation of the past is also a contribution to our own future.

TOWARDS A MORE CRITICAL APPROACH TO BIO-EVENTS

WALLISER, Otto H. *)

A contribution
to Project
GLOBAL
BIO-
EVENTS

IUGS
UNESCO

Abstract: Global bio-events are manifold in respect to causes and extent. There can be recognized the following patterns of global bio-events: (1) innovation-events; (2) radiation-events; (3) spreading-events; (4) extinction-events with stepwise or/and contemporaneous, rapid extinctions.

Probable causes for global bio-events are (1) cosmic causes, such as (1a) changes caused by the revolution of the solar's system within the Galaxy and (1b) impact of cosmic bodies; (2) earth-born causes, namely (2a) biological causes (mostly biological innovations) and (2b) abiotic = geological causes. The latter imply, among others, sea-level changes, changes of the physical and chemical composition of the ocean and the atmosphere, changes of climate, changes of oceanographic parameters. Impacts of cosmic bodies may then have a catastrophic effect, if one or several of the affected systems are already near to instable conditions, the latter caused independently by other geologic processes.

Holostratigraphy, that is the combination of all available stratigraphies, such as bio-stratigraphy, event-stratigraphy, chemo-stratigraphy and others, provide us with an extreme high time-resolution.

In next future, the main task in order to answer open questions in respect to causes and processes in connection with global bio-events, will be to elaborate more and more detailed and precise data. Therefore we need interdisciplinary cooperation.

Introduction

Why are global events such fascinating happenings; why, since several years, do they have permanent space in journals such as Nature or Science; why are global events enormously important in several respect, as for example for a better understanding of evolutionary and of geological processes? Parts of these questions might find an answer by a short historical reminiscence.

Events, that means drastic or at least important changes in the biotic or abiotic documentation of earth history, are well known since the very beginning of our science, i.e. since about the late 18th century. There, in the pre-Darwin time, the events have been interpreted as catastrophes, according to the scriptural Deluge (e.g. G. de Cuvier). Of course, it was believed that the catastrophes were followed by new creations of a "higher", more "progressive" and more "perfect" standard. In modern terms: extinctions, especially mass extinctions, are connected with a macro-evolutionary pattern.

As a reaction to Darwin's theory, gradualism (or micro-evolution) dominated all considerations, thus denying sudden changes. The latter

*) Institut und Museum für Geologie und Paläontologie der Universität, D-3400 Göttingen, F.R.G.

Lecture Notes in Earth Sciences, Vol. 8
Global Bio-Events. Edited by O. Walliser
© Springer-Verlag Berlin Heidelberg 1986

have been explained by a lack of information and/or of geological record (e.g. Ch. Lyell).

In the early decades of this century, and into the fourties, fundamental biological changes have been discussed again. O.H. Schindewolf, e.g., emphasized major genetic transformations in an early ontogenetic phase. Orthogenetic processes, besides of geological events, have been used as causal explanations for rapid biological changes. But at the same time, also more gradualistic views (e.g. G.G. Simpson) have been discussed and subsequently preferred.

Since a few years, the pendulum swung back again to some kind of catastrophism. The old discussion was revived, but at the beginning only under the labels of new terms or/and definitions, such as punctuated equilibrium versus gradualism, phyletic versus phylogenetic, macro-evolution versus micro-evolution. Since a few years this discussion became intensified on account of the impact hypothesis and the assumption of a certain periodicity in respect to the occurrence of major bio-events. The combination of these two latter ideas lead to several hypotheses, such as that about the dark dwarf Nemesis. The hypotheses have been enormous by stimulating as well as the modelling of such spectacular happenings as also new basic investigations in order to receive new and more precise basic data.

At the 5th ALFRED WEGENER-Conference at Göttingen 1986, which has also been the first international meeting of the IGCP Project 216, the state-of-the-art in respect to hypotheses about the causes of bio-events have been discussed. But then, these hypotheses became tested by the available data, especially new ones, presented at the conference. The present paper attempts to combine the authors standpoint with a short review of the conference.

How to handle the terms event, global event, bio-event

Each happening is an event. Therefore, in order not to wishy-washy this term, we should restrict it to extraordinary happenings. Then, in most cases an event has a time-span which is remarkably shorter than the intermediate times of relatively stable conditions or slow changes and developments, respectively. This necessary but intended weak definition is independent of the question whether this final event is caused by another, initial or ultimate event or whether it is the result of one or several processes.

Global event means, that it is detectable worldwide. Thereby it is not important whether the object under consideration (such as lithology, ocean chemistry, biosphere, etc.) has been affected in a great extent or only in a small sector.

The term biological event, shortened: bio-event, stands for an extraordinary change within the world of organisms (faunal and/or floral
change, Faunen- and/or Floren-Schnitt).

Pattern of global bio-events

It can be taken as a rule that a change of abiotic parameters lead to
a change of the ecosystem. This means that with such a change the disappearance or appearance of taxa is always connected. In so far, bio-
events can be recognized indirectly by a change of the abiotic parameters, such as lithology (including petrography as well as sedimentology) or chemical composition.

After having recognized an event in a section or a region, it needs
to examine its dimension: is it only local or basin-wide or global;
is it synchronous in all regions; did it affect only one or a few taxa
or did it cause mass extinction? These questions imply already the
different pattern of bio-events.

In the following, some categories of global bio-events are listed.
Thereby, bio-event categories of local or restricted regional extent are
not discussed, even if they are very important for a regional event-
stratigraphy (e.g. mass-mortality or regional population burst, regional
emigration and immigration of taxa, replacement of ecosystems by one
another, which existed already contemporaneously).

In our definition, a global bio-event always affects the total
population of one or more species. Thereby it is well possible that the
extinction of only one species, which is an important member of the
general food-chain, triggers the subsequent extinction of other species.
If I mention here the species level, we should be aware that taxa of
a higher rank, such as genera, families, etc., consist of species.

Another aspect, which is included in the following considerations,
is that of background-extinction and background-creation of species.
This aspect is especially important in those cases of global bio-events,
where only one taxon or a few species have been eliminated. In order to
distinguish background- from event-extinctions we need (a) a careful
cause-effect analysis and (b) an entire analysis of the relevant ecosystem, as pointed out in the key-note of A.J. Boucot (this volume).
With other words: it has to be proven that the creation or extinction
of a species is really unusual and extraordinary, caused by the event.

Considering all these mentioned aspects, we come to the following
categories of global bio-events

(1) **Innovation-event.** Biological innovations may lead to major changes in the world of biota. As an example may serve the coiling of certain cephalopodes, leading to the long-termed domination of the ammonoids. It has to be suggested that biological innovations played an important role especially in Precambrian and early Phanerozoic times (compare contributions of H.D. Pflug & E. Reitz, B.S. Sokolov & M.A. Fedonkin and M. Brasier; this volume). The first foundation of protein, of DNA, the creation of metazoan structures, the innovation of skeletal mineralization, etc., opened totally new worlds of ecosystems. This surely influenced and affected also the pre-existing ecosystems in a dynamic feedback system.

(2) **Radiation-event.** As pointed out below and in Fig. 1, radiations of certain taxa appear subsequently to extinction-events, mostly after short interval. A radiation-event implies an unusual high percentage of creations of taxa.

   Radiation-events also occur subsequently to a biological innovation-event. Nevertheless, in some cases such an innovation-caused radiation occurs only after an extinction-event. The most evident example for this kind of dependencies exists in the late Triassic innovation of mammals and their main radiation after the K/T mass extinction-event.

(3) **Spreading-event.** Under this term several kinds of sudden spreading out of species are united. But it has to be mentioned, that it is very difficult to clearly distinguish the normal, but also sudden spreading out of new species in times between bio-events on the one hand, from unusual spreading-events on the other hand. As an example may serve "Dictyonema" flabelliforme, which occurs worldwide at and together with the global transgression at the beginning of the Ordovician. Its unusual spreading out is obviously due to the innovation of a floating life habit. Another example among graptolites (compare also the contribution of B.-D. Erdtmann; this volume) is Monograptus uniformis. After the transgrediens Event in the latest Silurian, this species shows a sudden burst at the beginning of the Devonian System, but well after its first occurrence.

(4) **Extinction-events.** The most evident bio-events are the extinction-events, especially when they increase to mass extinctions, as for example at the O/S Event, the Kellwasser Event or the K/T Event. During a certain event, extinctions may occur in a very short time, thus producing a real catastrophe within the affected ecosystems or groups of organisms. In other cases, the extinctions may occur in several steps, even if these happen in a relatively short time

(compare the contributions of R. Th. Becker and of P. Schäfer & E. Fois-Erickson; this volume). Both types, catastrophic ones and stepwise ones may occur in different groups of organisms during one and the same event. As examples may serve the end-Palaeozoic and end-Cretaceous regressions, during which stepwise extinctions have been caused by several effects of the regressions (e.g. loss of biotopes, change of albedo, change of climate, and of oceanic parameters,etc.). Examples for the stepwise extinctions at the K/T Event are given in the contributions of J. Wiedmann and of H.J. Hansen et al., in this volume). At the K/T boundary, this stepwise mass extinction then has been superimposed by an additional event, i.e. the iridium-event. Another example is the Kellwasser Event (see below).

The above mentioned patterns of bio-events reflect the influence of global events to the evolution. Thus, for example, all known major bio-events show the following sequence: extinction -- short interval -- radiation. The latter might be composed of several radiations in different groups, which may happen within a certain time-span. This regularity may be explained by the following hypothesis (compare Fig. 1):

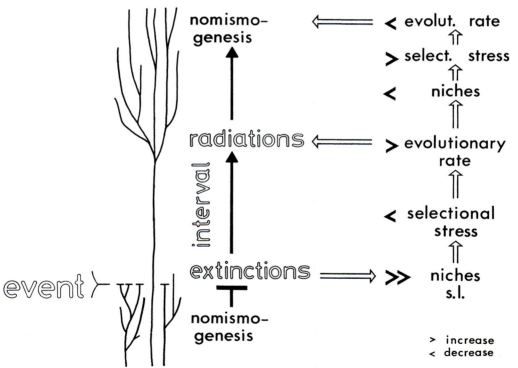

Figure 1.   General pattern of the extinction -- interval -- radiation sequence (for explanation see text).

In an undisturbed ecosystem, the selectional stress is relatively high. This means that only a few of the newly occurring mutations and recombinations successfully contribute to the evolution. Therefore the evolutionary rate is low. However, as soon as there occur niches, caused by preceding extinctions or by the creation of new biotopes, the selectional stress is reduced -- at least with respect to those forms which possess the potential ability of evolving into the direction of the niche. The fewer potential competitors for conquering the niche are present, the less will be the selectional stress. A lowering of the selectional stress means that relatively more newly occurring mutations and recombinations may be used for further evolution. Thus, the evolutionary rate becomes relatively high, but without any necessity for a change of the quantity or quality of mutations and recombinations, or, with other words, without any change of the genetic mechanisms. The high evolutionary rate is expressed by radiation. The latter leads to an occupation of the niches and through that to a normalization in respect to selectional stress and evolutionary rate.

This regular sequence also may bear the solution of the old problems micro-evolution versus macro-evolution, gradualism versus punctuated equilibrium. With other words: although global events may lead to a perturbations of ecosystems the evolution continues in a gradualistic manner to catastrophes in the world of organisms or to interruptions of certain evolutionary lines. Thereby the main parameters, these are mutation rate, selectional stress and evolutionary rate, are always in balance to each other. Only the ratio has changed. Thus, the punctuated equilibrium does not concern the evolutionary mechanisms, but the affected ecological systems. This sequence also demonstrates quite clearly a certain evolutionary pattern which we could call the **parado of evolution: extinctions accelerate evolution** . In so far, global biological events represent important and very interesting aspects for the understanding of evolutionary processes.

Causes

During the last few years, the impact of extra-telluric bodies has mainly been taken as an explanation for the initial cause of the main events. During the 5th ALFRED WEGENER-Conference, a more critical approach has been reached. Thereby the importance of the impact hypothesis has well been appreciated, but it became supplemented by the assumption, that als other causes, independently of impacts or sometimes enhanced by impacts lead to important bio-events.

As the author already pointed out formerly (Walliser 1984a, 1984b) different global bio-events may have different causes, and different causes may lead to similar events. An ultimate or initial cause may directly influence the biosphere, thus being also the final cause for a bio-event. On the other hand, an ultimate cause may only trigger processes, which then produce the final cause.

These processes may be very complex, with a network of actions, reactions, interactions, feed back systems etc. A further possibility is that such a process leads then to an event, when it overlaps with another process or several other processes, thus amplifying each other. Furtheron we should be aware that comparable final causes may produce events of quite different extent, because other parameters have changed in the course of earth history.

Figure 2 is a simplified flow-chart, showing that in spite of the mentioned possible multicausality, in most cases of global events sea-level changes, changes in sedimentology (often represented by black shales), and changes of the physical and chemical conditions (inclusively climatic ones), play an important role. This is also documented in the key-notes of W.T. Holser et al. and of P. Wilde & W.B.N. Berry, in this volume. But also most other contributions show that the mentioned parameters are of decisive importance. In addition, we should be aware that the impact of extra-telluric bodies trigger geologic processes which then lead to the final cause of a global event.

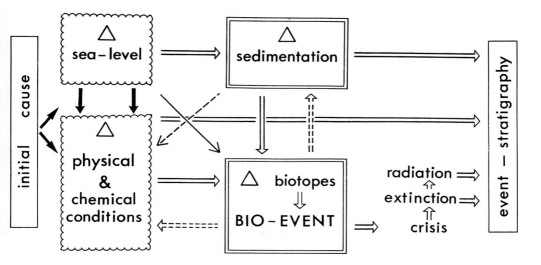

Figure 2.   Simplified flow-chart, illustrating the complexity of a selected example of geological processes (Δ = change).

In order to obtain a general view of the mentioned aspects, we may categorize the causes in the following way:

(1) **Extra-telluric = cosmic causes**

   (1a) **Changes** caused by the revolution of the solar's system within the Galaxy (see the following chapter).

   (1b) **Impact** of cosmic bodies, such as comets or asteroids. In this category we should try to distinguish two cases. In case one the impact is as thorough as to cause an unusual event in any case. In the other case the impact may lead only then to an event, if certain systems have already reached critical, weak conditions by other processes.

(2) **Earth-born causes**

   (2a) **Biological causes.** In most cases these are biological inno-vations, as already mentioned above.

   (2b) **Abiotic = geological causes.** These might be manifold, and only some important ones are mentioned in the following.

   S e a - l e v e l   c h a n g e s . These may be caused by climatic fluc-tuations or by tectonic processes. An already classical example is the glaciation at the outgoing Ordovician (see key-note of C.R. Barnes and further contributions of Rong & Chen and of P. Štorch in this volume). The relation between orogenesis and regression is shown in the contribution of R. Ingavat-Helmcke & D. Helmcke. But also other processes in plate tectonics, such as the formation of mid-ocean ridges, are responsible for sea-level changes. But even the hypothesis of an undulating expan-sion of the earth can not yet be excluded as one of several causes for sea-level changes. With other words: also sea-level changes can have different or multiple causes.

   C h a n g e s   o f   p h y s i c a l   a n d / o r   c h e m i c a l   c o m p o s i t i o n of sea-water or atmosphere. Thereby, the long-termed developmen of the chemical composition of both, ocean and atmosphere, played an important role in the development of the biosphere, at least in pre-Phanerozoic times. But also shorter fluctuation of the ocean chemistry are well correlated with major Phane-rozoic events.

   C h a n g e s   o f   c l i m a t e . Here we surely have also to distin-guish between long-term processes and short-term fluctuations. As an example for the latter, the Milankovitch-curve may be mentioned. In this connection, also the influence of volcanic activity to short-termed climatic perturbations has to be con-sidered. Of great interest in this respect is the contribution

of J. Besse et al. (this volume), in which the formation of the Indian Deccan Trapps is considered as a possible cause for the K/T Iridium Event. About long-term processes only few facts are already known. Here a self-perpetuating system has to be considered, in which the interdependency of biological productivity on the one hand and of O_2 and CO_2 content in the atmosphere on the other hand leads to a cyclic sequence of greenhouse effect and general cooling.

Changes of oceanographic parameters. These are obviously decisive for many of the global bio-events (see the key-note of P. Wilde & W.B.N. Berry; this volume). Oscillations of the anoxic layer, destratification, overturn, changes of the major ocean currents etc., certainly play an important role.

According to the statements given before, all these mentioned causes may itself trigger global events or may lead to those events on account of their combined effects. As an example may serve the Kellwasser Event at the natural boundary (but only close to the defined boundary) of the Upper Devonian stages Frasnian and Famennian. Since long time it has been recognized that this Fr/Fa boundary marks one of the most important bio-events in earth history. But only since a few years it became evident that this event -- with the extinction of biohermal coral reefs -- is connected with the globally traceable Kellwasser Horizon at the top of the Uppermost <u>Palmatolepis</u> <u>gigas</u> Zone ("Upper" KW Horizon). This latter indicates a typical anoxic or black shale event. Black shales occur already below the Upper KW Horizon, of which the "Lower" KW Horizon marks the uppermost part of the Lower <u>gigas</u> Zone. The number of black shales between the two KW Horizons seems to depend on the palaeogeographical position. This supports the author's opinion that each of the mentioned black shale horizons, occurring between pelagic cephalopode limestones, represents a more or less short-termed rise of the upper part of the anoxic layer. Palaeogeographically relatively deep positions have been reached even by a slight upward movement of the anoxic layer, whereas high positions have been reached only by an extremely extended rise. This oscillation of the anoxic layer may have caused one or the other minor bio-event, thus contributing to a stepwise disappearance of certain taxa.

The final, rapid bio-event happens at the very end of the Upper KW Horizon. There, e.g., the manticoceratid ammonites and the homoctenids became extinct. This time most probably coincides with the extinction of the whole biocoenosis of bioerhermal reefs, upon the shelfs as well as in the basins. Nevertheless, this catastrophe was selective. The

best explanation seems to be an ocean overturn, which happened at a
time of a critical destabilization, indicated by the globally occurring
Upper KW black shale. The nature of this additional factor, which
triggered then the final overturn, is still in debate. As seen
in the contributions of J. Lottman et al. and of J. Hladil et al. (this
volume), a regression or sea-level fall seems to be connected with the
top of the Upper KW Horizon. In addition, a climatic change is indicate
(see contributions of J. Kalvoda and of G.R. McGhee et al.; this volume
Probably, each of these factors could have triggered the ocean overturn
Theoretically this also could have happened through the impact of an
cosmic body.

The question of periodicity

Speculations about periodically occurring processes, changes or even
catastrophes can be traced back far into the last century. But only the
investigations of Raup and Sepkoski (see key-note of J.J. Sepkoski;
this volume) provided us with better data. They calculated the periodic
occurrence of extinction events every 26 Myr. They got this cycle on
account of a best-fit curve. With a Fourier analysis (Rampino & Stother
1984) a dominant periodicity of 30 Myr has been calculated. This latter
evidence is near to a palaeontological calculation of Fischer & Arthur
(1977), assuming a 33 Myr periodicity. Own calculations for the Devonian
show an interval of 8 Myr between major events. This could indicate a per
odicity of 24 Myr or -- even better -- of 32 Myr. This coincides approx
mately with the dominant cyclicity of 31 Myr for cratering and also wit
the time needed for the solar system to oscillate about the plane of th
Galaxy (Rampino & Stothers 1984). In addition, 8 times 32 Myr is close
to the solar system's revolution around the centre of the Galaxy, as
calculated by Trumpler & Weaver (1962).

In both cases, the assumption of a companion star with a highly
excentric orbit, as well as counting with an oscillation of the solar
system, provides us with a periodic maximum of impacts of cosmic bodies
But we should be aware, that sometimes, even in times of maxima, cosmic
bodies miss the earth and, on the other hand, that in the intervals
exceptional impacts may happen.

Further investigations and calculations in cooperation with colleagues of the astrophysics, should also consider cosmic influences to
other telluric and geologic phenomena, such as gravity and magnetism.
Perhaps we shall find in the future some kind of a cosmic rule, which
explains many of our biological and geological events.

High-resolution event-stratigraphy

As shown by the key-notes of C.R. Barnes and E.G. Kauffman (this volume), the combination of biostratigraphy and event-stratigraphy provides us with a high time-resolution. If we add all other stratigraphies. such as magneto-stratigraphy, chemo-stratigraphy, tectono-stratigraphy, volcano-stratigraphy (especially tephro-stratigraphy), etc., we may call it holostratigraphy. In addition to the high resolution, the event-stratigraphy has the advantage, that many of the recognized events can serve for global correlation even in the field.

   In spite of the high time-resolution, the event-stratigraphy comprises another important aspect: recognition and subsequent analysis of events leads to a much better understanding of the dependencies and interdependencies of events and processes in both, the biosphere and the geosphere.

Conclusions

Global biological events show a broad variety of causes and effects. Several categories of causes as well as of patterns can be recognized. Some kind of periodicity is most probable.

   Many of the open problems can only be solved, if further precise and detailed data from field and laboratory investigations are elaborated. For this aim multidisciplinary cooperation is necessary.

REFERENCES

BARNES, C.R. (1986): The faunal extinction event near the Ordovician-
    Silurian boundary: a climatically induced crisis.- This volume.
BECKER, R. Th. (1986): Ammonoid evolution before, during and after
    the "Kellwasser-event" - review and preliminary new results.- This
    volume.
BESSE, J.; BUFFETAUT, E.; CAPPETTA, H.; COURTILLOT, V.; JAEGER, J.-J.;
    MONTIGNY, R.; RANA, R.; SAHNI, A.; VANDAMME, D. & VIANEY-LIAUD, M.
    (1986): The Deccan Trapps (India) and Cretaceous-Tertiary boundary
    events.- This volume.
BOUCOT, A.J. (1986): Ecostratigraphic criteria for evaluating the
    magnitude, character and duration of bioevents.- This volume.
BRASIER, M. (1986): Precambrian-Cambrian boundary biotas and events.-
    This volume.
ERDTMANN, B.-D. (1986): Early Ordovician eustatic cycles and their
    bearing on punctuations in early nematophorid (planktic) graptolite
    evolution.- This volume.
FISCHER, A.G. & ARTHUR, M.A. (1977): Secular variations in the pelagic
    realm.- Soc. Econ. Paleont. Mineral., Spec. Publ. 25, 19-50.
HANSEN, H.J.; GWODZ, R.; HANSEN, J.M.; BROMLEY, R.G. & RASMUSSEN, K.L.
    (1986): The diachronous C/T plankton extinction in the Danish Basin.-
    This volume.

HLADIL, J.; KESSLEROVÁ, Z. & FRIÁKOVÁ, O. (1986): The Kellwasser event
    in Moravia.- This volume.
HOLSER, W.T.; MAGARITZ, M. & WRIGHT, J. (1986): Chemical and isotopic
    variations in the world ocean during Phanerozoic time.- This volume.
INGAVAT-HELMCKE, R. & HELMCKE, D. (1986): Permian fusulinacean faunas
    of Thailand - event controlled evolution.- This volume.
KALVODA, J. (1986): Upper Frasnian and Lower Tournaisian events and
    evolution of calcareous foraminifera - close links to climatic
    changes.- This volume.
KAUFFMAN, E.G. (1986): High resolution event stratigraphy: regional
    and global Cretaceous Bio-events.- This volume.
LOTTMANN, J.; SANDBERG, Ch.A.; SCHINDLER, E.; WALLISER, O.H. & ZIEGLER,
    W. (1986): Devonian events at the Ense area (Excursion to the
    Rheinisches Schiefergebirge).- This volume.
McGHEE, G.R., Jr.; ORTH, Ch.J.; QUINTANA, L.R.; GILMORE, J.S. & OLSEN,
    E.J. (1986): Geochemical analyses of the Late Devonian "Kellwasser
    Event" stratigraphic horizon at Steinbruch Schmidt (F.R.G.).- This
    volume.
PFLUG, H.D. & REITZ, E. (1986): Evolutionary changes in the Proterozoic
    - This volume.
RAMPINO, M.R. & STOTHERS, R.B. (1984): Terrestrial mass extinctions,
    cometary impacts and the Sun's motion perpendicular to the galactic
    plane.- Nature 308, 709-712.
RONG Jia-yu & CHEN Xu (1986): A big event of latest Ordovician in China
    - This volume.
SCHÄFER, P. & FOIS-ERICKSON, E. (1986): Triassic Bryozoa and the
    evolutionary crisis of Paleozoic Stenolaemata.- This volume.
SEPKOSKI, J.J., Jr. (1986): Global bioevents and the question of
    periodicity.- This volume.
SOKOLOV, B.S. & FEDONKIN, M.A. (1986): Global biological events in the
    late Precambrian.- This volume.
TRUMPLER, R.J. & WEAVER, H.F. (1962): Statistical Astronomy.- (Dover,
    New York).
WALLISER, O.H. (1984a): Geologic Processes and Global Events.- Terra
    cognita 4, 17-20.
-- (1984b): Global Events and Evolution.- Proc. 27th Internat. Geol.
    Congr. Moscow, Palaeontology 2, 183-192.
WIEDMANN, J. (1986): Macro-invertebrates and the Cretaceous-Tertiary
    boundary.- This volume.
WILDE, P. & BERRY, W.B.N. (1986): The role of oceanographic factors
    in the generation of global bio-events.- This volume.

DEVONIAN EVENTS AT THE ENSE AREA (EXCURSION TO THE RHEINISCHES SCHIEFERGEBIRGE)

A contribution to Project GLOBAL BIO-EVENTS

LOTTMANN, Jan *), SANDBERG, Charles A. **),
SCHINDLER, Eberhard, *), WALLISER, Otto H. *) &
ZIEGLER, Willi ***)

During the 5th ALFRED WEGENER-Conference, held as the first international meeting of the IGCP Project 216, two Devonian outcrops at the Ense near Wildungen (Rheinisches Schiefergebirge) have been visited.

During the Devonian, the Ense area belonged to the pelagic, i.e. basinal realm of the relatively shallow miogeosyncline. The basin has been morphologically modified by stable rises with socalled cephalopod limestones. In Givetian time, basaltic volcanics (spilites) could reach the sea level and gave rise for the settlement of reefs.

Locality 1: Blauer Bruch

800 m southeast of the railway station of Wildungen, at the Wenzigeröder Weg (GK 25 Bad Wildungen, 4820; r 10060, h 64240)

The former quarry "Blauer Bruch" is one of the famous localities of the Ense area. Already Waldschmidt (1885), Holzapfel (1895) and Denckmann (1893, 1901) described ammonoids from here. Since this time, the quarry has repeatedly been mentioned, mainly in connection with stratigraphical investigations. Thus, e.g. Bischoff & Ziegler (1957) and Wittekindt (1965), in connection with the establishment of a conodont chronology for the Middle Devonian. The exposed sequence comprises mainly cephalopod limestones and intercalated shales of Upper Emsian to early Frasnian age. It is complicated by faults and folding.

The field-trip was mainly concerned with the **pumilio** Event , represented in two layers. Both are separated of each other by nearly 3 m of cephalopod limestones. The lower pumilio layer is about 10 cm thick, the upper one about 15 cm. They are dark limestones within the discoides Limestone, situated in the Middle varcus Zone. They are composed of approximately 1-3 mm small, lenticular brachiopods, since long time known as "Terebratula pumilio". Roemer (1855) described them from the Hartz Mountains, where they also occur in 2 layers. Already Beushausen (1900) and Denckmann (1901) correlated these layers with those from the

---
*)   Institut und Museum für Geologie und Paläontologie der Universität, D-3400 Göttingen, F.R.G.

**)   US Geological Survey, Denver, Colorado 80225, U.S.A.

***) Naturmuseum und Forschungsinstitut Senckenberg, D-6000 Frankfurt a.M. 1, F.R.G.

Ense area and emphasized their stratigraphical significance. Later on, the two pumilio layers have also been recognized in other parts of the Rheinisches Schiefergebirge and in Morocco.

The systematic position of pumilio is still uncertain. Denckmann (1901) doubted the assignment to the terebratulids. H. Schmidt (1960) pointed out, that pumilio might be a free-swimming, juvenile stringo-cephalid. Ongoing investigations by J. Lottmann intend to clarify the taxonomic position and to solve the "pumilio problem": if preliminary datings can be verified, that each of the two horizons has always the same stratigraphical position, what then does that mean? Do all aspects (stratigraphical position, sedimentological structure, microfacies, character of the under- and overlying sediments, palaeogeographical position, aspects of functional morphology etc.) allow the assumption that the pumilio layers represent seasonal events? In this case, two time-horizons, representing an extremely short event, could be traced at least from Mid-Europe to North Africa.

Besides of the pumilio Event, some features of time-specific facies (TSF) have been presented for discussion. A characteristic example is the sequence near and through the Middle/Upper Devonian boundary: the lower part consists of a very distinct kind of nodular limestone, charac-teristic for the uppermost Maenioceras Stage. The overlying solid cephalopod limestone represents the Pharciceras Stage. Then follow dark and black shales with intercalated thin layers of limestone, which either yield or nearly consist of styliolinids and/or homoctenids. In one of these layers, the styliolinids are surrounded by a seam of ra-diate calcite.

The same sequence with the same very peculiar characteristics of the sediments is also known from Morocco. In addition, the "styliolinit with radiate calcite is identically represented in certain localities in South China. In so far, the TSF can be used as an additional para-meter for event-stratigraphy.

Locality 2: Steinbruch Schmidt

NNE of Braunau near Wildungen (GK 25 Armsfeld 4920; r 3509275, h 5661275

This quarry is also known as "Braunauer Kalkofen". Formerly, Frasni and Lower Famennian cephalopod limestones have been quarried and burned The two Upper Frasnian black Kellwasser Horizons (Kellwasserkalke), re-presenting the **Kellwasser Event** (Walliser 1980, 1984), are well recognized since long time. In the eighties of the last century, the quarry already has been known to v. Koenen and Waldschmidt. Especially the descriptions of the arthrodire fauna made the quarry well known all over the world. This fauna occurs in limestone nodules in the basal part of the Upper KW Horizon. To mention are the publications of v. Koenen, Jaekel (in the beginning of this century until 1928), Gross (in the early thirties) and Stensiö (1922-1963).

First mapping of the quarry and its surrounding area was done by Denckmann already in 1894. Denckmann and Wedekind worked on the goniatite fauna. The locality was also known to Schindewolf (1921), H. Schmidt (1928) and Pusch (1935).

1968 Lange descsribed about 70 conodont assemblages. They have been found in both KW Horizons. However, the bulk occurrence is in the "fish layer" of the Upper KW Horizon.

Buggisch (1972) in his geochemical work about the "Kellwasserkalke" as well as Tucker in his micro-facial investigations on pelagic lime- stones used samples from the Schmidt quarry. New geochemical data are now presented by McGhee et al. (see this volume).

The conodont chronology in the sequence of this quarry is mainly based on Ziegler (1958, 1962, 1971). Most recent data are from Sandberg & Ziegler (see below). Those colleagues cooperate with Schindler, who is studying the faunal distribution and the development of the micro- facies.

The Lower KW Horizon is about 45 cm thick and ends ca. 260 cm below the Upper KW Horizon. In addition to the KW Horizons, there still occur several black shales above the Upper KW Horizon. However, the main KW event is connected with the Upper KW Horizon. This horizon is about 40 cm thick and consists of black shales and intercalated black limestones, both yielding abundant homoctenids and entomozoan ostracods, besides of other fossils, such as goniatites, orthocone cephalopods and Buchiola. The uppermost intercalated limestone can be subdivided into two parts, layer 64 (5 cm) and layer 65 (7 cm). Then follow 7 cm black shales (layer 66) which finish the KW Horizon. The overlying layer 100 con- sists of a partly nodular grey limestone.

Remarkable is the fact, that the last observed occurrence of homoc- enids and manticoceratid goniatites is already in layer 64. Then, in the following layer 65 occurs a large pelecypode, thus indicating a major change between these two layers.

These observations coincide with the results of Sandberg and Ziegler: they have been studying the changes in conodont faunas across the Fras- ian-Famennian boundary in platform, reefal, slope, and basinal facies in Western Europe and Western North America. Their interim findings show a marked eustatic fall in sea level in all regions beginning near the top of the Uppermost gigas Zone and extending through the Lower and

Middle _triangularis_ Zones. This eustatic fall is marked by a marked increase in abundance of _Icriodus_ coupled with a decrease of _Palmato-lepis_. Within the Uppermost _gigas_ Zone at Steinbruch Schmidt the abundance of _Icriodus_ first drops from about 8 percent down to 0 percent at the base of the Upper Kellwasser Limestone, reflecting an eustatic rise, and then increases to a few percent within the lower limestone beds of this interval. However, in the upper part of layer 65 (= 16 of Ziegler), the abundance of _Icriodus_ increases markedly and proportionately to similar increases observed in other palaeoenvironmental settir in Belgium, Utah and Nevada.

REFERENCES

BEUSHAUSEN, L. (1900): Das Devon des nördlichen Oberharzes.- Abh. Preu Geol. L.-A., N.F. 30, 383 p.

BISCHOFF, G. & ZIEGLER, W. (1957): Die Conodontenchronologie des Mitte devons und des tiefsten Oberdevons.- Abh. hess. L.-Amt Bodenforsch. 22, 136 p.

BUGGISCH, W. (1972): Zur Geologie und Geochemie der Kellwasserkalke un ihrer begleitenden Sedimente (unteres Oberdevon).- Abh. hess. L.-Am Bodenforsch. 62, 68 p.

DENCKMANN, A. (1893): Schwarze Goniatiten-Kalke im Mitteldevon des Kel lerwaldgebirges.- Jb. kgl. preuß. geol. L.-A. f. 1892 13, 12-15.

-- (1895): Zur Stratigraphie des Oberdevon im Kellerwalde und in einig benachbarten Devon-Gebieten.- Jb. kgl. preuß. geol. L.-A. f. 1894 1 8-64.

-- (1901): Der geologische Bau des Kellerwaldes.- Abh. preuß. geol. L. A., N.F. 34, 88 p.

-- (1902): Erläuterungen zur geologischen Specialkarte von Preussen un benachbarter Bundesstaaten.- Berlin, 84 p.

GROSS, W. (1932a): Die Arthrodira Wildungens.- Geol. Paläont. Abh., N. F. 19, 61 p.

-- (1932b): Ein Wildunger Arthrodire in Nord Amerika.- Paläont. Z. 14, 46-48.

-- (1933): Die Wirbeltiere des rheinischen Devons.- Abh. preuß. geol. A., N.F. 154, 83 p.

HOLZAPFEL, E. (1895): Das obere Mitteldevon (Schichten mit _Stringoce-phalus_ _burtini_ und _Maeneceras_ _terebratum_) im Rheinischen Gebirge.- Abh. preuß. geol. L.-A., N.F. 16, 459 p.

JAEKEL, O. (1903): Über _Rhamphodus_ nov. gen., ein neuer devonischer Ho cephale aus Wildungen.- Sitz.-Ber. Ges. naturforsch. Freunde 7.

-- (1904): Über neue Wirbeltierfunde im Oberdevon von Wildungen.- Z. d geol. Ges. 56, 159-167.

-- (1906): Neue Wirbeltierfunde aus dem Devon von Wildungen.- Sitz.-Ber. Ges. naturforsch. Freunde Berlin, 73-86.

-- (1928): Untersuchungen über die Fischfauna von Wildungen.- Paläont. Z. 9, 329-339.

KOENEN, A. v. (1880): Vorlage von Fischresten aus dem Oberdevon von Bicken und Wildungen.- Z. dt. geol. Ges. 1880, XXXII, 673-675.

-- (1883): Beitrag zur Kenntnis der Placodermen des norddeutschen Ober devons.- Abh. kgl. Ges. Wiss. Gött. 30.

LANGE, F.G. (1968): Conodonten-Gruppenfunde aus Kalken des tiefen Ober devons.- Geologica et Palaeontologica 2, 37-57.

McGHEE, G.R., Jr.; ORTH, Ch. J.; QUINTANA, L.R.; GILMORE, J.S. & OLSEN E.J. (1986): Geochemical analyses of the Late Devonian "Kellwasser

Event" stratigraphic horizon of Steinbruch Schmidt (F.R.G.).- (this volume).

PUSCH, F. (1932, 1935): Beobachtungen im Devon und Kulm der Wildunger Gegend.- 1. + 2. Ber. an die Geol. L.-A. zu Berlin, 9 + 16 p.

ROEMER, F.A. (1855): Beiträge zur geologischen Kenntnis des nordwest-lichen Harzgebirges. III.- Palaeontographica 5, 1-46.

SCHINDEWOLF, O.H. (1921): Versuch einer Paläogeographie des europäischen Oberdevons.- Z. dt. geol. Ges. 73, 137-223.

SCHMIDT, H. (1928): Exkursion bei Wildungen (Devon, etwas Silur und Kar-bon).- Paläont. Z. 9, 5-8.

-- (1960): Die sogenannte "Terebratula pumilio" als Jugendformen von Stringocephaliden.- Paläont. Z. 34, 161-168.

STENSIÖ, E. (1922): Über zwei Coelacanthiden aus dem Oberdevon von Wil-dungen.- Paläont. Z. 4, 167-210.

TUCKER, M.E. (1971): Aspects of pelagic sedimentation in the Devonian of Western Europe.- Thesis, Reading Univ., 405 p. (unpubl.).

-- (1973): Sedimentology and diagenesis of Devonian pelagic limestones (Cephalopodenkalke) and associated sediments of the Rhenohercynian Geosyncline, West Germany.- N. Jb. Geol. Paläont., Abh. 142, 320-350.

-- (1974): Sedimentology of Palaeozoic pelagic limestones: the Devonian Griotte (Southern France) and Cephalopodenkalk (Germany).- Spec. Publ. internat. Assoc. Sediment. 1, 71-92.

WALDSCHMIDT, E. (1885): Über die devonischen Schichten der Gegend von Wildungen.- Z. dt. geol. Ges. 37, 906-927.

WALLISER, O.H. (1980): The geosynclinal development of the Variscides with special regard to the Rhenohercynian Zone.- in: CLOSS, H. et al. (eds.): Mobile Earth. Research Report Deutsche Forschungsgemeinschaft, 185-195 p., Boppard (Harald Boldt Verl.).

-- (1984): Geologic processes and global events.- Terra cognita 4, 17-20.

WEDEKIND, R. (1917): Die Genera der Palaeoammonoidea (Goniatiten).- Palaeontographica 62, 85-184.

WITTEKINDT, H. (1965): Zur Conodontenchronologie des Mitteldevons.- Fort-schr. Geol. Rheinld. u. Westf. 9, 621-646.

ZIEGLER, W. (1958): Conodontenfeinstratigraphische Untersuchungen an der Grenze Mitteldevon-Oberdevon und in der Adorfstufe.- Notizbl. hess. L.-Amt Bodenforsch. 87, 7-77.

-- (1962): Taxonomie und Phylogenie oberdevonischer Conodonten und ihre stratigraphische Bedeutung.- Abh. hess. L.-Amt Bodenforsch. 38, 166 p.

-- (1971): Symposium on Conodont Taxonomy - a Field Trip Guidebook, Post Symposium Excursion to Rhenish Slate Mountains and Hartz Mountains.- 47 p., Marburg.

General Aspects

ECOSTRATIGRAPHIC CRITERIA FOR EVALUATING THE MAGNITUDE, CHARACTER AND DURATION OF BIOEVENTS

BOUCOT, Arthur J. *)

Abstract: In order to have the capability for recognizing as many of the extinction and adaptive radiations in the fossil record as possible we should take advantage of the ecostratigraphic approach in our work. This means that we will carefully collect, stratum by stratum, data about the stratigraphic ranges of the individual taxa within individual community groups, biofacies narrowly construed, as opposed to the all too customary habit of lumping taxa from varied community groups together indiscriminately. Following this procedure enables one to far more easily recognize as well, those brief intervals when portions of the ecosystem were restructured, which is important owing to the fact that such restructuring commonly coincides with extinction and adaptive radiation events. It must be recognized that major changes in supra-specific abundance are fully as useful in pin pointing extinction and adaptive radiation events as are mere taxonomic compilations. The ecostratigraphic approach also emphasizes the fact that so-called "known" stratigraphic ranges are commonly far less than "true" ranges except for the small number of abundant genera and their species. Awareness of this last relationship makes it clear that there is no such thing as a "Background Extinction Rate" within any one community group, i.e., biofacies, because the species to species name changes within the genera of each community group are merely evidence of phyletic evolution, not the termination of a lineage. Emphasis is placed on the importance of separating out the major ecosystem components, such as the level bottom from the reef complex when trying to recognize event horizons, i.e., compilations that lump taxa from such ecosystem components together tend to blur the actual nature of the units being mixed together, giving rise to an artifactual background extinction (and adaptive radiation) rate. We now need to far more carefully sample beds above and below suspected event horizons, community group by community group, in order to discover whether or not the taxa involved in radiations and extinctions undergo a sigmoidal change in abundance or not. All of this requires that we carefully evaluate our data against a sound knowledge of classical biostratigraphy, based on the evolutionarily useful data developed during the past century and more.

Introduction

The concept of bioevents has been with us for a long time in one form or another. One need only mention the name of Cuvier to conjure up catastrophism affecting the organic world during the past, as well as the systematic work which D'Orbigny summarized in his Prodrome de Paléontologie (1850-52). During this century there has been intermittent discussion of major extinction events such as that separating the Permian from the Triassic, or the Cretaceous from the Tertiary. Additional to these relatively obvious, major extinction events, most of which were

---

*) Departments of Geology & Zoology, Oregon State University, Corvallis, Oregon 97331, U.S.A.

Lecture Notes in Earth Sciences, Vol. 8
Global Bio-Events. Edited by O. Walliser
© Springer-Verlag Berlin Heidelberg 1986

first recognized in the last century, we now have increasing concern about a number of important, although smaller scale, additional events of concern to Event Stratigraphy (Walliser 1984a, b).

The purpose of this paper is to point out the utility, the necessity for emplying ecostratigraphic *) methods in the evaluation of the biologic effects of such events. Of great concern here is whether or not the affected taxa, at all levels from the species on up, were eliminated from the fossil record "instantaneously" or "gradually". The correlative problem of whether or not new taxa appeared "abruptly", following an event, or "gradually", must also be addressed.

For well over a century there has been discussion of extinction events. One can define an extinction event loosely as a "time" when a large number of taxa disappear from the fossil record, without leaving any descendants in younger strata. Most of the evaluation of the time significance of such events has been done by graphically plotting the presence or absence of varied taxa during the geologic time intervals preceding and succeeding the event in question. The geologic time intervals involved have commonly been no larger than a geologic Period, commonly a geologic Series, and in recent years a paleontologic Stage or Zone. These time units have an absolute time duration ranging from a few million years to many millions of years. In this decade it has become increasingly common practice to collect fossils, sometimes centimeter by centimeter, adjacent to the suspected event boundary. This more detailed collecting has been carried out because of the obvious need to determine whether or not the extinction event occupied a relatively "brief" interval, as judged by thickness of sedimentary rock and zonally significant fossils, or whether it might not have involved a greater time interval. It has been commonplace to display this data (Fig. 1) in time-diversity plots which commonly indicate either abrupt disappearance (Fig. 1A) of many, varied taxa at a single horizon adjacent to the suspected event's position in the stratigraphic column, or, alternatively, to note that the many, varied taxa are only gradually (Fig. 1B) eliminated from the record prior to the event horizon's position. Similar plots may also be provided for what can sometimes be interpreted as subsequent adaptive radiations (Fig. 1) following the event horizon. Kauffman (1984) has provided ample data of the gradual elimination sort, while considering the nature of the terminal Cretaceous event.

---

*) Ecostratigraphy is based on the observation that fossils collected in various time intervals do not occur in a random manner. The fossil flora and fauna are neither a homogenized, uniform mixture occurring everywhere in the world, nor a set of non-repetitive occurrences in which every fossil locality provides a unique mixture.

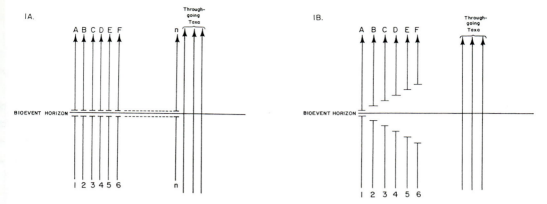

Figure 1. Time-Diversity Diagram indicating (1A) an instantaneous interpretation of a bioevent horizon. Taxa (below the bioevent horizon) that became extinct are numbered from one to "n", whereas (above the bioevent horizon) taxa that subsequently adaptively radiated are lettered "A" through "n".
Time-Diversity Diagram indicating (1B) a gradual interpretation of a bioevent horizon. Taxa are lettered and numbered as in Fig. 1A (above). Note that the taxa, both above and below the bioevent horizon, are miscellaneous taxa belonging to a variety of community groups, i.e., most of them do not co-occur in any individual community.

How should one view these "gradual" versus "instantaneous" versions of bioevents? Is it possible that some bioevents feature gradual extinction of varied taxa, whereas others are characterized by instantaneous extinction of varied taxa? Is it possible that some bioevents are followed by a gradual coming in of varied taxa, whereas others are characterized by an abrupt coming in of varied taxa? Or, are the terms gradual and instantaneous merely sampling artifacts? In the light of our available samples these are difficult questions to answer. But, if the term bioevent is to have any meaning these questions must be evaluated!

How to evaluate the rapidity of a bioevent

Introduction: It is critical that the rapidity of bioevents be evaluated if we are to have any hope of trying to understand causation. Bioevents taking place within a geologically instantaneous time interval (fractions of a year to a few decades for younger events in the Quaternary; a few thousand to a few tens of thousands further back in the record, should be distinguished from those which took place over a few million or even a few hundred thousand years). Once we are in the pre-C^{14} domain our capability for precisely measuring time in years or even a few thousand years is poor in most instances (except for such uncommon cases as laminae shown to be varves with annual value). Even within the Cenozoic where

our capabilities for isotopic absolute age determination, magnetic reversal stratigraphy, chemostratigraphy, and paleontologic dating are impressive for one whose attentions have been directed towards the older Paleozoic, there is great difficulty trying to work reliably at the very
short, human time scale. Therefore, it is probable that for some time we
will be restricted to estimates no better than a few thousands of years
at best, and more commonly working in the hundreds of thousands of years
range to even a few millions of years. But,even here it is useful to be
able to reliably discriminate between bioevents that took place over a
few hundred thousand years as contrasted with those which occupied a few
millions of years. There has been much useful discussion of these matters
during the past six years due to concern over the nature of the K-T bio-
event -- I need not remind you of the mass of data involved. What I will
attempt to do now, is to review and outline the possibilities for better
time resolution and recognition of bioevents in an ecostratigraphic context. By "ecostratigraphic context" I mean taking advantage of the fact
that organisms do not evolve solely within their own higher taxonomic
units as discrete entities. Rather, they evolve together with other,
taxonomically unrelated organisms, in communities, biofacies, associations, assemblages, or whatever other term you happen to prefer. We
should take advantage of this fact for purposes of bioevent recognition
and definition.

Biostratigraphers have been aware of this biofacies continuity of
the fossil record since the middle of the last century (D'Orbigny 1850-
52). I first became aware of this pattern as a graduate student working
in the Somerset County, northern Maine, where I mapped Late Silurian and
Early Devonian strata while also collecting and studying their fossils.
I learned later, while employed by the Paleontology and Stratigraphy
Branch of the U.S. Geological Survey, for whom I identified fossil collections of Silurian-Devonian age for field geologists, that this biofacies continuity in time was a routine character of fossil collections,
not a feature unique to the mid-Paleozoic of Somerset County, Maine. My
colleagues at the Survey had the same experience in other parts of the
stratigraphic column. Our experience merely reaffirmed that of genera-
tions of biostratigraphers. I also noted, as did my colleagues, that
within individual biofacies the species of the less common genera commonly had shorter stratigraphic ranges, which made them more useful for
dating and correlation, whereas the species of the more abundant genera
commonly had very long stratigraphic ranges which made them much less
useful for this purpose. Again, this conclusion was not unique to me or
my colleagues at the Survey, but reflected the experience had by gene-
rations of biostratigraphers. It also accounted for the many comments

we made to hopeful geologists "that you had better go back to the field and make a bigger fossil collection in order to have a better chance of obtaining some of the rarer, more short ranging fossils that would enable us to give you a more precise age determination". During this period, however, I did not spend any time pondering the evolutionary significance of these data -- their biostratigraphic utility was enough for me, as it has been for my professional forebears. Beginning in 1970 (see also, 1975, 1978, 1982, 1983, 1984a, b, c) I started to ponder these matters, and will now discuss their significance, particularly as it bears on the definition and recognition of bioevents.

Community evolution: I have previously (1978, 1982, 1983) discussed the pattern followed by organisms during community evolution (Fig. 2). It has been clear since the beginning that organisms do not evolve in some kind of ecologic vacuum. They evolve in intimate association with other, commonly unrelated taxa, in what may be termed a community. We are presently unable to evaluate the level, or levels, of coevolution present within communities. Whether or not one prefers the biologists term community for an evolving facies, or prefers to term it an association, or to consider other terms, is beside the point. The critical thing is that the fossil record is characterized by distinct sets of contemporary biofacies (beginning in 1975 I termed them community groups), which have definite, coincident time ranges (Boucot 1983, ecologic-evolutionary units).

How to recognize and define bioevents ecostratigraphically: In a series of papers (1978, 1982, 1983, 1984a, b) I discussed varying aspects of how to recognize and define community groups, i.e., the commonly, long ranging biofacies of the fossil record. I will briefly summarize the problem here. It is largely a sampling problem, as is so commonly the case in science. The critical item is to obtain a series of large samples of the same community type (= my term community group, or the narrowly defined biofacies of many authors, including such things as the Stringocephalus Community or Biofacies (not broadly defined biofacies such as graptolitic, shelly, pelagic, etc.)) through as long an interval of geologic time as possible. When the samples have been prepared, and counts made of the number of individuals belonging to each genus and species the following conclusions are commonly made: 1) The numerically abundant genera are usually represented by a single species (see Fig. 2 for a diagrammatic summary of these relations); 2) The numerically less abundant genera commonly are represented by a time sequence of species, with the number of such species being inversely proportional to the abundance of each genus; 3) The numerically more abundant genera also tend to be more eurytopic, with eurytopy being estimated in terms of such things as occur-

rence across a greater or lesser shelf width, and with the number of com
munity groups in which a genus occurs; 4) The numerically more abundant
genera also tend to be far more cosmopolitan geographically, and the les
abundant genera far more provincial.

Level bottom marine environment samples of megabenthos which include
several thousand specimens commonly include the species of many rare
genera.

How to recognize bioevents: The easiest way to recognize bioevent horizons is to follow the time honored biostratigraphic path. By this I mean
carefully noting those unique, globally distributed horizons where there
are major ecologic-evolutionary unit and subunit changes made manifest
by marked community group changes. To put it in traditional terminology
to note where there are major, global biofacies changes w i t h i n   the
same part of the ecosystem, i.e., remaining within the level bottom, cri
noid thicket, or other part of the environment -- not mixing units from
varied parts of the ecosystem. Such a procedure immediately, relatively
easily, points the way to the major bioevent horizons. In fact, these
horizons have been discussed at some length, beginning long ago, in that
extensive, dull biostratigraphic literature so familiar to practising
biostratigraphers. Its really very easy if one is only willing to become
familiar with this routine, readily available data.

How not to recognize and define bioevents: Raup (1986) has summarized
his views on the recognition and definition of extinction events. His
approach is a "statistical" one. He makes a basic assumption for which
the biostratigrapher can provide little support. Namely, that there is
a b a c k g r o u n d e x t i n c t i o n r a t e   operative during lengthy intervals of geologic time (essentially what I have called (1983) ecologic-
evolutionary units, which correspond in large part to D'Orbigny's (1850
1852) Étages). Extinction type bioevents are then weighed against
this so-called b a c k g r o u n d e x t i n c t i o n r a t e . In his 1986 pape
Raup provided a statistical treatment for family level taxonomic units
an example of how to recognize major extinction events. Raup's approach
fails to recognize that such a statistical approach is a very dull tool
with which to define and recognize extinction events as well as other b
events. This approach fails to recognize that evolution does not occur
some kind of ecological vacuum, and that sample quality (as in all statistical treatments) must be evaluated and taken account of. Purely sta
tistical treatments of paleontologic data, unless accompanied by a soun
natural history understanding of what is being dealt with, as well as a
reasonable attempt to evaluate and then weight varied parts of the samp
can be not only misleading, but can actually obscure significant matter

Specifically, I am concerned with the failure to separate out those parts of the marine ecosystem (in this example) which were evolutionarily largely decoupled from each other (such as pelmatozoan, algal, and bryo-zoan thicket communities, pelagic ammonoid communities, as well as reef complex communities), and which have significantly shorter stratigraphic ranges (commonly resulting from a later initiation and in some cases an earlier extinction than the co-occurring level bottom communities on which the bulk of the record is based). This mixing of taxa from unlike ecolo-gic units gives rise to a fallacious view about the times when family level taxa enter the geologic record by providing part of the data leading to the conclusion that there is a background rate of family entry into the record. Additionally unfortunate from the sampling viewpoint is that those families, probably the majority in most higher taxa, represented by rare genera and species, commonly have far shorter known ranges than would be the case were our samples of more equal quality (Would that the known ranges of fossils indeed equalled their true ranges! We are, unfortunately, far from this ideal for the great majority of taxa, the less common taxa.). The known ranges of rare genera and their species are being regularly extended for most groups as our sample size increases. Although this sampling problem is obvious for such excruciatingly rare taxa as the soft bodied animals, it was ignored when dealing with the well skeletonized megafossils. For example, the fossil record and known stratigraphic ranges of many pelmatozoan and stellaroid families is obvi-ously far short of what must be their true ranges. Raup's (1986) treat-ment dealt with families. Similar "statistical" treatments that deal with genera fall into the additional error of assuming that the generic and specific level evolution within an ecologic-evolutionary unit is largely, or at least significantly, of the cladogenetic type (see Jablonski 1986, Fig. 13, for a diagram making this assumption). Following the community evolution concept (Boucot 1975, 1978, 1982, 1983) which I have discussed as the basic model fitting the data of biostratigraphy (long-ranging bio-facies; rare genera and species having short stratigraphic ranges; abun-dant genera and species having far longer stratigraphic ranges; fixed num-ber of community groups/ecologic-evolutionary unit; stable number of com-munity group) it is clear that cladogenesis does not occur to any extent during the community evolution of the genera and species within each evol-ving community group (Figs. 2, 2A) during each ecologic-evolutionary unit; anagenesis, phyletic evolution is the norm. Jablonski (1986, Fig. 13) even introduces the concept of "stochastic, background extinction" which implies family level extinction within ecologic-evolutionary units, a concept for which there is no evidence if one analyses the data in an ecostratigraphic manner.

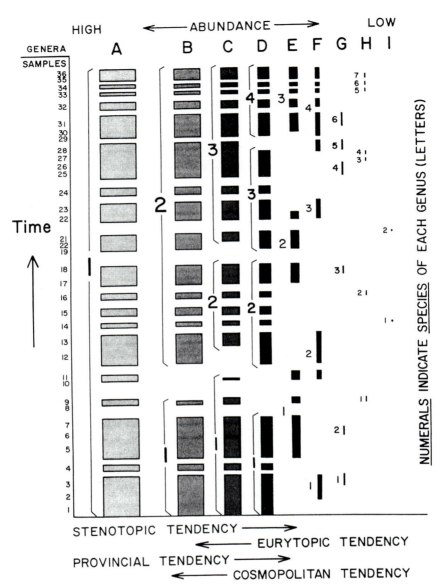

Figure 2.   Diagrammatic representation of the fossil record of a community group through an interval of time measured in a few fives or ten of millions of years. The genera (lettered) and their described species (numerals) are assigned line widths based on their relative abundance as individual specimens. Note that the species of the rarer genera tend to be both more rapidly evolving, and also have a much poorer fossil record correlating with their numerical abundance (rarer) as individual specimens (from Boucot, 1984, Fig. 1).

Note that this figure predicts that it will be the less abundant genera, which are also more stenotopic, and more provincial, which will be subject to the earliest extinction, whereas the most abundant genera and their species, which are also more eurytopic, and most cosmopolitan will be the last to go.

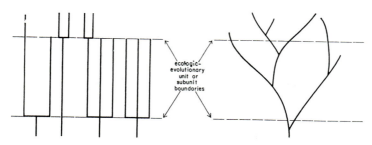

Figure 2A.   (Left) Cladogenetic pattern concluded to be characteristic
of the organisms belonging to individual ecologic-evolutionary units and
subunits. Note that cladogenesis is restricted to that brief moment in
time when new community units first appear. Cladogenesis here refers to
Metacladogenesis, i.e., quantum evolution mediated phenomena. Biogeographically mediated Diacladogenesis can, of course, occur anywhere within
an ecologic-evolutionary unit. (see Boucot, 1978, for definitions of
Metacladogenesis and Diacladogenesis)
      (Right) Cladogenetic pattern of the standard, random through time
type, which ignores the constraints imposed by what we know about community evolution. Note that this view permits cladogenesis to occur at any
time within an ecologic-evolutionary unit, or subunit, and is also consistent with important changes in species level diversity within any
ecologic-evolutionary unit or subunit as contrasted with the view outlined in "Left". Such random, within ecologic-evolutionary unit changes
in diversity do not occur. It is only by "superimposing" family trees
derived from ecologically unrelated, major parts of the global ecosystem
(such as level bottom, reef complex of communities, pelmatozoan thickets,
sponge forests, bryozoan thickets, pelagic cephalopod units, freshwater,
etc.) that one can simulate the unnatural random cladogenetic pattern.

So much for how   n o t   to recognize and define bioevents. The way to
recognize them is to take full advantage of the known characteristics of
community evolution as detailed above.

Suitability of different taxonomic levels for the recognition and definition of extinction events

I cannot recall having read a serious discussion about the suitability of
differing taxonomic levels for the recognition of extinction events, or
adaptive radiations. Clearly, though, the use of syntheses prepared at
different taxonomic levels does afford differing levels of precision and
ability to discriminate bioevents. Earlier (1978, pp. 621-624) I said
something about the taxonomic rate significance of summaries prepared for
the familial and lower levels. It has been clear for some time that family level introductions and removals from the stratigraphic record are not
randomly distributed in time if one remains within the same environment. Figure 4 diagrams some of the rationale for this conclusion. Figure
2 makes it clear that syntheses of special level changes/time interval
will be relatively dull statistical tools for the recognition of bioevents
as contrasted with higher taxonomic levels. This conclusion about species

level bioevents refers of course to the less common genera and their spe
cies (Fig. 2) which phyletically speciate at a fairly high rate. This
fairly high rate will make it difficult in the absence of a very large
sample to see above background rates of taxonomic extinction that might
reflect an extinction or adaptive radiation. It is entirely proper to
employ the term background rate of species level change here because of
the continuing levels of speciation going on in different generic lineages by means of phyletic evolution. The large measure of species change
that is diachronous makes it reasonable to discuss background rate if o
is measuring units of time in a few millions of years. Such would not, o
course, be true were one measuring time on a human scale operating unde
natural conditions (of which the present, modern situation is obviously
an extinction event that can hardly be viewed as natural). If one emplo
the species of the more abundant genera it will clearly be seen (Fig. 2
that there is such a low rate of change that bioevents will be very har
to recognize or define. This is particularly true not just within ecolo
gic-evolutionary units, but also crossing from one to the other due to
the fact that such genera and their species tend to be far more resista
to terminal extinction.

Turning to the use of the generic level one finds that things are n
far different from the specific level. The more abundant genera tend no
to give rise to other genera within any ecologic-evolutionary unit,
whereas the rare genera may more commonly undergo enough change in morphology to justify the setting up of phyletically evolved subgenera and
even genera in some instances. This being the case it will be difficult
although somewhat better than when dealing with species, to recognize a
define bioevents.

The basic reason for the failure at the specific and generic levels
is that they do not result from the quantum evolution adaptive radiatio
into new adaptive zones that the generation of new families and higher
taxa does, or the loss of such adaptive zones during extinction.

What little utility the generic level does possess for the recognition and definition of extinctions derives from some basic relations.
First of all, most families contain a small number of genera (data cull
from the Treatise on Invertebrate Paleontology and Romer
(1966) Vertebrate Paleontology indicate that half of the biva
ve families contain less than 9 genera, half the nautiloid families 8
genera, half the ammonoid families 5 genera, half the echinoid families
4 genera, half the trilobite families 5 genera, half the articulate bra
chiopod families 3 genera, and half the eutherian mammal families 4
genera). In all groups, such as those listed here, the mean number of

35

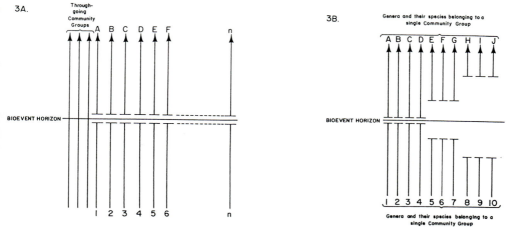

Figure 3A.   Time-Diversity Diagram indicating an instantaneous interpretation of a bioevent horizon employing data from a single community
group. Taxa are lettered and numbered as in Fig. 1. Note here that the
taxa within each community group have the same upper and lower stratigraphic ranges.

Figure 3B.   Time-Diversity Diagram indicating a gradual interpretation
of a bioevent horizon employing data from a single community group. Taxa
are lettered and numbered as in Fig. 1. Note here that the taxa within
each community group have the same upper and lower stratigraphic ranges.
Taxa, belonging to these community groups, which fail to reach the bio-
event horizon may reasonably be expected to have truly become extinct
prior to that boundary, rather than possibly being the result of environment not present immediately adjacent to the boundary, as is possible
when employing the miscellaneous type data diagrammed in Fig. 1B. The
same is true for diachronous appearances of new taxa above the bioevent
horizon.

genera occurs in families that contain more genera (9, 8, 7, 10, 9, 7,
and 22 respectively). Secondly, the stratigraphically long ranging, cosmopolitan, eurytopic forms, a minority of the genera, will tend to belong
to families present in more than one ecologic-evolutionary unit. It
follows from this that the majority of the genera, those which tend to
be stenotopic, provincial, and short ranging, will be reasonable indicators of extinctions if they are lumped together as family units.
Therefore, those families containing a small number of genera will be
more useful for this purpose. But, this is needless work, since one can
achieve the same result merely by plotting family ranges directly, which
bypasses the problem of mistaking phyletic, intra-ecologic-evolutionary
unit, phyletic evolution at the generic level as evidence of "background
extinction rate" instead of recognizing it for what it actually is --
merely routine phyletic extinction rate within evolving community groups,
and lacking any utility for recognizing or defining true extinctions.
Because of the large percentage of genera which belong to families that

include a large number of genera, and consequently have a better chance of including a few long ranging forms, it makes better sense to plot the ranges of families directly.

Going to the family and higher taxonomic levels au contraire! It has long been recognized, following Simpson's lead, that the origination of families and higher taxa, as well as their extinction, involves distinct adaptive zones, quantum evolution, the truly major evolutionary changes in the history of life, as contrasted with the trivial generic and specific level changes taking place within family units (I use the term family here to also include such suprageneric terms as subfamily and eve tribe). But, it has long been recognized that each of the major extinction events affecting the marine fossil record has its own flavor from a statistical point of view. The end of the Lower Cambrian sees the extinction of a very large number of higher taxa -- items that are in the subclass to phylum level, which because of their unique (and extinc character are very hard to categorize as to subclass to phylum. The archaeocyathids are a fine example, although Rowell (oral comm., 1985) points out that they survived into the Upper Cambrian in Antarctica as judged from associated trilobites. Extinctions below the subclass to phylum level at this time are not very notable. There is no similar Phanerozoic extinction event. The terminal Cambrian event seems to be chiefly a family and superfamily event. The terminal Lower Ordovician (Arenigian) event features the loss of many families and superfamilies, followed by adaptive radiations from the survivors plus the introduction of some new classes in abundance, such as the corals, plus majo relative abundance changes which include such things as bivalves, bryozoans, ostracodes, and articulate brachiopods. In terms of adaptive radiations and extinctions it is fully as important to discuss their overall effects on the relative abundances of higher taxa, and of commu nity groups, i.e., ecologic units, as it is to discuss pure statistical compilations at varied taxonomic levels. At the risk of appearing tedious, I would emphasize again that evolution does not occur in an ec logic vacuum. The terminal Ordovician event features familial level extinction very prominently, although a large number of superfamilies and some orders are also eliminated. The Frasnian-Famennian seems to be most prominent at the superfamilial level, although a fair number of orders, and some classes are also involved, as is also true for the Per mian. The Permian sees the extinction of an unusually large number of orders and lower categories down to and including species, plus some classes,which distinguishes it from the others. The terminal Triassic is not too well defined at present in these terms, but certainly does no seem to feature anything important at the subclass to phylum levels in the marine environment. The end Cretaceous event seems to differ from most of the others (see Kauffman 1984, and Jablonski 1986, for discussions). My point here is that major bioevents are most easily recognized and defined at the family and higher levels. And, I would like to again emphasize that the major terminal extinction events do not appear to coincide significantly with those in the nonmarine, especially the terrestrial environment.

I pointed out earlier (1983) that the defining, characterizing and ranking of extinctions demands that more than mere statistics of taxonomic units be considered. For example, the fact that all of the hippuritid bivalve genera and species disappear during the later Cretaceous, and their higher taxa automatically with them, is a fact. But, the persistence of the nautiloid cephalopods to the present time -- a single genus remaining, and that one biogeographically restricted, as is also the case with the rhynchocephalians and the rhipidistians, is not truly very different at the generic and specific as well as familial levels, similar in terms of pure abundance of specimens belonging to the four groups, but very different at suprafamilial levels. One should, when trying to characterize and recognize extinctions pay attention to changes in numbers of taxa at different taxonomic levels and also to relative abundance of individuals as well. The study of changing community structure is one good way to approach this aspect of the problem. Ecostratigraphiy requires that fossils be studied as members of evolving ecologic and biogeographic units occurring within a set of environments, not only as mere evolving taxa. Changing abundance relations are fully as important when recognizing, defining, ranking and describing extinction events as are taxonomic statistics. Practising stratigraphic paleontologists have dealt with these matters since the time of William Smith. We should do our best to interpret the mass of biostratigraphic information in ways conducive to a better understanding of both extinctions and adaptive radiations. We should not ignore it if we wish to truly comprehend the mysteries of extinctions and adaptive radiations.

These brief comments also should make it clear that in terms of taxonomic level the major extinction events differ from each other in many ways. When one adds to this the additional data of co-occurring changes in the relative abundances and community groups present (Boucot 1983, brief summary) then it becomes clear that no two major extinction or adaptive radiation events even begin to possess the same characteristics. The evolutionary and environmental significance of this unlikeness is uncertain.

Also to be reckoned with is the problem concerning why, after some of the major extinction events we have a geologically instantaneous adaptive

radiation affecting many groups, whereas after others there is a geologically lengthy delay before and adaptive radiation. In the latter category I am thinking of the level bottom environment's delayed rebounds
after the terminal Ordovician (about two-thirds of the Llandoverian),
Frasnian (the Famennian as far as continental shelf depth equivalent
benthos are concerned), an terminal Permian (all of the Scythian) extinction events. The reason(s) for this phenomenon are unknown. Yet they
need to be comprehended into any theory that attempts to explain bio-
events, or just how evolution operates. Also worthy of comment is the
question concerning whether or not the initiation of the many non-level
bottom complexes, such as the reef complex of communities, at times othe
than the major, ecologic-evolutionary units of the level bottom environment are merely stochastic events or involve unknown causes.

Dynamics of bioevents

During the course of the Alfred Wegener Conference on Bio-Events I was
struck by the fact that some of the participants, particularly Messrs.
Becker, Kauffman, and Ziegler, presented evidence strongly supporting
the concept that both the extinction of taxa prior to a bio-event horizo
and the subsequent adaptive radiation of descendant taxa from among the
survivors occur over a distinct time interval. There is the distinct
probability, in view of their evidence, that the decay and disappearance
of the taxa belonging to any higher taxon as well as their subsequent
adaptive radiation follows a sigmoid pathway. This is to say that the
initial growth or decay procedes slowly at first, builds up to a maximum, and then tapers off. This is the type of growth curve shown by many
populations belonging to a single species, as well as that character-
ising those specimens whose growth ceases when they reach adulthood. How
ever, no evidence was presented at the Alfred Wegener Conference concern
ing whether or not this type of extinction and subsequent adaptive radiation affects all of the taxa within a particular community or only the
more susceptible taxa within each community. It is essential, therefore
that additional research be carried out to determine whether or not this
sigmoid pattern affects the genera and species of each higher taxon independently, or whether certain communities are eliminated in toto
before other communities, with these communities containing a mixture o
higher taxa. The answer to this community extinction and subsequent
adaptive radiation consequent to the formation of new community types
may tell us something as well about possibilities for coevolution.

Conclusions

In view of the basic nature of the fossil record, viewed in community
terms, it is obvious that an understanding of bioevents requires the
following: 1) fossiliferous samples from immediately below and above the
suspected horizon must be obtained from localities representing as many,
varied environments as possible. 2) these varied samples should be obtained in order to get as many distinctive community groups immediately
adjacent to the suspected event horizon as possible. 3) these samples
should, naturally, be obtained from localities where zonal data indicates no disconformable possibilities. 4) after, and only after, a significant sample of the community groups known to exist during the time
immediately preceding and succeeding the event horizon, has been obtained
will it be possible to evaluate the question of "gradual" or "instantaneous".

Only after such samples have been obtained from the strata immediately underlying and overlying the suspected bioevent horizon will it be
possible to determine whether or not there is, indeed, a gradual or an
abrupt disappearance of taxa from within each of the community groups
present, as well as an abrupt or gradual appearance of new taxa from
within the overlying, post-bioevent horizon. Such sampling will then
make it clear whether older and younger horizons, respectively, from the
sampled community groups need to be considered in determining whether
gradual changes are involved or not. This type of sampling is far more
comprehensive than that thought suitable previously for obtaining the
necessary answers. But, as pointed out earlier, the sampling done previously is inadequate for providing definitive answers to the questions
which need answering. In the interests of a more reliable answer, more
labor must be expended. It is also clear that the obtaining of an adequate community group sample will probably involve many localities within any one biogeographic unit. It is unlikely, when dealing with level
bottom, marine organisms, which tend to occur as widely distributed
community types covering a large area of former sea floor, that sampling
of only a few carefully measured, closely spaced sections will even begin
to satisfy the sampling requirement discussed here. Again, this may pose
a real problem in terms of time and money needed, but if one is to defi-
nitively answer a major question it is reasonable that in some cases
major effort will have to be made. This type of major effort is prefer-
able to sterile arguments back and forth, based on inadequate samples.
It is unlikely, for example, that data from one region of the world will
be adequate for our purposes. Figure 3 outlines what, in principle, can
be obtained through careful sampling of varied community groups.

It is also clear that if some community groups are not found imme diately adjacent to the suspected bioevent horizon, either above or below, one cannot be certain whether their unique taxa, any not shared with other community groups that do occur immediately adjacent to the suspected horizon, became extinct earlier or adaptively radiated later than those known from strata immediately adjacent to the suspected bio-event horizon. Only a very large sample of those community groups which do not occur immediately adjacent, from as widely scattered a set of localities as possible, might provide one with a basis for suspecting that their unique taxa became extinct earlier, or appeared later, than those known from the immediately adjacent position. In other words, it is crucial that we make every sampling effort possible to dissociate mere sampling artifact influence on our thinking and conclusions from data that could materially mislead our thinking into error of one kind or another.

For example, Raup (1986, Fig. 4) cites data from a very limited Danish area that incorporates the Cretaceous-Tertiary boundary region. These data showed that certain articulate brachiopod genera and species are absent for a brief interval (he refers to these as "Lazarus taxa", using Jablonski's 1986, term). This gap in the brachiopod record prob-ably represents a mere sampling effect caused by a slight, local enviro mental shift that removed the environment necessary for the existance of the appropriate community type (community group). Raup then goes on to compare this minor generic and specific level, local omission, prob-ably a gap that will be filled in with expanded sampling elsewhere in the K-T region, with the major ordinal and even class level gap repre-sented by the absence in the Scythian (Lower Triassic), globally of varied taxa such as corals. For the Lower Triassic example the global sampling is entirely adequate to show that many high level taxa (fami-lies and higher) are truly absent. More importantly, these same taxa ar represented in many, but not all cases by totally distinct families and higher taxa in the younger Permian, where they last occur, as contraste with the beginning of the Middle Triassic where distinctly different, although potentially (but not certainly) descendant taxa are present (a typical example being Permian tetracorals replaced by Middle Triassic hexacorals, with no corals of any kind known in the Lower Triassic glo-bally) (think also of the high level changes affecting the articulate brachiopods, crinoids, echinoids, stromatoporoids, and bryozoans). The Danish example probably represents a sampling artifact that could be remedied if ecologically suitable localities immediately adjacent to th K-T boundary are found and sampled, whereas the Lower Triassic anomaly has been well enough sampled to convince most workers of its reality.

It is also clear that each of the major terminal extinction events of the Phanerozoic has its own distinctive character in terms of the statistics provided by varied taxonomic levels from the subclass and phylum on down, in the changing relative abundances of major taxa, and in the changing community groups present. It may well be, in view of this data, that no two major extinctions of the record resulted from the same mixture of causes.

In the previous discussion I have restricted myself to a consideration of the continental shelf depth equivalent marine fauna. But, it is obviously crucial to a better understanding of bioevents that we learn whether or not they simultaneously affect both nonmarine (terrestrial and aquatic) and marine environments in order that extra-terrestrial causation be considered. Present knowledge of the higher land plant fossil record strongly supports the conclusion that major innovations among the higher land plants do not correlate very well with similar adaptive radiations in the marine realm, nor that major extinction events in the marine realm correspond in time with events affecting the higher land plants. The diachronism of the paleobotanical time terms Palaeophytic, Mesophytic and Cenophytic with the paleozoological time terms Paleozoic, Mesozoic and Cenozoic testifies to this fact. The Palaeophytic-Mesophytic boundary is in the later Permian, the change from the Coal Swamp Flora to the Mesozoic Flora, and the Mesophytic-Cenophytic boundary is in the mid-Lower Cretaceous, representing the appearance of the flowering plants -- these are not times of major extinctions followed by subsequent adaptive radiations in the marine world. The terminal Ordovician extinction event in the sea has no parallel on land, where the Ordovician-earlier Silurian, higher land plant spore flora (Gray 1985) shows no evidence for extinction of any type. Jablonski (1986) has made similar comments about the later extinction events in the marine world, and their non-correspondance with major, higher land plant events. A truly zonal, global biostratigraphy for nonmarine animals, at a confidence level similar to that for the marine biota is not yet available, which makes nonmarine animal with marine animal bioevent comparisons difficult.

Because of the far lower level of biostratigraphic synthesis currently available for non-level bottom marine organisms it is crucial that their records be treated separately from those of the level bottom animals when trying to recognize and define bioevents. We still have a long way to go in the definition of most pelmatozoan thicket, algal thicket, sponge forest, bryozoan thicket, reef community complex, and other non-level bottom communities, as well as for their upper and lower stratigraphic ranges. Mixing data on the known stratigraphic ranges of their taxa in with that for the far better known level bottom taxa for pur-

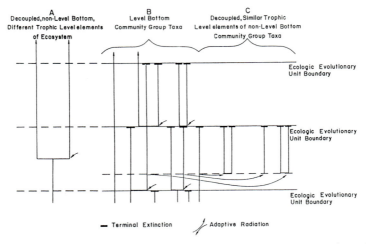

A
Decoupled, non-Level Bottom,
Different Trophic Level elements
of Ecosystem

B
Level Bottom
Community Group Taxa

C
Decoupled, Similar Trophic
Level elements of non-Level Bottom
Community Group Taxa

Ecologic Evolutionary
Unit Boundary

Ecologic Evolutionary
Unit Boundary

Ecologic Evolutionary
Unit Boundary

━ Terminal Extinction   ⟋ Adaptive Radiation

Figure 4.   Diagram outlining the timing of cladogenetic-adaptive radiation events affecting decoupled portions of the ecosystem. Column "A"
recognized the presence of ecologically, evolutionarily decoupled portions of the ecosystem which are taxonomically unaffected by significant
extinction and adaptive radiation events which materially affect other
portions, such as columns "B", and "C". Note too, that within SIMILAR
trophic units adaptive radiations significantly affecting one portion of
the similar trophic level need not correspond in time to that affecting
a parallel part. This applies to such things as adaptive radiations
giving rise to the reef complex of communities, or pelmatozoan thicket
community complexes on the one hand as contrasted with the standard level
bottom community groups.
    A real life example similar to this diagram might be the adaptive
radiation of high trophic level Silurian-Devonian vertebrates that in no
way corresponds to that of the level bottom or reef complex of communities. Recall that a significant number of the vertebrates sail right
through the Frasnian-Famennian, mid-Upper Devonian extinction (McGhee,
1982, p. 492) event with no taxic curtailment of any consequence, and
that the real adaptive radiation giving rise to the Silurian-Devonian
reef complex of communities is well after (Late Wenlockian, mid-Upper
Silurian) the appearance of the level bottom community groups from which
the reef taxa presumably were derived by adaptive radiation.

poses of bioevent recognition can only lead to confusion (Fig. 4). The

diachronous extinctions and adaptive radiations of these non-level bottom

ecologic units may represent another class of bioevents incapable of

affecting the more stable level bottom units, or possibly these are

merely stochastic events -- we need to compile the data and critically

examine the available evidence.

    In considering different bioevent horizons through time we must be

prepared to find more than one type of event. It may be that both gra-

dual and instantaneous events are present in the record. It is crucial

that we not fall into the trap of thinking that only one type of bio-

event timing can occur in the fossil record. There are probably many

controls over bioevents. Some of these controls may have acted in a geo logically instantaneous manner, such as the effects ascribed to the cometary-asteroid impacts that have been much discussed in the recent past. Others, such as geologically slow, but major changes in global climatic gradient may have behaved in the gradual manner; one immediately thinks of Pleistocene glacial and interglacial impacts on the European flora here. In all of this we must continually be wary about the effects of different extinction and adaptive radiation rates. If we are careful about collecting this rate data we may, ultimately be in a much better position to consider the causes of bioevents, rather than being in the present quandary (see Jablonski's 1986, extensive review, as well as my own brief summary, 1983). In all honesty we must admit that a true understanding of the causes behind extinction and adaptive radiation, bioevents, will come only through a most careful analysis of as much geologic and paleontologic data as possible, tempered with all the understanding that modern biology can provide for us. We find only correlations from our geologic materials. In order to directly translate these correlations from the possibility of mere chance coincidence, into the area of probable cause and effect, we must be provided with large, reliable, well studied samples from as large an area of the world as possible. This is an area of science where conclusion reliability is in largest part determined by sample adequacy. Sloan et al.'s (1986) account of post-Ir anomaly dinosaurs in the very early Paleocene of Montana serves as still one more warning about sample adequacy, as well as the need for far greater biostratigraphic precision in both fossil collecting and section measuring. Mere logic, no matter how penetrating, is no substitute here for sample adequacy. There is much work to be done; the Promised Land of understanding is still veiled in morning fog.

Acknowledgements

I am most grateful to Professor Otto H. Walliser for having arranged the Alfred Wegener-Conference, and for having patiently husbanded it through to the ultimate publication. I learned a great deal while participating in the Conference, and wish to thank my fellow participants for their assistance. I am also indebted to Dr. David Jablonski, Department of the Geophysical Sciences, University of Chicago, for his constructive criticism of an earlier version of the manuscript, although he should not necessarily be held responsible for all of the conclusions. Dr. J.G. Johnson, my long term colleague at Oregon State University, was most helpful with several items.

REFERENCES

BOUCOT, A.J. (1970): Practical taxonomy, zoogeography, paleoecology, paleogeography and stratigraphy for Silurian and Devonian brachiopods.- N. Amer. Paleont. Convention, Chicago, 1969, Proc. F., 566-611
-- (1975): Evolution and Extinction Rate Controls.- Elsevier, 427 p.
-- (1978): Community Evolution and Rates of Cladogenesis.- Ev. Biol. 11, 545-655.
-- (1982): Ecostratigraphic framework for the Lower Devonian of the Nort American Appohimchi Subprovince.- N. Jb. Geol. Paläont., Abh. 163, 81-121.
-- (1983): Does evolution take place in an Ecological Vacuum? II.- J. Paleont. 57, 1-30.
-- (1984a): The Pattern of Phanerozoic Community Evolution.- Proc. 27th Internat. Geol. Congr. 1, 13-21, Palaeontology, VNU Press.
-- (1984b): Constraints provided by ecostratigraphic methods on correlation of strata and basin analysis, by means of fossils.- Proc. 27th Internat. Geol. Congr. 1, 213-218, Stratigraphy, VNU Press.
-- (1984c): Ecostratigraphy.- in: SEIBOLD, E. & MEULENKAMP, J.D. (eds.): Stratigraphy Quo Vadis?. AAPG Studies in Geology 16, IUGS Spec. Publ. 14, 55-60.
D'ORBIGNY, A. (1850-52):Prodrome de Paléontologie.- Masson, 394, 427, 197, 190, 99 p.
GRAY, J. (1985): Microfossil record of the higher land plants: Advances in understanding of early terrestrialization, 1970-1984.- Phil. Trans Roy. Soc. London B 309, 167-195.
JABLONSKI, D. (1986): Causes and consequences of mass extinctions: A comparative approach. in: ELLIOTT, D.K. (ed.): Dynamics of Extinction Wiley, 183-229.
KAUFFMAN, E.G. (1984): The fabric of Cretaceous marine extinctions. in: BERGGREN, W.A. & VAN COUVERING, J.A. (eds.): Catastrophes and Earth History.-Princeton Univ. Press, 151-246.
McGHEE, G.R., Jr. (1982): The Frasnian-Famennian extinction event: A preliminary analysis of Appalachian marine ecosystems. Geol. Soc. Amer., Spec. Pap. 190, 491-500.
RAUP, D.M. (1986): Biological extinction in earth history.-Science 231, 1528-1533.
ROMER, A.S. (1966): Vertebrate Paleontology.- Univ. Chicago Press, 468 p
SLOAN, R.E.; RIGBY, J.K., Jr.; VAN VALEN, L.M. & GABRIEL, D. (1986): Gradual dinosaur extinction and simultaneous ungulate radiation in the Hell Creek Formation.- Science 232, 629-633.
WALLISER, O.H. (1984a): Geologic Processes and Global Events.- Terra Cog nita 4, 17-20.
-- (1984b): Global Events, Event Stratigraphy and "Chronostratigraphy" within the Phanerozoic.- 27th Internat. Geol. Congr., Abstracts 1, 208.

APPENDIX

In the body of this paper I have strongly warned against the unthinking use of taxonomic compilations in the recognition and definition of extinction events. By unthinking I have included the acceptance of "known" ranges of taxa as equivalent to "true" ranges. I have insisted on the us of ecostratigraphic principles in order to avoid this sampling error, which is a very serious one.

I would now like to briefly discuss a typical example. In 1971 I described a new, unique genus of Ludlow age, Upper Silurian brachiopod from the Klamath Mountains of northern California and the Roberts Mountains of Eureka County, Nevada. The genus was named Aenigmastrophia. The shells are excessively rare; two specimens known from a single locality in the Klamaths, and three specimens from two localities in the

Roberts Mountains. The collections of Silurian brachiopods from western
North America made by myself and others since WWII number in the many
tens of thousands of individual specimens. The communities within which
Aenigmastrophia occurs are relatively widespread and extend beyond the
Ludlow in many places. Aenigmastrophia represents a unique family level
taxon without doubt, and probably represents a unique superfamily as
well. If , however, one were to use the available data for this exceptionally rare genus one would probably restrict it to a small part of
Ludlow time for both time of origination and extinction. In view of its
rarity this would, obviously, be most misleading. The key to more specimens of originally very rare taxa is larger collections from ecologically
favorable locales.

J.G. Johnson (oral comm., 1986) comments that one of the Nevada
collections (loc. 4435) belongs to the Ludlovian age siluricus Zone
whereas the second collection (loc. 917) belongs to the top of the underlying ploeckensis Zone, i.e., the two occurrences could be very close in
time. Johnson also commented that his genus Antistrix (Johnson 1972) is
a terebratulid "from a four-foot interval of beds, mid-Middle Devonian
of Lone Mountain, Nevada - is known nowhere else". He also reminded me
that the Lochkovian notanopliid genus Callicalyptella from Carlin Nevada
is known from about 12 specimens collected on a single bedding plane
(Boucot & Johnson 1972). It is also worth pointing out that the Early
Devonian terebratulid genera Prorensselaeria, Mendathyris, Cloudothyris,
Lievinella, and Cloudella are all single locality, very rare taxa, and
that there are many, many such single locality taxa; fully enough to
make for many problems if one insists on employing the term background
extinction rate. Sepkoski (oral comm., 1986) at the Alfred Wegener-Conference mentioned to me that he now is eliminating from his diversity
through time compilations those taxa restricted in the reported record
to a single stage; this is a worthwhile and necessary step, but I would
encourage people making similar compilations to further consider that
there is nothing sacred about the single stage in this regard, particularly if there is any foundation for my contentions about the nature of
community evolution with their indications that   all   of the supra-
specific taxa within an evolving community group should have about the
same stratigraphic range -- both rare and abundant taxa -- with the
possible exceptions mentioned earlier about sigmoid curve disappearance
of some of these taxa within a community prior to an extinction event,
as well as the subsequent gradual, sigmoid introduction of these taxa
during an adaptive radiation.

BOUCOT, A.J. (1971): Aenigmastrophia, New Genus, a difficult Silurian
    brachiopod.- Smithsonian Contributions to Paleobiology 3, 155-158.
-- & JOHNSON, J.G. (1972):Callicalptella, a new genus of notanopliid
    brachiopod from the Devonian of Nevada.- J. Paleont. 46, 299-302.
JOHNSON, J.G. (1972): The Antistrix brachiopod faunule from the Middle
    Devonian of central Nevada.- J. Paleont. 46, 120-124.

GLOBAL BIOEVENTS AND THE QUESTION OF PERIODICITY

SEPKOSKI, J. John Jr. *)

Abstract: The hypothesis of periodicity in extinction is an empirical claim that extinction events, while variable in magnitude, are regular in timing and therefore are serially dependent upon some single, ultimate cause with clocklike behavior. This hypothesis is controversal, in part because of questions regarding the identity and timing of certain extinction events and because of speculations concerning possible catastrophic extraterrestrial forcing mechanisms. New data on extinctions of marine animal genera are presented that display a high degree of periodicity in the Mesozoic and Cenozoic as well as a suggestion of nonstationary periodicity in the late Paleozoic. However, no periodicity is evident among the as yet poorly documented extinction events of the early and middle Paleozoic.

Introduction

Extinction events are brief intervals of geologic time, normally less than several million years long, during which unusual numbers of taxa disappear. The idea that these events might be regularly spaced in time was introduced by Fischer & Arthur (1977) and supported by Raup & Sepkoski (1984) in a statistical analysis of extinctions of marine animal families. They found that extinction events were significantly nonrandom and could be described as displaying a 26-Ma periodicity (see also Raup & Sepkoski 1986, Sepkoski & Raup 1986). This conclusion has proven very controversial for a variety of reasons (Hallam 1984, Kerr 1985, Maddox 1985), including questions regarding the validity of some small extinction used in the analyses and the necessity of catastrophic forcing agents to explain the periodicity.

In this paper, I attempt to clarify some of these questions. I first discuss what is meant by periodicity of extinction and what kinds of causal agents are implied. I then review several criticisms of the idea and present new data and analyses at the genus level. These support the hypothesis of periodicity in the Mesozoic and Cenozoic but leave the pattern in the Paleozoic unclear.

Meaning of periodicity

Random versus periodic series

The hypothesis of periodicity in extinction events is basically an empirical claim based on statistical assessment of pattern in the fossil

*) Department of the Geophysical Sciences, University of Chicago, Chicago, Illinois 60637, U.S.A.

Lecture Notes in Earth Sciences, Vol. 8
Global Bio-Events. Edited by O. Walliser
© Springer-Verlag Berlin Heidelberg 1986

Figure 1. Examples of random ("independent") and periodic series of events.

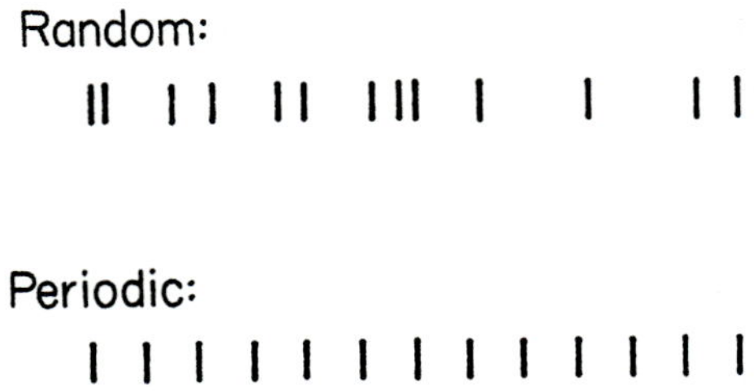

record. It is a claim that time intervals between events are too regular to be strictly random and that they approach constant length. Fig. illustrates the contrast between random and periodic time series of events. The upper, random series was generated by flipping a pair of coins and recording when both landed as heads. The expected frequency for this event is one in four trails, but the intervals between events are very irregular; some events can occur together while others may be separated by long intervals. This results in an irregular series charac terized by some loose clusters and some variably long gaps. Time series of this nature can be expected when events are independent of one another, as when coins are flipped.

In a strictly periodic time series, each event is dependent on the timing of the previous one and is separated from it by a constant inter val. The periodic time series in Fig. 1 has the same frequency of even as the random series (i.e. one in four) but has a very different appear ance because of the imposed serial dependency.

Mixed series

The two time series in Fig. 1 are end members of a spectrum of pattern reflecting relative dominance of independent versus dependent factors. Various constraints can make the random series appear more regular or periodic series less regular. For example, the operation of a recovery time after each event, during which subsequent events cannot occur, te to break up clusters in a random series. This can make intervals betwe events appear more regular, although there would still be long, irregu lar gaps between some pairs. At the other extreme, a nonstationary, or variable, period length would make intervals between periodic events appear less regular. This situation could occur when the period length is a function of time or contains a stochastic component. In the analy sis of Raup & Sepkoski (1986), for example, the time intervals between identified periodic events have a standard deviation of 2.4 Ma; this might reflect some "wobble" in the period length, although it could ju as well result from errors in the chronometric time scale (or both).

Obviously, if the standard deviation is large, a periodic series can be difficult to distinguish from a random one in statistical analysis.

Other situations that might degrade a periodic series include missing events and intermixed aperiodic events. Missing events may result when the periodic process "skips a beat" or when an event is so subdued that it cannot be identified in the data. Sepkoski & Raup (1986) found two gaps in their periodic series of Mesozoic familial extinction events, one in the Early Cretaceous and the other in the Middle Jurassic. The Early Cretaceous gap appears to be filled when better data are analyzed (see below), but no candidate for a Middle Jurassic event can yet be identified.

Time series that result from interference of two or more periodic signals with different wavelengths may contain "missing events" as well as clusters of events. However, so long as the observed time series is sufficiently long, most statistical techniques (e.g. Fourier analysis) can identify the separate periodicities. It should be noted, however, that multiple, interfering periodicities have not been observed in the history of extinction events.

Aperiodic events in a basically regular time series may reflect operation of independant causal agents that produce outcomes similar to those produced by the periodic forcing. An example in the fossil record is the terminal Pleistocene extinction of large terrestrial mammals. This event does not fit the periodic series for marine animals and thus appears an independent event with perhaps a separate cause (as suggested by the human predation hypothesis; see Martin & Klein 1984). The occurrence of aperiodic events makes a time series appear less regular but does not necessarily preclude discovery of periodicity. Many statistical techniques can still distinguish regular signals so long as the rogue events are not too frequent.

Implications for cause

Observation of periodicity in no way indicates what caused the regular series of events, but it does limit the range of possible causes. It suggests either that some single, external causal agent with clocklike behavior produces each event or that some internal dynamic (autocorrelated processes) of the system causes each event to occur after some set time interval. (There is no theoretical justification for a 26-Ma interval cycle in evolutionary systems.) Both implications are very different from that of a random series which suggests either multiple, independent causes of events or a single cause varing irregularly through time.

Another implication of periodicity, although not a necessary one, is that a causal hypothesis developed for one event may be applicable to all others. It is largely this implication that ties the hypothesis of extinction periodicity to hypotheses of extraterrestrial catastrophism. There is now persuasive evidence associating the terminal Cretaceous mass extinction with the impact of a large extraterrestrial body (Alvarez et al. 1980, 1982, 1984). Since the terminal Cretaceous event is a member of the periodic series, it has been inferred that the other events might also be associated with impacts, and several speculative hypotheses of clocklike ultimate forcing agents have been suggested (Davis et al. 1984; Rampino & Stothers 1984; Whitmire & Matese 1985). However, an impact at the end of the Cretaceous does not necessitate that all extinction events were caused by impacts. It remains possible that the end-Cretaceous impact was a rogue event that was coincidental with the operation of another, periodic agent of extinction, amplifying its effects. In order to test this possibility, important empirical efforts are needed, including (1) further search for signatures of large or multiple impacts at other extinction events (and between events); (2) further investigation of the detailed character of the periodic extinction events to determine if their patterns of taxonomic selectivity, temporal duration, and environmental and geographic distribution are similar; and (3) investigation of long time series of other environmentally sensitive geologic variables, such as stable isotope ratios, to determine if these display periodicities congruent with that of extinction.

Previous work on extinction periodicity

Summary of work on familial data

Raup & Sepkoski's (1984) initial claim of a statistical 26-Ma periodicity in extinctions was based on an analysis of marine animals families in the Late Permian to Miocene interval. These data were selected because a data base on fossil families was available and because the time scale for this part of the geologic record is more accurate and the 39 sampling units (stages) are shorter than in the earlier Paleozoic. It was implicitly recognized that families dampen the signal of extinction events since all constituent species must be eliminated before a family will register the extinction. Therefore, all extant families were culled from the data in order to enhance the signal, especially over the younger portions of the record. This culling left a residue of 567 families out of an initial sample of 2160. The metric chosen to measure extinc-

tion intensity was percent extinction, computed as the number of extinctions divided by the total number of families in each stage in the culled
data. This metric scales extinctions to the number of families at risk
and avoids potentially inaccurate estimates of stage durations.

In order to minimize possible bias, Raup & Sepkoski (1984) treated
all local maxima in their time series as if they were extinction events,
even though some most certainly were not (see Sepkoski & Raup 1986).
Several statistical analyses were performed on the data. The principal
analysis involved a bootstrap (or, more properly, "randomization") test
that measured the fit of periodic functions to the observed maxima relative to randomized maxima. This text was designed to assess only the
timings of events and not their magnitudes (which were assumed to vary
unpredictably) and was constructed to permit missing events. Also, because maxima had to be separated by at least one stage, a de facto recovery time of about 12 Ma (reasonable for rebound times after small
events; see Sepkoski 1984) was built into the randomization procedure.
The test showed that the observed data fit a 26-Ma periodicity significantly better than the randomized versions. Subsequent tests using only
significant extinction maxima in unculled familial data (Sepkoski & Raup
1986) and maxima in preliminary generic data (Raup & Sepkoski 1986) have
corroborated this result.

Criticisms

These analyses have been subjected to a number of criticisms. Questions
have been raised concerning (1) the culling of the familial data
(answered by demonstrating comparable results in unculled data); (2) the
taxonomic level employed (answered by demonstrating comparable patterns
in generic data); (3) the accuracy of the chronometric time scales
employed (answered in part by demonstrating significant results when only
the most accurately dated last 100 Ma are analyzed; see Raup & Sepkoski
1986); and (4) the use of nonstandard statistical techniques (see below).

The most extensive critique published to date has been by Hoffman
(1985a). He argued that paleontologic and chronometric data are very
noisy and that different treatments can generate very different patterns
of extinction. To demonstrate this, he computed 20 time series from the
familial data using various time scales and metrics of extinction intensity. He assessed which local maxima in the series were "minor" or "major" and presented a tabular summary that he contended displayed little
consistency in the timing of extinction events and therefore no periodicity.

I agree with Hoffman's conclusion that the familial data are noisy
but would contend that noise in itself does not necessarily preclude

discovery of pattern. The proper question is whether the data are so
noisy that no pattern at all is discernible. A common approach to such
a question when multipe time series are available is to "stack", or add
together, the series. The presumption is that random noise will deviate
in both positive and negative directions and thus tend to cancel or add
together weakly whereas nonrandom signal, varying in a consistent direc
tion, will tend to be amplified.

I have stacked Hoffman's data by assigning values of 0 to stages
without maxima, 1 to stages with minor maxima, and 2 to stages with maj
maxima in each time series. Fig. 2 illustrates the result of this exer-
cise. As evident, intervals containing well-known mass extinctions, suc
as the Upper Permian and Maestrichtian, form strong peaks in the stacked
time series. But other intervals containing smaller extinction events,
such as the Pliensbachian, Thithonian, Cenomanian, Upper Eocene, and
Middle Miocene, also form peaks, indicating that the time series are nc

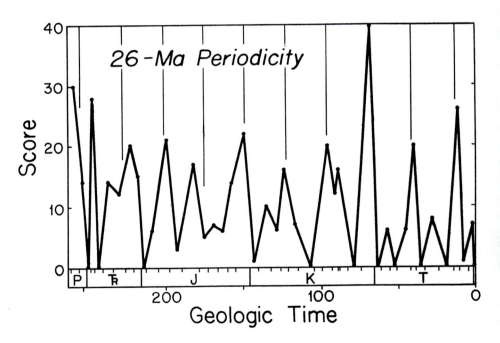

Figure 2. Time series of extinction in the Mesozoic and Cenozoic co
piled from data in Table 1 of Hoffman (1985a). The scores were compute
by assigning values of 0 to no maxima, 1 to minor maxima, and 2 to maj
maxima in each of Hoffman's 20 extinction time series and then summing
the values for each stage. The 26-Ma periodicity of extinction found b
Raup & Sepkoski (1984) is indicated by the vertical lines; this period
city fits the illustrated peaks rather well. The time scale is from Ha
land et al. (1982) with durations of Jurassic stages based on Westerma
(1984). Tics along the abscissa indicate stage boundaries.

entirely random. In fact, the pattern of peaks in the stacked time series fits a 26-Ma periodicity rather well, as indicated by the vertical lines in the figure. Thus, rather than falsifying the hypothesis of periodicity, Hoffman's time series would appear to lend support.

New data and analyses

Generic data

In attempt to circumvent problems in the familial data, I have been compiling a new data set on fossil marine genera with times of origination and extinction resolved to a finer stratigraphic framework. Preliminary reports on these data have been presented in Sepkoski (1986) and Raup & Sepkoski (1986). Below I present new analyses of a somewhat more refined version of the data.

Fig. 3 displays time series for the mid-Permian to Recent for four metrics of generic extinction (compare to Fig. 1 in Sepkoski & Raup

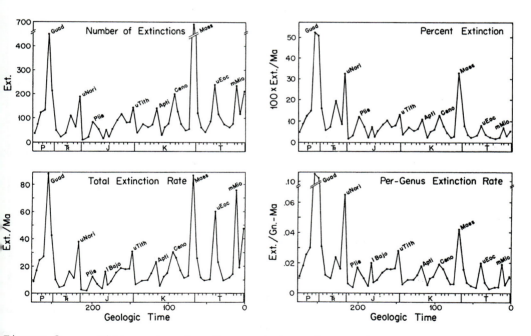

Figure 3.   Time series for four metrics of generic extinction intensity for the Permian (Sakmarian) to Recent, showing the occurrence and magnitude of extinction events. The time scale is the same as in Fig. 4. Tics along the abscissa indicate standard stages and not sampling intervals. Dots represent computed values and are placed over the centers of the sampling intervals. Abbreviations along the ordinates are "Ext." = number of extinctions, "Gn." = number of genera (standard diversity), and "Ma" = million years.

1986). The metrics are simple numbers of extinctions, percent extinction (number divided by standing diversity), total extinction rate (number divided by estimated duration of the sampling interval), and per-genus extinction rate (also called "extinction probability", which is the total rate divided by diversity). For all metrics, stratigraphic stages have been subdivided or amalgamated so as to make the durations of the sampling intervals as even as possible. The Sakmarian, Leonardian Norian (including Rhaetian), Bajocian, Tithonian, Campanian, and Lower Miocene are split into two intervals ("substages"), and the Albian into three; amalgamated stages are the Induan and Olenekian, the Turonian and Coniacian, and the Pliocene and Pleistocene. This manipulation has resulted in 51 sampling intervals with a mean duration of 5.5 Ma and standard deviation of 1.2 Ma.

A total of 9773 genera are represented in Fig. 3, of which 5594 are extinct. The times of extinction of 67 % are resolved to specific sampling intervals and another 8 % to stages that have been subdivided. All low-resolution data are distributed over the intervals in proportion to the high-resolution data.

The genera used in Fig. 3 actually represent only a subset of the 15,780 genera that have been compiled for the illustrated interval. Excluded genera are those confined to single stratigraphic intervals. When the total sample is used, there is a high correlation (r = .675) between numbers of extinctions and originations in the 105 sampling intervals over the entire Phanerozoic. Although this correlation might be interpreted as resulting from equilibrium processes governing generic diversity (cf. Mark & Flessa 1977, Hoffman 1985b), it more probably reflects monographic effects: extensive taxonomic studies of certain richly fossiliferous formation produce rare genera that appear to have their originations and extinctions confined to single stratigraphic intervals. When such single-interval genera are culled from the data set the correlation between extinctions and originations decreases to .286. It is largely this manipulation that produces differences in appearance between the generic extinction curves in Fig. 3 and those published by Sepkoski (1986) and Raup & Sepkoski (1986). Note, however, that these differences are quantitative; the positions of extinction maxima do not change between the complete and culled data sets.

Patterns of Mesozoic-Cenozoic extinction

The time series in Fig. 3 present a visual impression of fairly uniform nonrandom spacing of extinction maxima, coupled with considerable variation in their amplitudes. The two highest maxima are generally the Upper Permian (Guadalupian and Tatarian) and the Maestrichtian. These interval

contain 670 and 695 generic extinctions, respectively, corresponding
to 76 % extinction across the entire Upper Permian (compared to 57 %
at the familial level) and 33 % in the Maestrichtian (compared to 17 %
for families). Using the rarefaction curves of Raup (1979), these values
would result from a 92 % extinction at the species level in the Upper
Permian and 60 % extinction in the Maestrichtian. The upper Norian (which
includes the Rhaetian) also appears as a major extinction maximum, comparable to the Maestrichtian, when the number of extinctions is scaled
to the relatively low standing diversity (about 600 genera) of the time.

Distributed among the larger extinction maxima are six or more
lower peaks that appear faily evenly spaced and more uniform in amplitude. Five of these maxima have been recognized in time series based on
families (Sepkoski & Raup 1986): the Pliensbachian, upper Tithonian,
Cenomanian, Upper Eocene, and Middle Miocene. The Aptian also appears as
a maximum comparable to the better-documented extinction events; an
extinction event in this stage was predicted, but not observed, in the
analysis of periodicity in familial data.

The shape and relative amplitudes of the six lesser maxima vary somewhat depending on the metric. There is some tendency for the maxima to
appear as broad peaks extending over several sampling intervals rather
than as sharp spikes centered over a single stage. However, no consistency in shape is apparent from one maximum to the next, so that it is
not clear whether this pattern represents real, long-term variation in
extinction intensity or simply stochastic variation in extinction and
in the data. Despite this nonuniformity of shape, there is a remarkable
similarity in amplitude of the six smaller extinction maxima in some
metrics; five of the six range between .017 and .020 extinctions per
genus-Ma in the per-genus extinction rate (upper Tithonian = .026). If
this pattern is real, it suggests a uniformity of response to perturbations and, perhaps, a uniformity in the magnitude of most perturbations.

In addition to the total of nine fairly evenly spaced extinction
events identified above, there are several other maxima that are irregularly scattered through the time series. These include the Plio-Pleistocene at the right end of the series, which has been recognized by Stanley (1984, 1986) as containing a small extinction event. The lower Bajocian, labelled in two of the graphs in Fig. 3, forms a sharp peak in the
extinction metrics scaled to time; this may result, however, from under-
estimation of the duration of the sampling interval (estimated as half
the duration of the Bajocian) coupled with failure to sample extinctions
in the adjacent intervals, especially the preceding Aalenian. The Carnian, to the left of the upper Norian, also contains a maximum in all

extinction metrics. It is not clear whether this reflects a true extinction event as yet unrecognized in detailed biostratigraphic analyses or simply an artifact resulting from failure to samply many Upper Triassic genera in the succeeding Norian; ammonoids, which are the most thoroughl studied taxa in the Triassic, do not show a distinct peak of extinction in the Carnian.

Autocorrelation analysis

Because of persisting questions as to what does or does not constitute an extinction event, it is desirable to analyze the entire generic dat set rather than just selected extinction maxima. One statistical technique available for assessing the strength of periodicities in time series data is autocorrelation analysis. This technique involves correlating data in each sampling interval with data lagged some set number of intervals backward or forward in the series. The lag can be varied among analyses to produce what is essentially a spectrum of autocorrelation coefficients. The analysis assumes equal duration of sampling intervals and thus was precluded by the nature of the previous familial data (although see Kitchell & Pena 1984). The more equal sampling inter vals in the generic data permit its use, although results may be conservative since some variance remains in interval durations.

Autocorrelation analysis of a time series with periodic properties will produce varying results depending on the nature of the cycle. A perfectly cyclic signal, as in a sine curve, will produce autocorrelations that move from +1 at lag equals zero, to -1 at lag equals one-half wavelength, back to +1 at lag equals one wavelength. Factors that will decrease the magnitude of the autocorrelations include unequal sampling intervals, noise in the data, presence of aperiodic events, variation in the amplitude of the periodic events, and low-frequency variation or trends (such as long-term increase or decrease in the mean

Fig. 4 displays correlograms ("autocorrelation spectra") for the fo metrics of generic extinction in Fig. 3. All show a pattern that is con sistent with periodicity. The correlations vary from positive at a lag of 1, to negative around a lag of 3, and back to positive at a lag of 5 This cycle is then repeated at lags of 6 through 10 (although with a downward drift indicating some low-frequency trends, such as irregular temporal increase in amplitude of events or level of "background" extir tion between events). This pattern is thus evidence of a periodic signa with a wavelength between 26.1 Ma (5.5 Ma x 9.5/2) and 28.9 Ma (5.5 Ma x 10.5/2), which is consistent with the 26.2 $\pm$ 1 Ma period length previously computed from the familial data (Sepkoski & Raup 1986). Note th

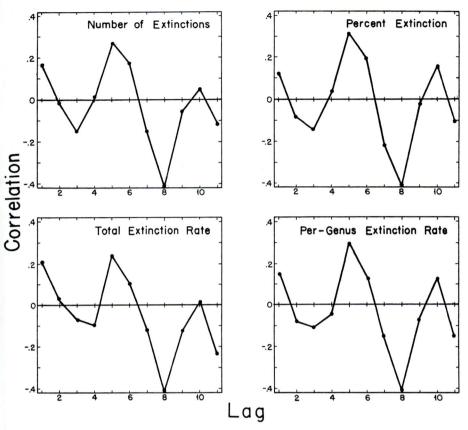

Figure 4. Correlograms for the four time series in Fig. 3. Autocorrelations were computed for 49 intervals (Leonardian to Plio-Pleistocene) with lags of 1 to 11. Prior to computation, data were transformed to logarithms to minimize variation in amplitude among events and recalculated as residuals from regressions on time to eliminate simple temporal trends.

this result is not consistent with the random walk model proposed by Hoffman & Ghiold (1985) and Hoffman (1985a) which predicts low peaks at lags of 4 and 8 sampling intervals.

Paleozoic extinction patterns

Although the autocorrelation analysis provides further support for periodicity in Mesozoic-Cenozoic extinction, it reveals no evidence of periodicity in the Paleozoic. Fig. 5 illustrates the per-genus extinction rate over the whole of the Paleozoic. Again, stages and series have been variously subdivided to produce sampling intervals averaging 5.5 Ma in duration (standard deviation = 1.4 Ma). Of the 6257 genera occurring in

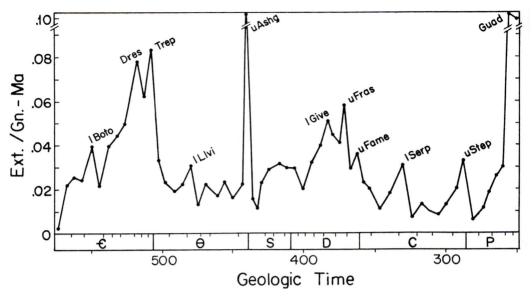

Figure 5. Per-genus extinction rate for the Paleozoic. Plotting conventions are the same as in Fig. 3. The time scale is from Harland et al. (1982), with stages in the Cambrian rescaled from Sepkoski (1979); Ordovician from the average of McKerrow et al. (1980), Gale & Beckinsale (1983), and Ross & Naeser (1984); Silurian from the average of Boucot (1975), Jones et al. (1981), Gale & Beckinsale (1983), and Harland et al. (1982); and Devonian from Boucot (1975).

more than one stratigraphic interval, 66 % of their extinction times have been resolved to one of the 60 sampling intervals and another 20 % to a stage or series that has been split into two or three subdivisions

Correlograms for this time series as well as for the other extinction metrics display no cyclic variation in magnitude. Generally, the autocorrelations are positive for lags of 1 to 7 and negative for lags of 8 to 11 when the entire Paleozoic is analyzed. However, some portion of the Paleozoic do exhibit some regularity in extinction events. The Carboniferous contains two extinction maxima, one in the lower Serpukhovian at the base of the Upper Carboniferous and the other in the upper Stephanian (Gzelian) at the end of the system. These maxima are nearly equal in amplitude and are rather evenly spaced between the extinction events at the end of the Devonian and Permian. However, the waiting time between these events is considerably longer than most in the Mesozoic and Cenozoic, averaging about 37 Ma (standard deviation = 3.1 Ma). This is actually compatible, though, with the waiting time between the Upper Permian and Upper Triassic events, which is somewhat in excess of 30 Ma and thus may be intermediate between the late Paleozoic and post-

Triassic Mesozoic. Such change in waiting times suggests that the extinction periodicity may be nonstationary with the period length decaying over time. However, more detailed statistical analyses are needed to confirm or reject this speculation.

Prior to the Carboniferous, there is no positive indication of periodicity. Fig. 5 displays high extinction intensities over most of the Devonian, with maxima in the lower Givetian, upper Frasnian, and upper Famennian. The last two of these maxima correspond to the six extinction events (the Kellwasser and Hangenberg) recognized by House (1985) for Devonian ammonoids. The high average intensities over the rest of the Devonian may reflect a combination of other, smaller events coupled with generally high provinciality and frequent extinction of endemic genera. Most of the Silurian also displays relatively high average intensities although without any definite peaks of extinction; some treatments of the data, however, suggest a low maximum in the Ludlovian.

The Ordovician contains two apparent maxima of extinction. The well-known terminal Ordovician mass extinction appears as a pronounced peak over the upper Ashgillian. This maximum is comparable in amplitude to the Upper Permian event but, because of the shorter duration, reflects extinction of fewer genera (435 out of approximately 1000). Earlier in the Ordovician there is a low maximum over the lower Llanvirnian which evidently corresponds to the event noted by Boucot (1983) around the top of the Lower Ordovician.

Extinction patterns in the Cambrian appear broadly similar to those in the Devonian. There are high average intensities, especially over the Middle and Upper Cambrian, with maxima in the lower Botomian, Dresbachian, and Trempealeauan. The last two reflect biomere extinction events among trilobites, documented by Palmer (1979, 1984) and others.

The seemingly chaotic pattern of extinction maxima in the early and middle Paleozoic certainly does not suggest any periodicity. However, it may be too early to conclude that periodicity is definitely absent. Fig. 5 represents at best a rough estimate of extinction patterns, and more highly resolved global stratigraphic and taxonomic data, coupled with detailed biostratigraphic analyses of critical intervals (e.g. House 1985), are needed before definitive data for testing the hypothesis are available.

Conclusions

Multiple lines of statistical evidence now exist for an approximately 26-Ma periodicity in extinction events over the last 250 Ma of the Phanerozoic. However, this periodicity cannot be easily traced back into

the Paleozoic. There is a suggestion of a nonstationary periodicity in
the late Paleozoic but no positive evidence for periodic events, or even
combinations of periodic and aperiodic events, in the early and middle
portions of that era. Future work needed to test these observations, in
addition to more sophisticated models and quantitative analyses of extinction patterns, include detailed comparative studies of the duration,
taxonomic selectivity, environmental and geographic distribution, and
geologic correlates of extinction events. Such studies will provide
needed information on precisely when extinction events occur and how
similar their effects are. From this information we might better determine their cause and the source of their clocklike behavior.

Acknowledgements

I thank James Quinn for suggesting autocorrelation analysis and Richard
Chappell for help in interpreting results. This work received partial
support from NASA grant 2-282.

REFERENCES

ALVAREZ, L.W.; ALVAREZ, W.; ASARO, F. & MICHEL, H.V. (1980): Extraterre
  trial cause for the Cretaceous-Tertiary extinction.- Science 208,
  1095-1108.
ALVAREZ, W.; ALVAREZ, L.W.; ASARO, F. & MICHEL, H.V. (1982): Current
  status of the impact theory for the terminal Cretaceous extinction.-
  in: SILVER, L.T. & SCHULTZ, P.H. (eds.): Geological Implications of
  Impacts of Large Asteroids and Comets on the Earth. Geol. Soc. Amer.
  Spec. Pap. 190, 305-316.
-- ; KAUFFMAN, E.G.; SURLYK, F.; ALVAREZ, L.W.; ASARO, F. & MICHEL, H.V
  (1984): Impact theory of mass extinctions and the invertebrate fossi
  record.- Science 223, 1135-1141.
BOUCOT, A.J. (1975): Evolution and Extinction Rate Controls.- Elsevier,
  Amsterdam, 427 p.
-- (1983): Does evolution take place in an ecological vacuum? II.- J.
  Paleont. 57, 1-30.
DAVIS, M.; HUT, P. & MULLER, R.A. (1984): Extinction of species by peri
  odic comet showers.- Nature 208, 715-717.
FISCHER, A.G. & ARTHUR, M.A. (1977): Secular variations in the pelagic
  realm.- in: COOK, H.E. & ENOS, P. (eds.): Deep-Water Carbonate Envi-
  ronments. Society of Economic Paleontologists and Mineralogists,
  Spec. Publ. 25, 19-50.
GALE, N.H. & BECKINSALE, R.D. (1983): Comments on the paper "Fission
  track dating of British Ordovician and Silurian stratotypes" by
  R.J. Ross and others.- Geological Magazine 120, 295-302.
HALLAM, A. (1984): The causes of extinction.- Nature 308, 686-687.
HARLAND, W.B.; COX, A.V.; LLWELLYN, P.G.; PICKTON, C.A.G.; SMITH, A.G.
  & WALTERS, R. (1982): A Geologic Time Scale.- Cambridge Univ. Press,
  Cambridge, 131 p.
HOFFMAN, A. (1985a): Patterns of family extinction depend on definition
  and geological timescale.- Nature 315, 659-662.
-- (1985b): Biotic diversification in the Phanerozoic: diversity inde-
  pendence.- Palaeontology 28, 387-391.

HOFFMAN, A. & GHIOLD, J. (1985): Randomness in the pattern of "mass extinctions" and "waves of origination".- Geol. Mag. 122, 1-4.
HOUSE, M.R. (1985): Correlation of mid-Palaeozoic ammonoid evolutionary events with global sedimentary perturbations.- Nature 313, 17-22.
JONES, B.G.; CARR, P.F. & WRIGHT, A.J. (1981): Silurian and Early Devonian geochronology -- a reappraisal with new evidence from the Bungonia Limestone.- Alcheringa 5, 197-208.
KERR, R.A. (1985): Periodic extinctions and impacts challenged.- Science 227, 1451-1453.
KITCHELL, J.A. & PENA, D. (1985): Periodicity of extinctions in the geologic past: deterministic versus stochastic explanations.- Science 226, 689-692.
MADDOX, J. (1985): Periodic extinctions undermined.- Nature 315, 627.
McKERROW, W.S.; LAMBERT, R.St.J. & CHAMBERLAIN, V.E. (1980): The Ordovician, Silurian, and Devonian timescales.- Earth and Planetary Science Letters 51, 1-8.
MARK, G.A. & FLESSA, K.W. (1977): A test for evolutionary equilibria: Phanerozoic brachiopods and Cenozoic mammals.- Paleobiol. 3, 17-22.
MARTIN, P.S. & KLEIN, R.G. (1984): Quaternary Extinctions: A Prehistoric Revolution.- Univ. of Arizona Press, Tuscon, Arizona, 892 p.
PALMER, A.R. (1979): Biomere boundaries re-examined.- Alcheringa 3, 33-41.
-- (1984): The biomere problem: evolution of an idea.- J. Paleont. 58, 599-611.
RAMPINO, M.R. & STOTHERS, R.D. (1984): Terrestrial mass extinctions, cometary impacts and the Sun's motion perpendicular to the galactic plane.- Nature 308, 709-712.
RAUP, D.M. (1979): Size of the Permo-Triassic bottleneck and its evolutionary implications.- Science 206, 217-218.
-- & SEPKOSKI, J.J., Jr. (1984): Periodicity of extinctions in the geologic past.- Proc. National Academy of Sci., U.S.A. 81, 801-805.
-- & SEPKOSKI, J.J., Jr. (1986): Periodic extinction of families and genera.- Science 231, 833-836.
ROSS, R.J., Jr. & NAESER, C.W. (1984): The Ordovician time scale -- new refinements.- in: BRUTON, D.L. (ed.): Aspects of the Ordovician System. Universitetsforlaget, Oslo.
SEPKOSKI, J.J., Jr.(1979): A kinetic model of Phanerozoic taxonomic diversity. II. Early Phanerozoic families and multiple equilibria.- Paleobiol. 5, 222-251.
-- (1984): A kinematic model of Phanerozoic taxonomic diversity. III. Post-Paleozoic families and mass extinctions.- Paleobiol. 10, 246-267.
-- (1986): An overview of Phanerozoic mass extinctions.- in: JABLONSKI, D. & RAUP, D.M. (eds.): Pattern and Process in the History of Life. Springer, Berlin.
-- & RAUP, D.M. (1986): Periodicity in marine extinction events.- in: ELLIOTT, D.K. (ed.): Dynamics of Extinction. Wiley, New York, 3-36.
STANLEY, S.M. (1984): Marine mass extinctions: a dominant role for temperature.- in: NITECKI, N.H. (ed.): Extinctions. Univ. of Chicago Press, Chicago, 69-117.
-- (1986): Anatomy of a regional mass extinction: Plio-Pleistocene decimation of the Western Atlantic bivalve fauna.- Palaios 1, 17-36.
WESTERMANN, G. (1984): Gauging the duration of stages: a new appraisal for the Jurassic.- Episodes 7, 26-28.
WHITMIRE, D.P. & MATESE, J. (1985): Periodic comet showers and Planet X.- Nature 313, 36-38.

CHEMICAL AND ISOTOPIC VARIATIONS IN THE WORLD OCEAN DURING PHANEROZOIC TIME

HOLSER, William T. *), MAGARITZ, Mordeckai **) &
WRIGHT, Judith ***)

Introduction

We ask the question as to whether there is any relation between major
extinction events and changes in ocean chemistry. One can imagine that
such a connection might be direct: mass extinction may be associated
with a dramatic drop in productivity in surface waters that will decrease
the fraction of organic to bicarbonate carbon, which would also lead to
changes in the cycles of nutrient elements as well as sulfur. Alternatively, a connection might be indirect: both mass extinction and changes
in the cycles of carbon and other elements might be common results of an
external factor, such as a major regression in sea level. We ask the
question, but as yet there are only partial answers, and many remaining
puzzles.

Evidence for secular changes -- both long-term trends and short
events -- of the composition of the world ocean has been reviewed by
Holser (1984) and Holland et al. (1986). After recapitulating some of
this background material, I will review some specific studies that have
recently been done (mainly in our research group) on trends and events
in the Paleozoic and early Mesozoic, with only passing mention of the
voluminous work done elsewhere on the Cretaceous and Tertiary and its
boundary.

Some general controls on ocean chemistry

Certain aspects of the world ocean are of particular importance to its
chemistry. It is a large reservoir that today is mixed horizontally by
currents on a time scale of hundreds of years, and vertically by the
downflow of cold dense polar waters on a scale of thousands of years.
The main input for most elements is eroded and carried by the world
rivers, although for some (e.g., Ca) mid-ocean ridge hydrothermal systems
are at least equally important. Their main outputs are into common sediments -- such as Ca and Sr into limey muds, or in sediments requiring
special geological circumstances -- the same pair of elements into eva-

---

*)   Department of Geology, University of Oregon, Eugene, U.S.A.

**)   Isotopes Department, Weizmann Institute, Rehovot, Israel.

***) Department of Geological Sciences, Arizona State University, Tempe,
      U.S.A.

Lecture Notes in Earth Sciences, Vol. 8
Global Bio-Events. Edited by O. Walliser
© Springer-Verlag Berlin Heidelberg 1986

porites. Those elements with residence times in the ocean that are long relative to mixing times (e.g., S, Sr) are both more homogenous and more difficult to modify than those with short residence times (e.g., C, rare earth elements (REE), trace metals). The vertical circulation and depth profile of many of these elements that have short residence times are controlled by their involvement with the biomass, particularly with primary productivity in the surface waters.

The relative inputs of organic carbon (Corg) and dissolved oxygen from this surface water to the underlying water masses and sediments may result in an anoxic or suboxic zone in the sediments or sometimes in the overlying water column. Thus extraordinary supplies of Corg that might have been generated by high productivity in surface waters may form black muds rich in Corg and sulfide. Alternatively, the supply of oxygen to bottom waters may be greatly reduced if the cold polar waters, that drive vertical circulation in today's ocean, are replaced by warm salty bottom waters generated by excess evaporation on shallow shelves or bordering basins at low latitudes. Shallow shelves are also conducive to both a high productivity of Corg and a high efficiency of its storage in the rapidly sedimenting shallow waters. This important parameter, the extent of shallow shelves in any geological period, is a function of eustatic (as well as local) rise of sea level, which in turn may be caused by an increase in mid-ocean ridge activity, melting of continental glaciers, or other geophysical processes that are less well understood. The exogenic chemical cycle is a complex fabric of interconnected chemical, biological and physical processes, including feedback loops that may either enhance or dampen an external forcing.

Changes from a steady state in this complex chemical system may be inferred from a variety of clues in the geological record. Mineralogical markers, such as the incidence of dolomite vs. calcite, or of aragonite vs. calcite öoids (Wilkinson et al. 1985) have suggested changes in atmospheric CO_2 pressure or cation ratios through time. Comparison of major element chemistry of fluid inclusions in halite crystals of marine evaporites have imposed limits of the variation of these elements since at least the late Paleozoic (Holland et al. 1986). Inventories through time of rock types, such as halite, anhydrite, black shales, and phosphorites tell us more than we have yet appreciated concerning the chemical history of the sea water from which they were deposited (Ronov 1980, Holser, Maynard & Cruikshank 1986). But some of the most useful indicators of change in the exogenic cycle are found by studying the ratios, sediments, of closely related chemical species: the rare-earth elements (REE), or isotopes of a particular element (S, C, Sr, Nd).

A chemical example: Rare Earth Elements

These elements follow one another very closely in their geochemistry.
They are taken up most readily in biogenic apatite, such as conodonts
or ichthyoliths, where in early diagenesis they record a "rare-earth
pattern" of the overlying sea water (Wright et al. 1984; Wright, Schrader & Holser 1986). A striking feature of this pattern in most modern
marine sediments is a glaring deficiency in the element cerium (Ce) when
compared with its REE neighbors lanthanum and neodymium, expressed as
Ce(anom) = log $\{3Ce(n)/[2La(n) + Nd(n)]\}$, where the n's indicate
that the analyzed values are normalized to standard shale levels. Thus
both the sea water and biogenic apatite in most of today's oceans exhibit a strongly negative Ce(anom) = -0.1 to -1.0, as shown in the top
panel of Figure 1. However regions where anoxic conditions are endemic,
as the Black Sea, have Ce(anom) near zero -- that is, all the REE including Ce are like those in average sediments. The Ce that is missing in
the deep-sea apatite is found adsorbed in fine authigenic metal hydrox -
ides -- ferric hydroxides coating shells and sedimentary grains, or in
manganese nodules. Apparently these carriers are re-dissolved in anoxic
seas, and Ce is re-circulated to remove the negative cerium anomaly.
Thus Ce(anom) is a measure of the oxidation/reduction system prevailing
in the surrounding seas.

We have measured these REE by neutron activation analysis in over
200 microsamples of biogenic apatite, mainly of Paleozoic conodonts. The
first result is a long-term trend (Fig. 1): in the early Paleozoic
Ce(anom) is close to zero, indicating a prevelance of anoxic conditions
in these oceans. Of course the samples of conodonts were actually taken
from shelf limestones, which are not a reducing environment, but this
only means that the Ce has been re-circulated into surface waters from
offshore anoxic deeps. Beginning in Devonian time we begin to see a
mixture of slightly negative Ce(anom), and in the Pennsylvanian and
Permian shelf samples we have Ce(anom) near -0,5 like today's well-
ventilated ocean. These relations confirm the biostratigraphic oceanographic model of Berry & Wilde (1978).

We don't yet have much good data to establish a trend for the Meso-
zoic/Cenozoic, but one set of detailed samples from the Solnhofen Basin
shows the negative Ce(anom) in the Fäulen layers, alternating with
Ce(anom) near zero in the Flinzen layers (Wright, Holser & Schrader
1986). Alternating reducing and oxidizing conditions are in accord with
current models for these unusual strata. In another detailed study
across the Cambrian/Ordovician boundary, Ce(anom) shows variations, even

Figure 1. Ce(anom) of REE in present sea water and biogenic apatite (upper panel), and in fossil apatite (lower histograms). Dashed line approximately separates oxidizing conditions on the left from reducing conditions on the right. After Wright, Schrader & Holser (1986).

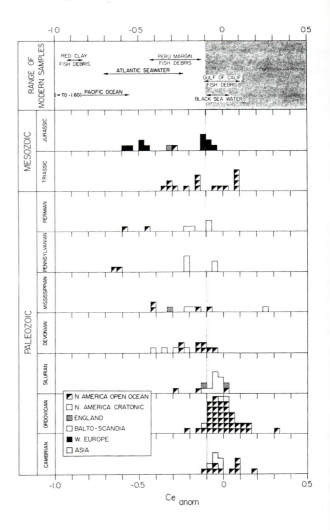

within the small range of +0.1 to −0.1 of Figure 1, that are consistent stratigraphically within a wide area of western North America, but different in China (Wright, Miller & Holser 1986). Thus this approach has possibilities for determining both trends and short-term variations in the oxidation system of sea water. We are now beginning a detailed study of sections in the Mississippian, and while the potential of this method is great, the costs in time and expense are large.

Isotopes

Ratios of isotopes of both stable and radiogenic species are one of the most informative markers of changes in the exogenic cycle. The devia-

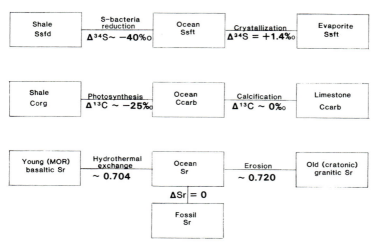

Figure 2.   Schematic representation of the relations of S, C, and Sr isotopes in the ocean.

tions from average crustal ratios are small; in order to more easily appreciate these deviations they are stated in the "del" notation in the cases of H, O, C, and S, where a del value is the deviation in parts per thousand from an international standard. There are many causes for changes in isotope ratios in sea water and its associated sediments: differing input sources (C, S, Sr, Nd), fractionations during evaporation (H, O) or deposition in sediments (O, C, S), temperature (O), specific biotic effects (C), or diagenesis in a milieu of differing isotope ratio (O). The main effects are laid out in Figure 2: for both C and S the major fractionation of isotopes occurs during biologically mediated reduction by photosynthesis or bacterial sulfur reduction. The reduced forms of carbon and sulfur, mainly in shales, are lighter isotopically by 20-25 ‰ and 35-40 ‰, respectively. The remaining carbonate and sulfate in sea water tend to be proportionally heavier, if the inputs have remained the same, and this change is recorded without much fractionation in the corresponding carbonate and sulfate sediments. Sr (as well as Nd) is a different story. Isotopes of these elements do not fractionate, so their isotope ratios in sea water are simple a balance of the input fluxes of heavier isotope -- Sr from old continental granites, or lighter isotopes -- Sr from young basalts like those on the mid-ocean ridges. S and Sr have such long residence times that they are probably well mixed in the ocean at most times, but Nd has such a short residence time that it should reflect more local inputs (Keto & Jacobsen 1985).

Carbon, although it has a substantial residence time in the ocean

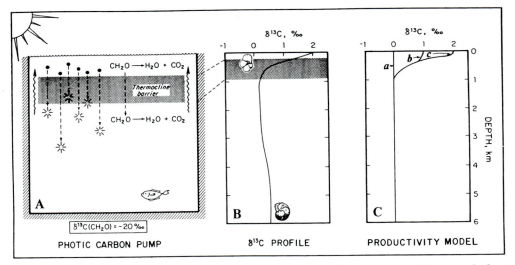

Figure 3. Fractionation of carbon isotopes between surface and deep waters, by downward transfer of Corg. (a) -- mechanism; (B) -- average profile of the present world ocean; (C) -- effect of decreasing productivity c>b>a on the carbon isotope gradient. After Berger & Vincent (1986).

as a whole, is presently differentiated into two reservoirs by intense fractionation that "pumps" Corg with light del 13C from a productive surface layer into deeper waters. This results in a gradient of del 13C of 1-2 ‰, between surface and deeps, which is a measure of this productivity (Fig. 3, Berger & Vincent 1986).

"Age curves" for isotope ratios, like those of Figure 4 (Holser 198 estimate variations through time of the mean isotope ratios in the mixe ocean. For long-term trends the curve represents a statistical series, which attempts to smooth out "noise" of various origins: incorrect age assignments, local fractionations, vital effects, and diagenesis. A pri mary criteria for confidence in any segment of a curve is the match found among separate basins worldwide.

The sulfur isotope age curve shows some of the best established long-term trends, which where recognized already more than 20 years ago The early Paleozoic exhibits high del 34S, indicative of a high net flux of sulfide from ocean to sediments; it drops to a minimum in the Permian, and returns to intermediate levels in the Mesozoic. Variations in carbon are less dramatic and tend to get lost in the noise. In the long term, variations in del 13C in carbonates tend to be inverse to those of del 34S in sulfates (Veizer et al. 1980); this was predicted o the theory that the oxygen of the atmosphere would only be maintained b a balance in these two oxidation/reduction systems. The mechanism that

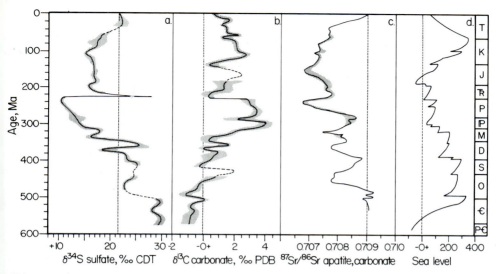

Figure 4. Age curves of S, C, and Sr in marine sediments, and Vail sea-level curve. Shading gives range of uncertainty; dashed lines, lack of data. After Holser (1984).

drives this balancing act is obscure, but may depend on the proportions of reduction that take place in continental regimes that are sulfur-poor to those that occur in sulfur-rich euxinic marine basins (Berner & Raiswell 1983).

Although the Sr isotope age curve is ostensibly generated by quite different proximate processes than are the curves for C and S, it's long-term trends resemble those of the S curve. This relation suggests some underlying common cause -- mid-ocean ridge activity has been suggested, but the system has not been convincingly modelled.

Isotopic events

The long-term trends of the isotope age curves are punctuated by sudden shifts. Although some of the sharp excursions in the S isotope curve were detected 10 years ago (Holser 1977), only in the last 5 years have important events in the C and Sr curves been revealed by very detailed profiles of limited time slices under maximum stratigraphic control. The close control that is available in the Cenozoic and Cretaceous from both microfossil and paleomagnetic stratigraphy has allowed the clear recognition of sharp changes of 1-2 ‰ in del 13C associated with the waning of Pleistocene glaciation, the Late Miocene recession of sea level, and earlier Miocene and mid-Cretaceous anoxic events (Arthur et al. 1985, Berger & Vincent 1986). Of particular interest is the very sharp but

short-lived negative excursion of del 13C associated with the iridium
anomaly at the Cretaceous-Tertiary boundary (Zachos & Arthur 1986).

Although stratigraphic control is less precise during the Paleozoic
some of the isotopic events are so large and extensive that they are
beginning to be well characterized. Some of these events are associated
with major biotic events.

Late Proterozoic time witnessed a prolonged high of del 13C > 5 ‰,
punctuated by episodic reversions to low values (Knoll et al. 1986). We
have mapped the final high in detail in a section on the Siberian platform, which shows a sharp decrease for 30 m across the Vendian/Tommotia
boundary (Fig. 5, Magaritz et al. 1986). The decline in biological productivity signalled by this drop continued into the lower Tommotian, wi
a temporary recovery shown by a moderate peak of del 13C in the middle
Tommotian. Two sections in China that are also designated as the Pre-
cambrian/Cambrian boundary, but which are probably not correlative with
our Siberian section, exhibit a sharp drop of del 13C associated with a
weak iridium anomaly. In the Chinese sections the del 13C anomaly is
mapped for only 20 m (~2000 yr) above the boundary. The scales of these
three investigations differ by orders of magnitude: 5000 m, 200 m, 1 m,
respectively.

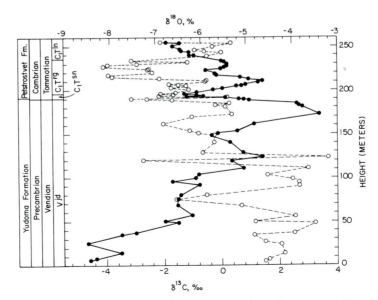

Figure 5.   Profiles of carbon (solid line) and oxygen (dashed line)
isotopes across the Vendian-Tommotian boundary on the Siberian Platforr
After Magaritz et al. (1986).

The Carboniferous was also a time of high average del 13C and
del 34S, as well as Sr isotope ratio (Fig. 4). We have measured one
section in New Mexico, which spans the late Carboniferous (~ Namurian-
Sakmarian), to compare an interval of background extinction with our
other profiles across major extinction events (Holser & Magaritz 1986a).
We were surprised to find that even here sharp excursions to low values
of del 13C occur; at least two of these correlate with regression/trans-
gressions, at the Morrowan-/Atokan and Atokan/Desmoinesian stage boundaries (in Westphalian A: Ross & Ross 1985). These anomalies remain to
be confirmed by del 13C profiles in distant basins.

We have investigated the Permian/Triassic transition worldwide; only
a very brief summary can be given here. Typical results are shown
schematically in Figure 6 (Holser & Magaritz 1986b). Del 34S is at its
minimum during Permian time, until it rises abruptly in the Lower Triassic (Skythian), falling again to moderate values in the Middle Triassic.
In carbon isotope profiles we find everywhere -- western USA, Zechstein
Basin, and in the waters of Tethys from the Alps to China -- a high of
del 13C of +3 to 7‰ that begins at the end of the Capitanian and
abruptly disappears just below the P/Tr boundary (Holser, Magaritz &
Clark 1986). Apparently this is the final manifestation of the high seen
in the late Carboniferous, although it has not been mapped in detail
through the early Permian. Strontium isotopes slide to a minimum in the
early Permian, and abruptly rise at the P/Tr boundary. We have also re-
evaluated sea-level changes during this interval. The regression,

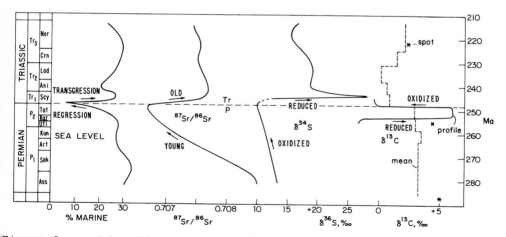

Figure 6.   Schematic representation of events across the Permian-
/Triassic boundary: sea level, strontium isotopes, sulfur isotopes, and
carbon isotopes (all in marine sediments). For carbon a stratigraphic
profile is shown solid, means of literature values dashed, and spot
values as stars. After Holser & Magaritz (1986b).

already well known to occur in the late Permian, seems to have happened
very fast, nearly 250 m in the last stage of the Permian, and the subsequent transgression was even faster in the first substage of the earl
Triassic.

Iridium has been detected in a few samples at the P/Tr boundary in
China (Sun et al. 1984, Xu et al. 1985), but there is some question as
whether these represent a significant anomaly (Asaro et al. 1983). Iridium was looked for but not found at the P/Tr boundary in the Transcauc
sus (Alekseev et al. 1983).

The isotope data adds complexity to our picture of Permian-Triassic
affairs, rather than furnishing a ready solution to the interrelations
of extinction, tectonics, salinity, and temperature that had previously
been suggested. What has emerged so far from this new analysis is that
both extinction and the sharp drop from earlier highs of del 13C may ha
had a common cause in a major regression of sea level. The reduced base
level also generated a pulse of heavy Sr isotopes eroded from the conti
nents. Concerning the Spathian high of del 34S, we can only speculate
that the same incident dissolved Permian salts to very effectively stra
tify some deeps in the early Triassic ocean, where excess sulfur reduction raised del 34S in the remaining marine sulfate.

Regression as a cause of a drop in del 13C is supported by the preliminary data from the Carboniferous, although the Carboniferous regres
sions, which may have been of glacial origin, were apparently not so

Figure 7.   Summary of extinction   events
and chemical events through the Phanerozoic.

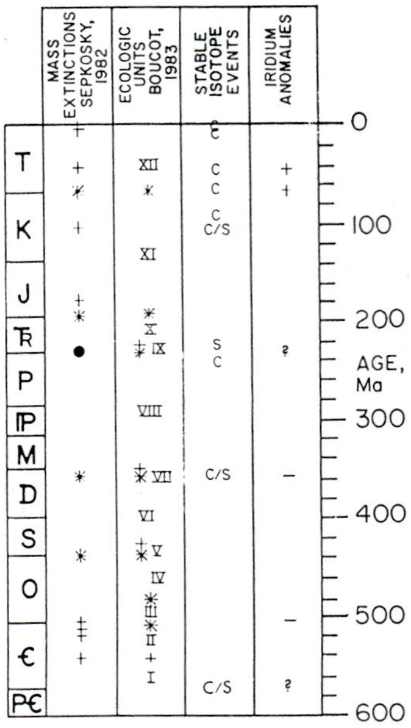

effective in extinguishing biota. Near the Precambrian-Cambrian boundary there was also a major regression (Brasier 1982), although this seems to have preceeded the final drop in del 13C. In the classical model regression would have decreased ecological niches, diversity, productivity, and burial of biota by reducing available shelf habitats. All of these work in the same direction to reduce previous highs of del 13C, and they were probably abetted by the re-oxidation of the excess Corg previously accumulated on shallow shelves or paralic basins. The relations among extinctions, regressions, drops in del 13C, and other chemical parameters, lasting some hundreds of thousands of years, are difficult to connect with an impact origin. Clearly much remains to be done in the geochemical investigation of mass extinction events.

Acknowledgements

The stratigraphic control and sampling that is an essential framework for this kind of geochemistry depends largely on a long list of biostratigraphers and paleontologists that have generously helped both us and other geochemists in attacking the problem of extinction events. We urge your continued collaboration. Supported by U.S. National Science Foundation Grants EAR 8115985, 8319429 and 8400222 to the University of Oregon.

SELECTED REFERENCES

ALEKSEEV, A.S.; BARSUKOVA, L.D.; KOLESOV, G.M.; NAZAROV, M.A. & GRIGO-RYAN, A.G. (1983): The Permian-Triassic boundary event: Geochemical investigation of the Transcaucasia section.- Lunar Planet. Sci.14, 7-8.
ARTHUR, M.A.; DEAN, W.E. & SCHLANGER, S.O. (1985): Variations in the global carbon cycle during the Cretaceous related to climate, volcanism, and changes in atmospheric CO_2.- Amer. Geophys. Un. Geophys. Monogr. 32, 504-530.
ASARO, F.; ALVAREZ, L.W.; ALVAREZ, W. & MICHEL, H.V. (1983): Geochemical anomalies near the Eocene/Oligocene and Permian/Triassic boundaries.- Geol. Soc. Amer. Spec. Pap. 190, 517-528.
BERGER, W.H. & VINCENT, E. (1986): Deep-sea carbonate: Reading the cargon-isotope signal.- Geol. Rdsch. 75, 249-269.
BERNER, R.A. & RAISWELL, R. (1983): Burial of organic carbon and pyrite sulfur in sediments of Phanerozoic time: A new theroy.- Geochim. Cosmochim. Acta 47, 855-862.
BERRY, W.B.N. & WILDE, P. (1978): Progressive ventilation of the oceans -- an explanation for the distribution of Lower Paleozoic black shales.- Amer. J. Sci. 278, 257-275.
BRASIER, M.D. (1982): Sea-level changes, facies changes and the Late Precambrian-Early Cambrian evolutionary explosion.- Precamb. Res. 17, 105-123.
HOLLAND, H.D.; LAZAR, B. & McCAFFREY, M. (1986): Evolution of the atmosphere and oceans.- Nature 320,27-33.

HOLSER, W.T. (1977): Catastrophic chemical events in the history of the ocean.- Nature 267, 403-406.
-- (1984): Gradual and abrupt shifts in ocean chemistry during Phanerozoic time.- in: HOLLAND, H.D. & TRENDALL, A.F. (eds.): Patterns of Change in Earth Evolution. Springer-Verl./Dahlem Konf., Berlin, 123-143.
-- & MAGARITZ, M. (1986a): A carbon isotope profile in marine Pennsylvanian carbonate rocks from New Mexico.- (in revision).
-- & MAGARITZ, M. (1986b): Events near the Permian-Triassic boundary.- (in prep.)
-- ; MAGARITZ, M. & CLARK, D.L. (1986): Carbon-isotope stratigraphic correlations in the Late Permian.- Amer. J. Sci. 286, 390-402.
-- ; MAYNARD, J.B. & CRUIKSHANK, K.M. (1986): Modelling the natural cycl of sulphur through geological time.- in: BRIMBLECOMBE, P. & LEIN, A.V. (eds.): Evolution of the Global Biogeochemical Sulphur Cycle. Wiley, New York (in press).
KETO, L.S. & JACOBSEN, S.B. (1985): The causes of $^{87}Sr/^{86}Sr$ variations in seawater of the past 750 million years.- Geol. Soc. Amer. Abstr. Progr. 17, 628.
KNOLL, A.H.; HAYES, J.M.; KAUFFMAN, A.J.; SWETT, K. & LAMBERT, I.B. (1986): Secular variation in carbon isotope ratios from Upper Protero/ zoic successions of Svalbard and East Greenland.- Nature (in revision).
MAGARITZ, M.; HOLSER, W.T. & KIRSCHVINK, J.L. (1986): Carbon-isotope events across the Precambrian/Cambrian boundary on the Siberian Platform.- Nature 320, 258-259.
RONOV, A.B. (1980): Osadochnaya Obolochka Zemli.- Nauka, Moscow, 80 p. (Transl. Inter. Geol. Rev. 24, 1313-1388 (1982)).
ROSS, C.A. & ROSS, J.R.P. (1985): Late Paleozoic depositional sequences are synchronous and worldwide.- Geology 13, 194-197.
SUN Yiyin; XU Daoyi; ZHANG Qinwen; YANG Zhengshong; SHENG Jinzhang; CHEN Chuzhen; RUI Lin; LIANG Xiluo; ZHAO Jiaming & HE Jiwen (1984): The discovery of iridium anomaly in the Permian-Triassic boundary clay in Changxing, Zhejiang, China and its significance.- Internat. Geol. Congr., 27th, Moscow, Abstr. 8, 309-310.
VEIZER, J.; HOLSER, W.T. & WILGUS, C.K. (1980): Correlation of $^{13}C/^{12}C$ and $^{34}S/^{32}S$ secular variations.- Geochim. Cosmochim. Acta 44, 579-587
WILKINSON, B.H.; OWEN, R.M. & CARROLL, A.R. (1985): Submarine hydrothermal weathering, global eustacy, and carbonate polymorphism in Phanero zoic marine oolites.- J. Sed. Petrol. 55, 171-183.
WRIGHT, J.; SEYMOUR, R.S. & SHAW, H.F. (1984): REE and Nd isotopes in conodont apatite: Variations with geological age and depositional environment.- Geol. Soc. Amer. Spec. Pap. 196, 325-340.
-- ; MILLER, J.F. & HOLSER, W.T. (1986): Conodont chemostratigraphy across the Cambrian-Ordovician boundary: Western United States and southeast China.- in: AUSTIN, R.L. (ed.): Conodonts: Investigative Techniques and Applications. Horwood, London (in press).
-- ; SCHRADER, H. & HOLSER, W.T. (1986): Variations in rare earth element distributions of Recent and fossil apatite and paleoredox of ancient oceans.- Geochim. Cosmochim. Acta (in revision).
ZACHOS, J.C. & ARTHUR, M.A. (1986): Paleoceanography of the Cretaceous/ Tertiary boundary event: Inferences from stable isotopic and other data.- Paleoceanography 1, 5-26.

THE ROLE OF OCEANOGRAPHIC FACTORS IN THE GENERATION OF GLOBAL BIO-EVENTS

WILDE, Pat & BERRY, William B.N. *)

A contribution to Project
GLOBAL BIO-EVENTS

Introduction

The oceanic environment although relatively stable with respect to salinity and volume over the past 600 million years is a dynamic system both physically and chemically. Variations over time in temperature, current patterns, water mass formation, and non-conservative element chemistry have occurred with depth and geographically. These physical-chemical changes have provided a background for biological evolution. The biological world has some interaction with chemical realm in the ocean as seen by the buffering of the chemical nutrients: phosphate, nitrate, and silica by phytoplanktonic organisms. Accordingly, it seems likely that oceanographic changes, particularly those that modify various marine habitats, could and would influence evolutionary development. In particular, episodic or long periodic oceanic events may be a factor in accelerating evolutionary changes and eliminating taxa that can not adapt to the change in the oceanic environment. The intent of this review is to identify various oceanographic phenomenon that have the potential to invoke sufficient global modifications to the ocean to effect biological change. It will be assumed that short term or local events such as tides, seasonal upwelling, tropical storms, or El Nino like events, because of their high frequency in terms of geologic time or evolutionary change provide noise to a basic environmental pattern to which organisms adapt without producing rapid evolutionary change.

Physical processes

The physical oceanographic processes are those that produce the mass properties of the ocean relating to the density and the flow of water at various depths and interactions between the atmosphere and the oceans. These include evaporation-precipitation (Salinity), water mass formation and flow, and mixing rates among water masses. In the present ocean, except at high latitudes where water masses are forming, the ocean is density-stratified into two major layers. The upper or mixed layer is homogenized by the wind to a depth of about 100 meters. The lower layer or deep water from about 1000 m to the bottom (average depth about 4000 m) forms at high latitudes. These two layers are separated by a

---

*) Marine Sciences Group, Department of Paleontology, University of California, Berkeley, California 94720, U.S.A.

Lecture Notes in Earth Sciences, Vol. 8
Global Bio-Events. Edited by O. Walliser
© Springer-Verlag Berlin Heidelberg 1986

Figure 1.   Temperature,
Salinity, Density Diagram.
Modern T-S curves after
Monin et al. (1977).

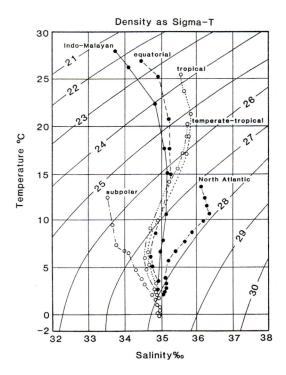

8604-003

Figure 2.   Global
Oceanic Volumes.
Data from Montgomery
(1958).

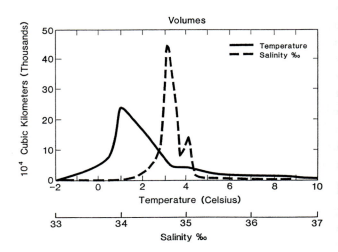

boundary layer in which temperature and density change rapidly known respectively as the thermocline and the pycnocline. A basic tool used by the physical oceanographer to study oceanic structure and mixing between various water masses is the T-S or Temperature-Salinity plot with contours of density as Sigma-T overprinted (Fig. 1). Thus on a T-S plot the variations in the pycnocline are exaggerated while the surface and deep layers plot as close clusters. The diagram is constructed so that when waters of different temperature and salinity are mixed, the resultant density of the mixture is found along a straight line connecting the two points proportional to the initial volumes of each. Fig. 2 shows the present distribution of volumes of water indicating the predominance of cold high latitude water and the uniformity of oceanic salinity.

Chemical processes

The chemical oceanographic processes are those that relate to the composition of the dissolved content of the ocean and the chemical interaction among sea water, the atmosphere and sea floor. The dissolved constituents can be classed (Fig. 3) as (1) conservative (composition related to salinity); (2) nutrient (composition buffered chiefly by biological activity); (3) non-nutrient gases (composition related to atmospheric content and solubility in sea water); and (4) non-conservative or not sufficiently studied to be assigned.

Processes that affect the nutrient composition of sea water obviously would influence biological processes through interactions within marine food webs. Such processes include introduction of dissolved material into the oceans by rivers, volcanoes, diagenesis; removal of material by precipitation, replacement, and sedimentation; modification of the chemical environment such as pH or redox potential. The major transfer mechanism, based on the bulk composition of marine biomass, between the dissolved content of sea water and the marine biological world is through the phytoplankton. Nutrient and nutrient related elements may be modelled by use of an ideal mole of a phytoplankton, representing marine primary productivity (modified from Redfield et al. (1963) and Richards (1963)):

$$[(CH_2O)_{106}(NH3)_{16}H_3PO_4] \{Ni, Zn, Se, Cd, Be, Cu, Ni, Rn, Ra\}_n \text{ to pmoles}$$

In the photic zone, these elements are bound in the living organism in their proper proportion.

In oxic waters, upon death, oxidative decay matter releases the elements back into sea water as the organic particle sinks through the water column:

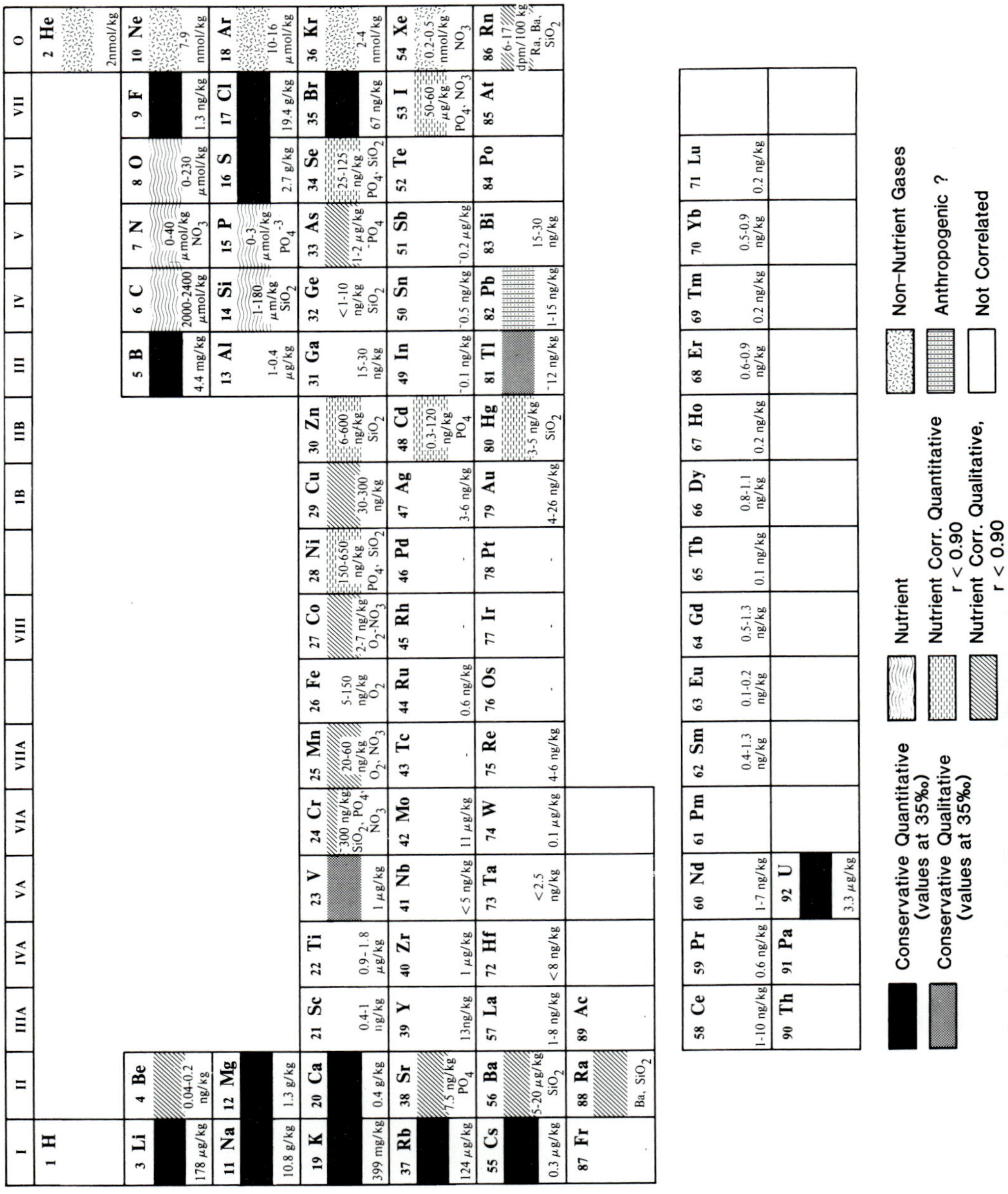

Figure 3. Periodic table of the dissolved elements in sea-water. Dat[a] from Quinby-Hunt & Turekian (1983).

$$\text{Mole Phytoplankton} + 138\ O_2 \rightarrow$$
$$106\ CO_2 + 132\ H_2O + HNO_3 + H_3PO_4 + \text{Trace Metals}$$

Thus, dissolved nutrients and nutrient-related metals are at a minimum in the photic zone, rapidly increase to the base of the pycnocline, then remain constant in deep water (Fig. 4).

In anoxic waters, initially, oxygen is replaced by nitrate and nitrite as the oxidant with ammonia being the reduced nitrogen end product:

$$\text{Mole Phytoplankton} + 53\ NO_2^- \rightarrow$$
$$106\ CO_2 + 16\ NH_3 + 53\ NH_4 + H_3PO_4 + \text{Trace Metals}$$

Trace metals would build up in such waters until sufficient ammonia as ammonium ion was available to produce soluble metal amine:

$$NH_4^+ + \text{Trace Metals} \rightarrow \text{Metal Amines}$$

As the concentration of ammonia in this zone far exceeds the concentrations of the dissolved trace metals, most of the metals would be complexed.

As anoxicity increases upon consumption of nitrate and nitrite, sulfate becomes the thermodynamically favored oxidant:

$$\text{Mole Phytoplankton} + 53\ SO_4^{2-} \rightarrow$$
$$106\ HCO_3^- + 53\ H_2S + 16\ NH_3 + H_3PO_4 + \text{Trace Metals}$$

However, dissociated H_2S would react with many of the released metals and metal amine complexes to form particulate sulfides; thus depleting the water column of dissolved metals:

$$\text{Trace metals (Ni,Zn,Cu,Cd,Hg, etc.)} + S^{2-} \rightarrow \text{Metal Sulfides} \downarrow$$

The decay of phytoplankton sinking into anoxic waters would proceed by reduction of nitrogen compounds to ammonia, then reduction of sulfate to sulfide with the precipitation of trace metals as sulfides. Thus contrary to the oxic situation, dissolved metals would decrease or remain constant with depth in anoxic waters rich in sulfides as seen in the modern Black Sea (Fig. 5).

Physical-chemical interactions

With the density and chemical stratification of the ocean generally in a steady state, any significant effect on marine biota must be caused by transfer of water from depth into the photic zone. Here, in the upper 50 to 100 meters, changes in species composition and phytoplankton productivity could occur. The process of vertical advection of water is given the general name of upwelling and occurs at various rates, at various locations, and from various depths.

Figure 4a.   Oxic Conditions
in modern open ocean: Nutrients
after Bainbridge (1979) and
Quinby-Hunt et al. (1981).

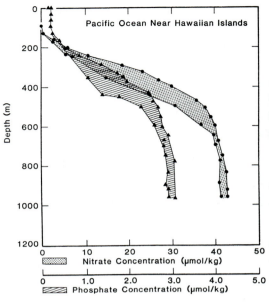

8604-002

Figure 4b.   Oxic
Conditions in modern
open ocean: Trace
metals data after
Bruland (1980).

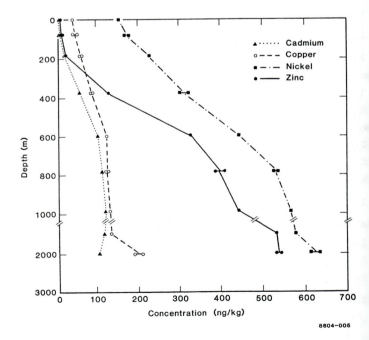

8604-005

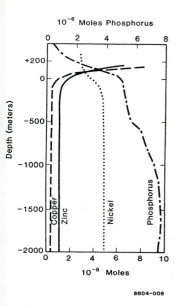

Figure 5.   Anoxic Conditions in the Black Sea. Phosphorous after Fonselius (1974). Trace Metals after Brewer and Spenser (1974).

8604-008

Types of upwelling

Three main types of upwelling may be identified functioning with various time scales:

I. Planetary

A. divergence of water by Ekman transport caused by winds. Effective depth relatively shallow as a function of the wind stress. Vertical rise: 10 to 80 meters/month (Wooster & Reed 1963).

B. displacement by continual renewal of water masses at source. Slow: millimeters/day. Present resident time is about 1000 years (Broecker & Peng 1982, p. 20) or one mixing cycle.

C. overturn of deep water to the surface (Wilde & Berry 1984). Rate unknown but presumably much less than one modern mixing cycle.

II. Regional

A. seasonal Ekman (high pressure off West Coast). Characteristics similar to planetary Ekman upwelling (Wooster & Reid 1963).

B. off-shore advection by current moving off shore (Peru-Chile Current near the equator).

III. Local

A. obstruction of current by a seamount producing Taylor columns (Owens & Hogg 1980) and uplift through continuity.

B. closed eddy circulation producing low pressure (Robinson 1983).

C. Bernouli uplift through constricted straits (Stommel et al. 1983).
D. breaking internal waves in shallow water (Hall & Pao 1971).

Duration of upwelling

The injection of chemically different water into the surface layer woul
have affects related to volume of water, depth of the source, and duration of the upwelling event. Fig. 6 indicates the volumes by latitude
for modern coastal upwelling. The volumes generated by overturn would
be significantly greater than those produced in coastal upwelling.

Changes in density stratification

Destabilization of the oceanic water column which would bring deep-wate
into the photic zone has the greatest potential for inducing rapid chan
ges in the chemistry of the upper ocean. Such changes would be a function of chemistry of the upwelled water in relation to the chemistry of
the oxidized surface layers. Modification of the chemistry of waters in
the surface layer has the potential for instituting evolutionary change
through (1) destroying or restricting the life range of existing taxa,
(2) opening of new niches or expanding of formerly restricted niches, (
opening or restricting food sources to neutrally affected taxa.

Generation of overturn

Table 1 shows the stability of the ocean for various climates demonstra
ting that glacial climates produces the most stabile conditions and the
thinnest pycnocline. Fig. 7 gives the energetics for displacement of
water from the base of the pycnocline to the photic zone for a transition from warm to cooler climates. Overturn can be initiated by massive
displacement by some physical event or when the oceans are at low stability by normal geophysical processes or by a change in the major sour
of water.

As shown in Fig. 8, an obvious occurrence of overturn can be cause
by climatic change where the source of the densest deep water shifts ge
graphically. During non-glacial climates where the temperature at high
latitudes does not go below 5° C (HLWa), the densest water is formed a
the cold edge of the salinity maximum (SMW) in mid-latitudes. Thus, th
relatively warm but salty water forms the bottom water overlain by colder but less saline high latitude water. At the onset of a cold period
where the high latitude water temperatures fall below 5° C, this wate
(HLWb) becomes denser than water from the salinity maximum, displacing

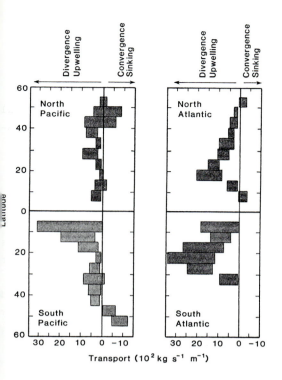

Figure 6.   Coastal Upwelling Volumes-Eastern Atlantic Ocean – after Wooster & Reid (1963).

8604-006

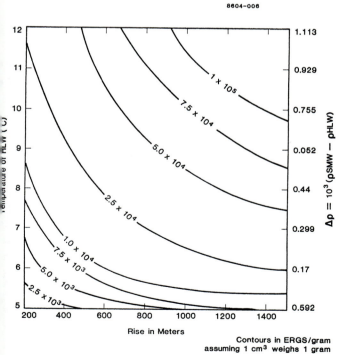

Contours in ERGS/gram assuming 1 cm^3 weighs 1 gram

Figure 7.   Potential Energy of Displacement for Overturn Conditions preceding cold climates. From Wilde & Berry (1984).

UNIT	Non-Glacial	CLIMATE Pre- or Postglacial	Glacial (Modern)
Z (thickmess m)	2900	1400	900
$10^8 E$ (m^{-1})	43	84	523
$10^4 N/2\pi$ (Hz)	3	4.5	11
2 /N X 60 (min.)	51	36.8	14.8
$10^4 \mu$ (sec.$^{-1}$)	6	8.9	21

Table I.   Stability in main pycnocline for various Climates - after Wilde & Berry (184, p. 148). Z = model thickness of pycnocline - after Wilde & Berry (1982); E = static stability (Hesselberg & Sverdrup 1915 N = buoyancy frequency (Gill 1982, p. 51-52); μ = vertical shear (Munk 1966, p. 710). Stability increases with higher values.

Figure 8.   Global Oceano-graphic Model as a function of climate - modified after Wilde & Berry (1984). Dashed lines encompass major water mass sources.

EQW = Equatorial Water
STW = Sub Tropical Water
SMW = Salinity Maximum Water
SSW = Shelf Sea Water
MW = Metiderranean Water
     (model for high sali-nity shelf sea)
HLW = High Latitude Water

Climates: Warm Climate (light pattern) - high latitude (HLWa) sea tem-peratures greater than 5° C and SMW densest and deep water. Pre or Post Glacial Climate (light pattern and intermediate pattern) - high latitude (HLWb) sea temperature less than 5°C but no sig-nificant sea ice formation.

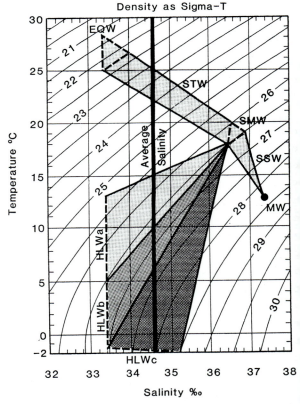

8604-004

that water in an overturn. If the cold trend continues, eventually, sea ice will form in sufficient quantities to produce even denser deep water (HLWc) as seen in the modern ocean. However no overturn is likely due to the increasing stability of the water column. Upon return to warmer climates the potential for overturn would occur as the oceans at high latitudes warms above 5° C and mid-latitude water again becomes denser than high latitude water.

Due to the change in the geographic source of deep water at the conditions of near zero stability required for an oceanic overturn, the sequence of overturn would depend on the direction of climatic change. For an overturn caused by the onset of cold climates, the high latitude source displaces the mid-latitude source. Accordingly, the overturn would be generated from high latitudes toward the equator. During the overturn, the tropical and subtropical surface would be only slightly colder as the deep water during warm climates was formed at mid-latitudes. For an overturn caused by the cessation of cold climates, the direction of the overturn would be from mid-latitude to both high latitudes and low latitudes. The major thermal effect would be in the tropics due to the upwelling of cold deep water. In either case, areas such as the eastern boundaries of oceans, which have significant wind produced coastal upwellings, would experience the effects of overturn first due to the enhanced vertical advection.

In general, an overturn is a special case of the usual displacement of less dense water upward by water of greater density. The global extent of a particular overturn depends on the volume and geographic locations of the various sources of the water masses. For the simple model discussed here, the density, and thus the volume and rate of formation of high latitude sources, varies as a function of climate. However, the convergence or downsinking can occur elsewhere. At such locations, as at the mid-latitude salinity maximum, rising overturn water will be suppressed or diluted with the downsinking water. Thus, deep water from an overturn will mix into the surface principally where divergence already occurs or where prior water mass formation has ceased or has become volumetrically insignificant. At convergences or other areas where water masses are forming, refuges would be found where the biota would be relatively unaffected by the overturn. Such areas could be "homes" for taxa that later repopulate areas or divergence decimated during the overturn.

Types of overturn

The effect on biota in the photic zone of such overturns would be a function of the chemistry of the deep water displaced to the surface.

Oxic overturn: In present well-oxygenated water, nutrients and trace metals increase with depth. In general, upwelling of additional nutrients causes favorable conditions for increased photosynthesis if the upwelling occurs when sufficient light is available. This is less a problem in the tropics. However, upwelling of trace metals may have toxic or inhibitory effects. Barber et al. (1971) noted that in the Peru-Chile area, initial rise of nutrients in surface waters is not immediately followed by increased productivity. The lag, they thought, was due to the fact that the deep water was not properly conditioned for use by photosynthetic organisms. Also, studies on the viability of deep water for the Ocean Thermal Energy Conversion programs (Terry & Caperon 1982) showed that deep water was initially inhibitory to increased photosynthesis. Experimental addition of metal chelators to upwelled water has increased productivity (see discussion in Provasoli 1963). Secondary effects on other organisms during upwelling have been found with "red tide" conditions. Dinoflagellates such as <u>Gonyaulax</u> and other algae produce potent neurotoxins which if ingested cause mass mortality in fish and in other higher organisms (Provasoli 1963). Such mass mortalities occur at the peak of the bloom suggesting a threshold level of nutrient enrichment before toxic effects occur. The major effect of an oxic overturn, which would occur during a transition from a glacial climate to a warm climate, potentially would reduce primary productivity initially due to the lack of available metals. If the rate of overturn was slow enough to permit chelation then the additional available nutrients could produce a toxic red-tide type bloom. This could have catastrophic effects on non-photosynthesizing taxa. As this water is cold and well oxygenated there may be some thermal effects, but there would not be any significant effect on oxygen-breathing organisms.

Anoxic overturn: Inhibitory responses by surface to near-surface dwelling organisms increase with the depth of the source of the upwelled water in the modern well-oxygenated ocean. However, if the water became anoxic with depth the effect on such organisms would differ as a function of how the redox potential would change the chemical equilibria. Wilde et al. (1984) and Wilde (1986) have proposed a chemical zonation of anoxic waters based on the sequence of dominant dissolved oxidants (Fig. 9). Benthic organisms would be effected most as the invasion of anoxic water of various compositions would change the composition of the water overlying their habitat. In simple upwelling or displacement, each zone shifts upward at the rate of upwelling. During relatively slow upwelling the chemical species may have a chance to convert to the local redox equilibrium composition during the rise. However, during overturn the

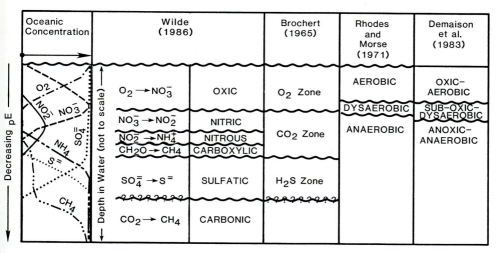

Figure 9.   Oceanic Redox Zones - after Wilde (1986).

whole pycnocline and surface layers are mixed rapidly. So there is time before a new equilibrium is established when, as a function of kinetics, the various ions in various oxidation states are present in the surface layer. This assumes that the rate of mixing is faster than the rate of chemical kinetics. Depending on the mixing and kinetics of reactions with oxidized waters, the various ions present in the source zone will reach the surface or into the photic zone unchanged or without being oxidized to higher states. The Wilde (1986) model predicts the following influx into surface layers as a function of anoxic zonation in the overturned pycnocline.

I.   Nitric to Oxic

         nitrate, high trace metals

II. Nitrous to Oxic

         nitrite, ammonia, ammonia trace metal complexes

III. Carboxylic to Oxic

         ammonia, ammonia trace metal complexes

IV. Sulfatic to Oxic

         sulfide, ammonia, no free trace metals

    In general, the productivity of photosynthetic organisms such as the green algae and the blue greens would be enhanced by upwelling from all zones above the Sulfatic due to their preference for reduced nitrogen compounds (Eppley et al. 1969). Simple non-sulfidic anoxic waters would not hurt such groups. The effect or inhibition caused by the metals and metal complexes would be diminished in anoxic overturn as their concentration would be lowered by precipitation as sulfides. Such enhanced

productivity may create vast areas of "red tide" conditions that are
extremely toxic for higher organisms consuming such algae. For respirin
organisms, any anoxic water would be toxic and anoxic overturn would ha
catastrophic effects on such groups.

Bio-events and overturns

The chemistry of the water column would be changed either by upwelling
or overturn from the various redox zones. The intrusion of these zones
onto the shelf at high stands of sea level would effect not only the
organisms in the water column but also the benthos and their living
space. Both epifauna and infauna would be restricted to bottoms with
overlying oxygenated waters. Fig. 10 shows the postulated status of
redox conditions since the Cambrian with intervals and type of possible
overturn noted. With the ventilation of the oceans by the end of the
Permo-Carboniferous glaciation, anoxic overturns became less likely.
The evidence for extinction events is derived from organisms that are
fossilized, generally ones with hard parts. These organisms are several
trophic levels above phytoplankton, which leave little or no fossil
record. Taxa in higher trophic levels were not necessarily affected
directly by the overturn. Motile forms such as nekton could avoid modified water. The demise of mobile species may have been caused indirectl
by changes in food supply, which may be directly influenced by changes
in the chemistry of sea water. The toxicities associated with a rapid
increase in productivity as "red tide" conditions would affect all high
organisms attracted to the bloom organisms if the toxin is released int
the water. It would seem that organisms with a short food web, that are
common in upwelling areas, would be affected most. Such groups also
would not be adapted to lower productivity generated by rapid oxic over
turn. During any extinction bio-event most taxa survive, so there must
be a refuge for survivors if the event is the result of an overturn. Op
ocean or near-shore groups living in convergence (downsinking) areas
generally have longer food webs due to generally lower primary product:
vity there. These groups may radiate into the upwelling areas after an
overturn. The following characteristics typifies generic groups with
respect to their ability to survive an overturn.

Most affected	Least affected
A. Eutrophic	A. Oligotrogphic
B. Benthic life stages	B. Euplanktonic
C. Benthic feeders	C. Near shore < 50 m depth
D. Short food web	D. Long or complex food web
E. Oxygen respirers/infauna	E. Anaerobes
F. Migratory in water column	

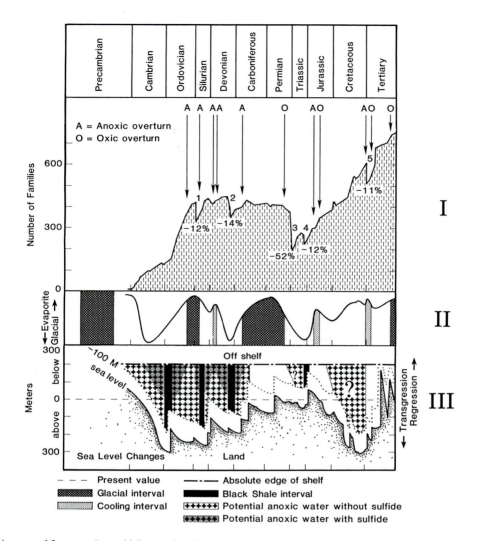

Figure 10.   Possible Relationship among Overturns, Mass Extinctions, Climate, and Shelf Conditions - modified from Wilde & Berry (1984). Row I: Distribution of Families after Raup & Sepkoski (1982). Overturns paired at onset and retreat of cold intervals. Absolute time of crossing of 5° C isoterm at high latitudes is not known.
Row II: Glaciation (dark pattern) and cold intervals (light pattern) after Frakes (1979). Climatic curve after Meyerhoff (1970).
Row III: Redox conditions on the Shelf. Sea level curve after Vail et al. (1977).

The causes of such complex events as mass extinctions, faunal replacements, and other bio-events in the geologic record also are likely to be complex and varied. Major oceanographic events such as overturns and modification of the chemistry of sea water are just some

of the probable contributory causes for changes in the pace of evolution, which should be factored into a complete investigation of any
particular biological event.

Acknowledgements

The authors wish to thank M.S. Quinby-Hunt for reviewing the manuscript
and offering many useful suggestions. M.A. Krup did her usual excellent
job in preparing and designing the manuscript and executing the illustra
tions. Partial support was provided by the Institute of Geology and Geophysics, University of California. This is contribution MSG-86-008 of th
Marine Sciences Group, University of California, Berkeley.

SELECTED REFERENCES

BAINBRIDGE, A.E. (1979): GEOSECS Pacific final hydrographic data report,
    22 August 1973 to 1 June 1974, R/V Melville: Geosecs Operations Group
    Publ. 32, np.
BARBER, R.T.; DUGDALE, R.C., MacISAAC, J.J. & SMITH, R.L. (1971): Vari-
    ations in phytoplankton growth associated with the source and condi-
    tioning of upwelling water: Investigacion Pesquera 35, 171-193.
BORCHERT, H. (1965): Formation of marine sedimentary iron ores.- in:
    RILEY, J.P. & SKIRROW, G. (eds.): Chemical Oceanography. Acad. Press
    London 2, 159-204.
BREWER, P.G. & SPENCER, D.W. (1974): Distribution of some trace metals
    in the Black Sea.- in: DEGENS, E.T. & ROSS, D. (eds.): The Black Sea
    - Geology, Chemistry, and Biology. Amer. Assoc. Petroleum Geologists
    Mem. 20, 137-143.
BROECKER, W.S. & PENG, T.-H. (1982): Tracers in the Sea.-Eldgio Press,
    689 p.
BRULAND, K.W. (1980): Oceanographic distributions of cadmium, zinc,
    nickel and copper in the north Pacific.- Earth and Planetary Sci.
    Letters 47, 176-198.
BRYDEN, H.L. & STOMMEL, H.M. (1984): Limiting processes that determine
    basic features of the circulation in the Mediterranean Sea.- Oceano-
    logica Acta 7, 289-296.
DEMAISON, G.J. & MOORE, G.T. (1980): Anoxic environments and oil source
    bed genesis.- Amer. Assoc. Petroleum Geologists Bull. 64, 1179-1209.
EPPLEY, R.W.; ROGERS, J.N. & McCARTHY, J.J. (1969): Half saturation
    constants for uptake of nitrate and ammonium by phytoplankton.- J.
    Phycology 5, 333-340.
FONSELIUS, S.H. (1974): Phosphorous in Black Sea.- in: DEGENS, E.T. &
    ROSS, D. (eds.): The Black Sea - Geology, Chemistry, and Biology.
    Amer. Assoc. Petroleum Geologists Mem. 20, 144-150.
FRAKES, L.A. (1979): Climates Throughout Geologic Time.- Elsevier,
    Amsterdam, 310 p.
GILL, A.E. (1982): Atmosphere-Ocean Dynamics.- Acad. Press, London, 662
HALL, M.J. & PAO, Y.-H. (1971): Internal Wave breaking in a Two-Fluid
    system.- Boeing Sci. Res. Lab. Doc. DI-82-1076, 141 p.
HESSELBERG, T. & SVERDRUP, H.V. (1915): Die Stabilitäts-Verhältnisse de
    Seewassers bei vertikalen Verschiebungen.- Berg. Mus. Arb. 15, 16 p
MEYERHOFF, A.A. (1970): Continental Drift, II. High-latitude evaporite
    deposits and geologic history of Arctic and North Atlantic Oceans.-
    J. Geol. 78, 406-444.

MONIN, A.S.; KAMENKOVICH, V.M. & KORT, V.G. (1977): Variability of the
    Oceans (translated from Russian).- Wiley, New York, 241 p.
MONTGOMERY, R.B. (1958): Water characteristics of the Atlantic and World
    Ocean.- Deep-Sea Res. 5, 134-148.
MUNK, W. (1966): Abyssal Recipes.- Deep-Sea Res. 13, 701-730.
OWENS, W.B. & HOGG, N.G. (1980): Oceanic observations of stratified Tay-
    lor columns near a bump.- Deep-Sea Res. 27, 1029-1045.
PROVASOLI, I. (1963): Organic regulation of phytoplankton fertility.-
    in: HILL, N.H. (ed.): The Seas. 2, Wiley-Intersci., New York, 165-219.
QUINBY-HUNT, M.S.; FANNING, K.; ZIEMAN; D.; WALSH, T.W. & KNAUER, G.A.
    (1981): Nutrient and Dissolved Oxygen Studies at OTEC sites.- in:
    Proc. 8th Ocean Energy Conf.: Marine Technol. Soc., Washington, D.C.,
    537-545.
-- & TUREKIAN, K.K. (1983): Distribution of the Elements in Sea-Water.-
    EOS 64, 130-131.
RAUP, D.M. & SEPKOSKI, J.J. (1982): Mass extinctions in the marine fossil
    record.- Science 215, 1501-1503.
REDFIELD, A.C.; KETCHUM, B.H. & RICHARDS, F.A. (1963): The influence of
    organisms on the composition of sea-water.- in: HILL, N.H. (ed.):
    The Seas. 2, Wiley-Intersci., New York, 26-77.
RHODES, D.C. & MORSE, J.W. (1971): Evolutionary and ecologic significance
    of oxygen-deficient marine basins.- Lethaia 4, 413-428.
RICHARDS, F.A. (1965): Anoxic basins and fjords.- in: RILEY, J.P. &
    SKIRROW, G. (eds.): Chemical Oceanography. Acad. Press., London 1,
    611-645.
ROBINSON, A.R. (ed.) (1983): Eddies in Marine Science.- Springer-Verl.,
    Berlin, 609 p.
STOMMEL, H.; BRYDEN, H. & MANGLESDORF, P. (1973): Does the Mediterranean
    outflow come from great depth?- Pure and Applied Geophys. 105,874-889.
TERRY, K.L. & CAPERON, J. (1982): Phytoplankton assimilation of carbon,
    nitrogen, and phosphorous in response to enrichments with deep-ocean
    water.- Deep-Sea Res. 29, 1251-1258.
VAIL, P.R.; MITCHUM, R.M., Jr. & THOMPSON, S. III (1977): Seismic strati-
    graphy and global changes of sea level, part 4: Global cycles of
    relative changes of sea level.- in: STANLEY, D.J. & KELLING; G. (eds.)
    Seismic Stratigraphy - Applications to Hydrocarbon Exploration. Amer.
    Assoc. Petroleum Geologists Mem. 26, 83-97.
WILDE, P. (1986): Model of Redox Zonation in the Late Precambrian-Early
    Paleozoic Ocean: submitted to Amer. J. Sci.
-- & BERRY, W.B.N. (1982): Progressive Ventilation of the Oceans-Poten-
    tial for Return to Anoxic Conditions in the Post-Paleozoic.- in:
    SCHLANGER, S.O. & CITA, M.B. (eds.): Nature and Origin of Cretaceous
    Carbon-Rich Facies. New York, Acad. Press, 209-224.
-- & BERRY, W.B.N. (1984): Destabilization of the oceanic Density struc-
    ture and its significance to Marine "Extinction" events.- Palaeogeo-
    gr., Palaeoclim., Palaeoecol. 48, 143-162.
WOOSTER, W.A. & REID, J.L. (1963): Eastern Boundary Currents.- in: HILL,
    M.N. (ed.): The Seas. 2, Wiley-Intersci., 253-280.

PRECAMBRIAN TO LOWER CAMBRIAN

EVOLUTIONARY CHANGES IN THE PROTEROZOIC

PFLUG, Hans D. & REITZ, Erhard *)

A review is presented of the currently available evidence of life in the Proterozoic, with special reference to the interrelationships between microbial activity and the first appearance of metazoans. Fossil organic microstructures can be detected in thin sections of the rock under the light microscope and examined in demineralized sections of the rock under the transmission electron microscope (TEM). They are chemically analyzed in microprobes and spectrophotometer microscopes. On the basis of such studies, the interaction of microorganisms with the formation of minerals in cherts, carbonates and iron sediments can be traced back to Archean times. Filamentous and coccoid cyanobacteria have the best chance of becoming lithified and structurally preserved in the sediments.

Microbiota

All known organisms utilize metal organic complexes in their cell organization. These metals may be captured in intracellular traps such as metallo-thionines or in extracellular macromolecular networks such as the cell wall and sheath. Certain microbes can immobilize large quantities of metals at their cell surface by forming aggregates of insoluble metal complexes. Such adsorption occurs when the positively charged ions are attracted to negatively charged ligands and biopolymers. After deposition and burial of the dead cell, the metals may react with H_2S produced by sulfate-reducing bacteria and form metallic sulfides. Such metal enrichment is of special interest because it leads to the formation of sedimentary ore deposits. A worldwide correlation between the geological record of these ore deposits and the evolutionary history of microbes is indicated.

The interaction of microbes with the formation of stromatolites and ore deposits, can be demonstrated on well preserved occurrences. Several of the facies are evidently time dependent and bound to periods that coincide with major steps in the evolution of the terrestrial atmosphere and biosphere (Fig. 1). The stratiform gold and uranium ores of the lower Proterozoic (2400 - 2800 MY) indicate anoxic conditions during deposition. Bacteria are apparently involved in the formation of the facies. The banded iron formations occur most often in the subsequent time span 2300 - 1700 MY ago. Structures resembling spherical or filamentous

*) Geologisch-Paläontologisches Institut, Universität Gießen, D-6300 Gießen, F.R.G.

Lecture Notes in Earth Sciences, Vol. 8
Global Bio-Events. Edited by O. Walliser
© Springer-Verlag Berlin Heidelberg 1986

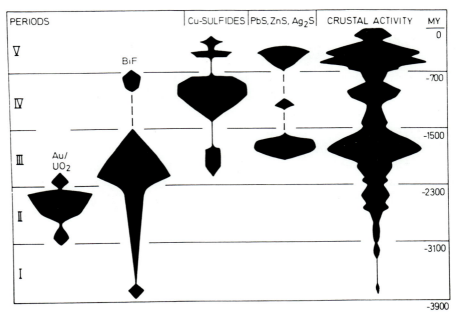

Figure 1. Chronological distribution of stratiform ore deposites and crustal activity. Ordinate: time in million years (MY); stages I-V: suggested increase of atmospherical oxygen (partly after Meyer 1985).

Plate 1. Microfossils of the Proterozoic and Cambrian

Figs. 1-12. Microfossils from the Gunflint Banded Iron Formation, Schreiber locality, Ontario. (1-5) <u>Gunflintia minuta</u> Barghoorn (Cyano-bacteria); (6-9) <u>Huronispora microreticulata</u> Barghoorn (Cyanobacteria); (10) cf. <u>Bavlinella</u> sp. Shepeleva (Cyanobacteria); (11) <u>Eomicrhystridium</u> sp. Deflandre (Algae); (12) <u>Eomicrhystridium barghoorni</u> Deflandre (Al-gae). (Bar 10 µm, see Fig. 12)

Figs. 13-15. Sphaeromorph acritarchs (Prasinophyta) from the middle Belt Super group (Montana/Idaho). (13) cluster of <u>Leiosphaerida</u> specimens; (14) <u>Nucellosphaeridium</u> sp. Timofeev; (15) <u>Montanella beltensis</u> Pflug. (Bar 10 µm, see Fig. 14)

Figs. 16-18. Middle Cambrian acanthomorph acritarchs from North-Eastern Bavaria, Germany. (Bar 10 µm, see Fig. 18)

Figs. 19-23. Cyanobacteria (cf. <u>Nostoc</u>) from cherts of the Spilitic Group, Bohemian Upper Proterozoic. (Bar 10 µm, see Figs. 22, 23)

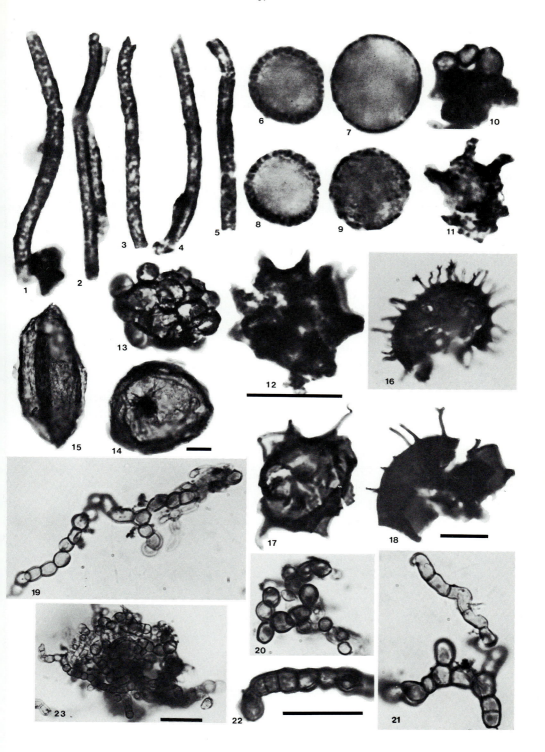

cyanobacteria are most common in Precambrian BIF (Pl. 1, Figs. 1-10). There are also many reports on the occurrence of iron precipitating bacteria, such as the <u>Metallogenium</u> type and others (Cloud 1984).

The main time for stratiform cooper formation coincides with that of the early diversification of the eukaryotic phytoplankton, 1100 to 650 MY ago (period IV in Figs. 1, 3). At present, most copper enzymes are found in eukaryotes. Only a few copper proteins, such as azurin and plastocyanins are also found in certain aerobic prokaryotes requiring an oxygen-rich atmosphere. In fact, redbeds and other fully oxidized sediments became widespread in period IV. At the same time, prokaryotes with a more elaborated morphology make their appearance in stromatolites (Fig. 3). A few red beds are known from the preceeding period III, which was dominated by the banded iron formation. They indicate that the atmosphere became increasingly oxygenated from that point onwards. Copper deposits can be thus interpreted as a marker of the younger Proterozoic 1200 - 650 MY ago as far as their biological significance is concerned. At this time, the first diversification of eukaryotic phytoplankton occurs. It is suggested that the evolution of eukaryotic algae was accompanied by an increase in copper, coupled with a decrease in iron in the cell (Ochiai 1983).

Oldest acritarches have been recently reported from the ca. 2000-MY-old Gunflint Formation of Ontario (Pl. 1, Figs. 11-12). About a dozen specimens belonging to four different morphological types ("form species") have been identified hitherto. The specimens occur in clear portions of a stromatolitic chert together with <u>Huronispora</u> sp. and <u>Gunflintia</u> sp., two common representatives of cyanobacterian affinity. The findings suggest that eukaryotic phytoplankton was in a diversified evolutionary stage in Gunflint times. Consequently their origin must go back to times earlier than 2000 MY ago.

In the younger Proterozoic, subsequent to the Gunflint, another group of eukaryotic organisms, the Prasinophyta, become the dominant elements in the phytoplankton spectra. These unicellular organisms are

---

Plate 2.   Fossils from Kuibis-quartzite, Nama system, Namibia

Figs. 1, 5. <u>Pteridinium simplex</u> Gürich 1930

Figs. 2, 3. <u>Rangea schneiderhöhni</u> Gürich 1930

Fig. 4.   <u>Ernionorma concretor</u> Pflug 1972

Figs. 6, 7. <u>Erniobaris</u> sp. 1972

Fig. 8.   <u>Kuibisia glabra</u> Hahn & Pflug 1986

            Bar = 10 mm

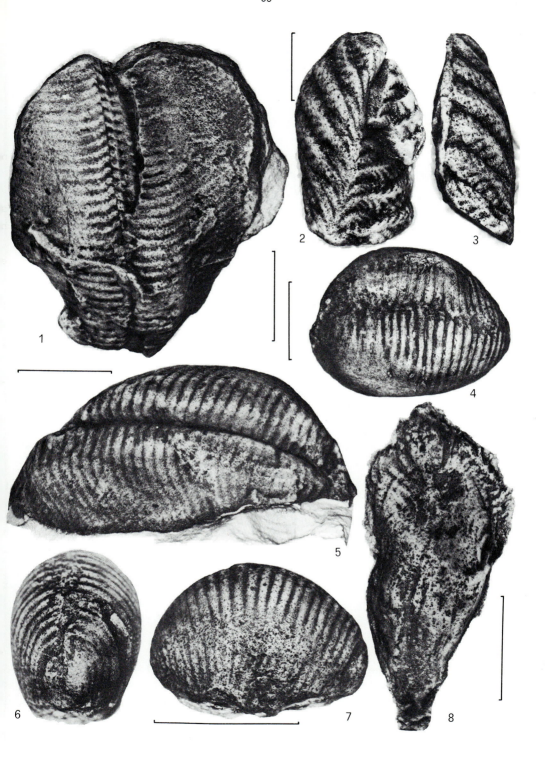

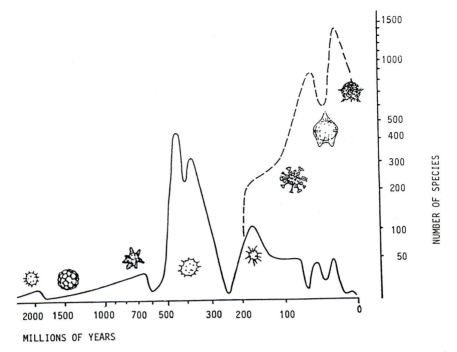

Figure 2. Diversity of eukaryotic phytoplankton species in the last 2000 MY plotted on logarithmic scales. Abscissa: time in MY, ordinate: number of known species, - acritarchs and prasinophytes, --- modern phytoplankton.

commonly placed among the green algae. The oldest representatives have been reported from the 1680-MY-old Daihongyu Formation of China (Lei-Ming 1985). Similar microbiota are also known from the Beltian series of Idaho/Montana (Pl. 1, Figs. 13-14), where they occur in the lower- and middle part of the sequence about 1400 to 1100-MY in age (Horodyski 1980, Pflug & Reitz 1985). It seems from the available evidence that the evolution of the eukaryotic plankton took a periodical course (Fig. 2). Two periods of low diversity are indicated in the record, one located at the end of the Paleozoic the other one at the closure of the Protero zoic.

Metazoa

The fossils from the Kuibis quartzite (Nama system) in South West Afric belong to the oldest metazoans known. The age of the stratum falls into the age bracket 610 - 650 million years (Early Vendium). The fossils ar often preserved in a detailed three-dimensional condition which is apparently the result of synsedimentary processes of silification. Most

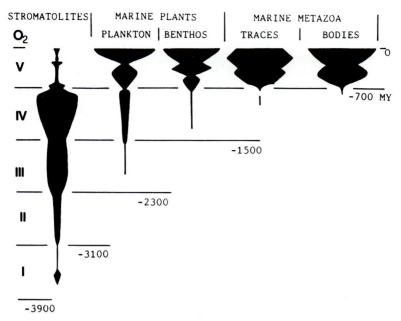

STROMATOLITES | MARINE PLANTS | MARINE METAZOA
PLANKTON | BENTHOS | TRACES | BODIES

O₂

V

IV

III

II

I

-700 MY
-1500
-2300
-3100
-3900

Figure 3. Chronological distribution of life. For explanation see Fig. 1 ("Stromatolites" after Walter & Heys 1985).

of the finds belong to the problematic group Petalonamae with the subgroups Pteridiniida, Rangeida and Erniettida (Pl. 2, Figs. 1-7). The petalonamian body is a thin-walled sack comprising a voluminous cavity. The general symmetry is bilateral, this can be seen from the pattern of exterior ribs and interior septa. The Petalonamae are uncomparable to any known phylum of metazoans. We believe that they lived as sessile plankton feeders on the bottom of the shallow sea. It is worthy to note in this context that their appearance coincides reasonably well with the commencement of late Proterozoic phosphogenesis. A massive increase in the biomass of the photic zone through phytoplankton blooms has also been postulated for this time, particularly in the late Riphean/early Vendian.

Another representative of the Kuibis is Cloudina Germs 1972, a calcareous tube that was probably produced by a worm-like organism of sessile, plankton feeding mode. The preserved structure is composed of several calcareous cones which are set one into the other comparable to segments of a telescope. The Cloudinidae are possibly related to ancestral annelides but evidently much more primitive than any known member of this phylum (Hahn & Pflug 1985a). The family comprises the Vendian genera Cloudina, known from Namibia and Brasilia and Acuticloudina from the Lower Cambrian of Argentina.

Additional finds of early metazoans have been described from the Kuibis recently (Hahn & Pflug 1985b). Ausia fenestrata closely resembles representatives of the recent genus Veretillum (Pennatularia), but this similarity is most probably the result of a morphological convergence rather than a relationship. Kuibisia glabra, a solitary "polyp" of approximately conical shape is comparable to recent Ceriantharia in certain features, but also this similarity must not necessarily mean a relationship (Pl. 2, Fig. 8). The genera Ausia and Kuibisia are interpreted as members of an early branch of "Coelenteroid" affinity in which nematocysts and a medusa generation were not yet developed. Medusae are unknown from the Kuibis quartzite, but they appear in an overlying stratum which is probably of late Vendian age.

Persimedusites is a medusoid fossil from the Young-Precambrian Esfordi formation of Central Iran (ca. 600 MY). Its subumbrella face is preserved in detailed condition showing remnants of oral tentacles radial lobes and body tentacles arranged in alternating position with the lobes. Some of the features resemble Scyphomedusae(Hahn & Pflug 1980).

At Corumbá-Ladário in Mato Grosso (Southwestern Brazil), Young Precambrian megafossils occur in a shaly horizon which according to radiometric data is of late Vendian age (ca. 600 MY). The fossils are preserved in a three dimensional condition by their chitinous periderm. All specimen hitherto found belong to the same kind of organism, named Corumbella werneri (Hahn et al. 1982, Hahn & Pflug 1985b). The structures resemble the living Scyphozoan Stephanocyphus in their morphology.

A critical comparison of the present fossil record shows that few of these Precambrian lineages can be traced with certainty into the lower Cambrian. A gap separates the Precambrian and Cambrian world of Metazoa. The discontinuity corresponds to a major faunal break in the sense of Schindewolf (1954) and Sokolov & Fedonkin (1984).

REFERENCES

CLOUD, P. (1984): The Cryptozoic biosphere: Its diversity and geological significance.- Proc. 27th Internat. Geol. Congr. Moscow 5, 173-198.
HAHN, G.; HAHN, R.; LEONARDOS, O.H.; PFLUG, H.D. & WALDE, D.H.G. (1982): Körperlich erhaltene Scyphozoen-Reste aus dem Jungpräkambrium Brasiliens.- Geologica et Palaeontologica 16, 1-18.
-- & PFLUG, H.D. (1980): Ein neuer Medusen-Fund aus dem Jung-Präkambrium von Zentral-Iran.- Senckenbergiana lethaea 60, 449-461.'
-- & -- (1985a): Die Cloudinidae n. fam., Kalk-Röhren aus dem Vendium und Unter-Kambrium.- Senckenbergiana lethaea 65, 413-431.
-- & -- (1985b): Polypenartigen Organismen aus dem Jung-Präkambrium (Nama-Gruppe) von Namibia.- Geologica et Palaeontologica 19, 1-13.
HORODYSKI, R.J. (1980): Middle Proterozoic shale-facies microbiota from the lower Belt Mountains, Montana.- J. Paleont. 54, 649-663.

LEI-MING, Y. (1985): Microfossils from Precambrian rocks of the Daihong-
    yu Formation of Jixian, North China.- Acta Palaeont. Sinica 24, 112-
    116.
MEYER, Ch. (1985): Ore metals through geologic history.- Science 227,
    1421-1428.
OCHIAI, E. (1983): Inorganic chemistry of earliest sediments.- in:
    PONNAMPERUMA, C. (ed.): Cosmochemistry and the origin of life. 235-
    276, Dordrecht, D. Reidel, P.C.
PFLUG, H.D. & REITZ, E. (1985): Earliest Phytoplankton of Eukaryotic
    Affinity.- Naturwiss. 72, 656-657.
SCHINDEWOLF, O.H. (1954): Über die möglichen Ursachen der großen erdge-
    schichtlichen Faunenschnitte.- N. Jb. Geol. Paläont., Mh. 457-465.
SOKOLOV, B.S. & FEDONKIN, M.A. (1984): Organic world of the Vendian
    period.- Proc. 27th Internat. Geol. Cong. 2, 159-170, VNU Sci. Press.
WALTER, M.R. & HEYS, G.R. (1985): Links between the rise of the Metazoa
    and the decline of stromatolites.- Precambrian Research 29, 149-174.

GLOBAL BIOLOGICAL EVENTS IN THE LATE PRECAMBRIAN

SOKOLOV, Boris S. *) & FEDONKIN, Michael A. **)

The Precambrian history of life on Earth constitutes just 6/7 of the whole history of life. However, classical paleontology developed over 200 years only on the basis of studies of the Phanerozoic, i.e. the latest stage of the evolution of biosphere. But it was the Precambrian, which saw indeed global cardinal events and processes connected with the formation of living systems, the evolution of the cell and its energetic mechanisms. These processes resulted in the pre-Phanerozoic rebuilding of ecological systems of our planet, drastic changes in the composition of its atmosphere, hydrosphere and lithosphere. To a considerable extent, all that are biogenic phenomena.

The second half of the current century is characterized by abruptly increasing knowledge of life in the Precambrian. Of great interest is the Late Precambrian stage of geological history, when the most important events took place and prepared a qualitatively new Phanerozoic eon.

The most vivid event of the Late Precambrian was evidently the appearance of multicellular animals of tissue organization (Metazoa). We cannot determine an exact moment of their appearance, but it seems that once having appeared, the metazoan organization could produce by several radiations for a short period of time (in geological sense) such a diversity, which we observe in the Vendian. Unlike metazoan organization, multicellularity could appear many times in various kingdoms depending primarily on environmental conditions.

We know about the appearance of Metazoa in the history of biosphere only from indirect data. This event evidently took place before the beginning of the Vendian period, as in the early half of the Vendian, i.e. directly above glacial deposits, known as the Laplandian or Varangarian tillite horizon, we can see the remains of a higly differentiated world of animals.

Some workers relate the appearance and wide expansion of Metazoa with an abrupt decrease in the quantity and diversity of stromatolites in the Late Riphean, Vendian and especially in the Early Paleozoic. An important global event is taking place during this time interval. Stromatolites as benthic communities, being the most typical of the shallow

---

*) Department of Geology, Geophysics, Geochemistry and Mining Sciences of the USSR Academy of Sciences, 117901 Moscow, U.S.S.R.

**) Paleontological Institute of the USSR Academy of Sciences, 117321 Moscow, U.S.S.R.

water basins of Earth during almost 3 billion years, became suddenly
reduced for a relatively short period of time.

The simplest explanation of this event is the appearance and expansion of Metazoa, which cropped cyano-bacterial mats, destroyed by bio-
turbations the stability of substrate, which is necessary for the initial
phases of bioherm formation, and then forced out stromatolites as
communities. The cause of the stromatolite reduction in the Late Riphean
and in the Vendian seems to be more complicated. We should refer to an
analysis of other biological and abiotic events in the Late Precambrian
in order to understand it.

Many workers, who studied the Upper Proterozoic, noticed one more
important event, that is an abrupt fall of taxonomic diversity of algal
microfossils and acritarchs in the Vendian, as compared with the Upper
Riphean. It is difficult to see whether the quantity of microfossils
and stromatolites and their diversity decreased parallel to each other
during Late Precambrian, because the systematics of these fossils is
still being formed, and the stratigraphic correlation of the Precambrian
units is not devoid of contradictions. But in the second half of the
Vendian, the diversity of stromatolites and microfossils slightly increases. Then, the Early Cambrian sees an outbreak of the diversity of
microfossils with organic wall, but this is not the case with stromato-
lites.Their habitats became occupied by other organisms, including reef
building groups (Archaeocyatha, calcareous algae and others).

It cannot be excluded, that a turning moment in the evolution of
ecosystems became the Varangarian glaciation, which is known to be the
largest one in the history of Earth. The interaction of glaciers with
atmosphere, hydrosphere and lithosphere entailed considerable climatic
and geographic changes, which stimulated sharp alterations in the character of habitat. Glacial-eustatic lowering of sea level led to regression
which resulted in the disappearance of vast epiplatform seas and shelf
seas. This circumstances as well as falling temperature, a high water
turbidity because of the increasing frequency of storms, and reducing
transport of material from the continents could have led to an abrupt
narrowing of the zone, where stromatolites are widespread, furtheron to
decreasing productivity and even the disappearance of some groups of
phytoplankton.

The Varangarian glaciation probably saw mass extinction of some
groups of invertebrates of which we know nothing. But the groups which
survived "a great cold" gave rise to radiation in the lower half of the
Vendian period, marked by becoming warm and gigantic transgression. Rapid
expansion of these animals, known now as the Ediacarian fauna, became an

obstacle on the way towards the previous domination of stromatolites.
The domination of Metazoa seems to begin since that time.

Rapid diversification of Metazoa took a relatively short period of
time. In the first half of the Vendian the growth of taxonomic diversity
slowed down. A period of a relatively slow evolution in the middle
Vendian was followed by an epoch of the extinction of many groups of
Metazoa, as only rare problematic forms of meduzoids and small trace
fossils occur in the second half of the Vendian (the Kotlin Horizon of
the Russian Platform).

The above mentioned phenomena, such as the disappearance of fossil
Metazoa and the extinction of phytoplankton in the Vendian seem to have
a causal relationship. But along with the extinction of megascopic groups
of soft bodied Metazoa in the second half of the Vendian we can speculate
about the decreasing dimensions of surviving groups of invertebrates.
This hypothesis seems more probable, if we compare small shelly fossils
of the Tommotian Stage of the Lower Cambrian and gigantic forms of soft
bodied animals of Ediacarian type. The decreasing dimensions or minia-
turization of Metazoa in the second half of the Vendian can be considered
as a factor of survival. But this event, if it indeed took place, could
lead to important ecological and evolutionary-morphological consequences.
An indirect confirmation of the hypothesis of decreasing sizes of Metazoa
in the second half of the Vendian can be an abrupt increase in the buried
biomass of macrophyte flora of vendotaenid algae in the deposits of the
Kotlin Stage of the Upper Vendian. This phenomenon is especially vivid
on the Russian Platform and can be indicative of the eutrophication of
shallow water marine basins with all negative results for the fauna.
This worked as selective factor in the favour of forms having small
dimensions.

The discovery of actinomycetes on the thallom of vendotaenid algae,
which has been recently made (M.B. Fnilovskaya), allows the assumption
about a gigantic outbreak of bacterial consumers at the Kotlin age,
which ate not only algal, but also animal tissues. This event can also be
considered as one of the causes for an abrupt fall of soft bodied Metazoa remains from the Late Vendian taphocoenoses.

The end of the Vendian (the Rovno Stage), again sees an abrupt increase in the dimensions of animals, primarily among representatives of
benthos, in particular of infauna. This is clearly seen from trace fossils, which become large, more complicated and deeper. The sea floor is
increasingly colonized by infauna.

At the end of the Vendian some groups of Metazoa begin to build a mineral skeleton, and this biological innovation becomes evident at the beginning of the Cambrian (the Tommotian Stage). Skeleton became the basis of radiation for some groups including such primitive forms as sponges and Archaeocyatha. The change in the geometry of trace fossils at the Rovno age was followed by that of the body at the Tommotian age. If we observe in the historical development of ichnofossils tendencies towards complication and deepening, then the development of skeletal forms (primarily of sedentary ones) is marked by the tendency towards complication and growth above the substrate. The Late Vendian and the Early Cambrian saw events which led to a sharp vertical expansion of the zone inhabited by benthos as well as to a change in general taphonomic situation: a period of the fossilization of soft bodied Metazoa was over since the second half of the Vendian. With the beginning of the Cambrian paleontologists concentrate mainly on skeletal remains.

Thus the Late Precambrian saw events which had considerable consequences for the future of the planet and its biosphere. The most important events of this interval were the following: 1) the appearance of Metazoa; 2) the reduction of the diversity of stromatolites, that started in the Upper Riphean; 3) the extinction of certain phytoplankton groups and probably of some groups of the Prevendian Metazoa during the Laplandian glaciation; 4) the rapid radiation and expansion of Metazoa in the first half of the Vendian; 5) the extinction of many groups of the Vendian fauna in the second half of the Vendian and possible decreasing sizes of organisms, which survived during this interval of history; 6) the increasing colonization of the sea floor and the increase in the sizes of infauna, furtheron the appearance of mineral skeleton in some groups of Metazoa at the end of the Vendian; 7) rapid radiation of invertebrates with mineral skeleton and their expansion at the beginning of the Cambrian. All these events took place against the background of abrupt changes of the geological, geochemical and geographical situation.

PRECAMBRIAN-CAMBRIAN BOUNDARY BIOTAS AND EVENTS

BRASIER, Martin *)

A contribution to Project
GLOBAL BIO-EVENTS

Geological and evolutionary events across the Precambrian-Cambrian boundary mark a major turning point in Earth history. Calibration and interpretation of these events, however, is hampered by the lack of an agreed biostratigraphic scale, especially in strata without trilobites. Attention has recently focussed upon the earliest 'small shelly fossils', representing a diversity of diminutive invertebrate skeletal remains. Conoidal phosphatic microfossils (CPM's) seem to be among the most promising elements for biostratigraphy since they are relatively resist-ant and widespread, while recent work indicates evolutionary series of species. Small molluscs and tubular problematica also provide important biostratigraphic and palaeoecological information.

Until recently, emphasis has fallen upon premier research into suc-cessions on the Siberian Platform, recently summarised by Sokolov & Zhuravleva (1983) and Rozanov & Sokolov (1984). Richly fossiliferous Precambrian-Cambrian boundary successions are becoming better known in other regions, though, as will be outlined below.

Avalon-Baltic successions

Research on the condensed but highly fossiliferous Hyolithes Limestone of Nuneaton and Comley Limestone of Shropshire, England, has elucidated successive species of conodont-like Rhombocorniculum, hyolithelminth Torellella, sclerites Sunnaginia, Eccentrotheca and Lapworthella, whose order of appearance is similar in Siberia and England (Brasier 1986). Of these fossils, undoubtedly the most valuable is the elaborately sculp-tured Rhombocorniculum, with three successive species in England and Siberia. Unpublished work also confirms a similar sequence of Rhombo-corniculum spp. in the Bornholm-Scania succession.

Rhombocorniculum stratigraphy has yet to be confirmed from comparable strata in southeastern Newfoundland to Massachusetts (Dr. E. Landing, pers. comm. 1986). In this region, however, the earliest assemblage contains molluscs Aldanella, Anabarella and Heraultipegma, tube Anabari-tes trisulcatus plus CPMs Fomitchella sp. and Lapworthella ludvigseni (Bengtson & Fletcher 1983, Landing 1984). This assemblage may be equiva-lent or older than the earliest assemblage bearing Sunnaginia neoimbri-cata and Torellella lentiformis in England. The third rich assemblage

*) Department of Geology, Hull University, Hull HU6 7RX, U.K.

Lecture Notes in Earth Sciences, Vol. 8
Global Bio-Events. Edited by O. Walliser
© Springer-Verlag Berlin Heidelberg 1986

contains monoplacophorans <u>Randomia</u> <u>aurorae</u> and <u>Prosinuites</u> <u>emarginatus</u>
and bivalve <u>Fordilla</u> sp., CPMs <u>R</u>. <u>insolutum</u> and <u>Torellella</u> <u>biconvexa</u>
(e.g. Brasier 1984) plus remains of the button-like <u>Mobergella</u> <u>radiolata</u>
gp. This assemblage occurs widely from Bornholm, through England to
Newfoundland (Brasier 1986). The fourth assemblage contains <u>R</u>. <u>cancel-</u>
<u>latum</u> and/or early <u>Callavia</u> Zone trilobites in England and Newfoundland.
A fifth assemblage has CPMs <u>lapworthella</u> <u>cornu</u> gp., <u>R</u>. <u>cancellatum</u>, <u>R</u>.
<u>exsolutum</u> n. sp., and <u>Microdictyon</u> <u>effusum</u> gp. with late <u>Callavia</u> Zone
trilobites of the <u>Serrodiscus</u> <u>bellimarginatus</u> assemblage. The gastropod
<u>Pelagiella</u> and brachiopod <u>Botsfordia</u> <u>caelata</u> also become common from
about this level (e.g. Rushton 1966).

This succession compares well with that found on the Siberian Platform. The first appearance of gastropod <u>Aldanella</u> <u>attleborensis</u> provides
the lowest datum for possible correlation between these two main areas.
Successive assemblages with <u>Torellella</u> <u>lentiformis</u>, Rhombocorniculum
<u>insolutum</u>, <u>R</u>. <u>cancellatum</u> and <u>Lapworthella</u> <u>cornu</u> provide higher points
of correlation. <u>R</u>. <u>insolutum</u> notably occurs in strata bearing the first
trilobites in Siberia (e.g. <u>Fallotaspis</u>; Rozanov & Sokolov 1984) and
Bornholm-Scania (Brasier, unpublished) whereas trilobites first appear
with <u>R</u>. <u>cancellatum</u> in the Anglo-Avalon region.

Tethyan successions

Biostratigraphic correlation of southern Asiatic ('Tethyan') boundary
sections is strengthened by the wealth of new data coming out of China
(e.g. Xing et al. 1984) and also from the Lesser Himalayas of India (e.g
Bhatt et al. 1985) plus new data on sections from the Elburz mountains
of northern Iran (Dr. B. Hamdi, in prep.). The Himalayan and Elburz sequences show a succession of faunas and lithologies very comparable with
those of Meishucun of Yunnan province and Maidiping of Sichuan province
of China while the Elburz sequence also compares with the Salany-Gol
sequence in western Mongolia (e.g. Voronin et al. 1982). Bio-, litho-,
and event-stratigraphy allow further correlation with the <u>Torellella</u>
and <u>Rhombocorniculum</u>-bearing strata of Kazakhstan (e.g. Missarzhevsky &
Mambetov 1981) linking the stratigraphy of Siberia and Avalon-Baltic
with the Tethyan belt.

Clastic strata bearing large algae of the <u>Chuaria</u> and <u>Vendotaenia</u>
groups plus sapropelic material occur low down in the successions of
China and Iran, the latter occurrence associated with early phosphatic
tubes of <u>Hyolithellus</u> (data of Dr. B. Hamdi). At the western limit of
the Tethyan zone in Spain, a similar algal assemblage was found associated with the first arthropod traces of <u>Monomorphichnus</u> and diminutive,

complex deposit-feeding traces of 'Cambrian-type' (Brasier et al. 1979). This suggests the existence of a 'transitional epoch' between Vendian and Cambrian, here informally called Zone 0.

Across much of Asia, this macroalgal assemblage passes up into dolomites with distinctive late Vendian stromatolites and calcareous algae. Zone I skeletal assemblages often appear towards the top of such dolomites and usually range through phosphatic-clastic strata deposited during a major phosphogenic event. Typical elements of Zone I include trimerate tubes of Anabarites trisulcatus, protoconodont Protoherztina anabarica gp., simple circothecid hyoliths (e.g. Conotheca), conulariids (e.g. Hexangulaconularia) and wiwaxiid-type sclerites (e.g. Lopochites). Assemblages of this type occur across Asia and range into northwestern Canada (Nowlan et al. 1985).

Zone II assemblages differ in the addition of diverse gastropod and monoplacophoran-type shells (e.g. Latouchella korobkovi gp., Aldanella attleborensis gp.) plus numerous wiwaxiid-type sclerites. The simple cambroclavid CPMs of Zhijinites are also common in China. The change from Zone I to II faunas may occur either within or above phosphorite-chert units and the upper part of the zone reverts to shallow calcareous facies on the Yangtze Platform. The datum marked by the appearance of the Latouchella korobkovi gp. suggests correlation with the basal Tommotian sunnaginicus Zone of Siberia and the Aldanella attleborensis assemblage of Newfoundland to Massachussetts. In Maly Karatau, Kazakhstan, Torellella cf. biconvexa in the upper part of this lithological cycle suggest a late Tommotian or low Atdabanian age, so it seems that Zone II may span the Tommotian-lowest Atdabanian. Condensation of this zone may have obscured this in many places.

The base of Zone III is a widespread ferruginous disconformity across the Yangtze Platform, followed by iridium-enriched clays with anomalous carbon isotopes (e.g. Hsu et al. 1985) then by black shales bearing the chancelloriids Allonia tripodophora and Archiasterella pentactina. This assemblage also occurs in India and Pakistan and may be traced through the redlichiid faunal province to France and Spain (e.g. Dore & Reid 1965, Linan 1978). Here, Allonia tripodophora first occurs below the trilobite Bigotina and in Spain with low-mid Atdabanian archaeocyathids, while Archiasterella pentactina appears with trilobites of Callavia Zone aspect (Sdzuy 1969). Correlation of this level beyond the Tethyan area is problematical but suggested by the appearance of Bigotina with early Rhombocorniculum cancellatum in Siberia (Rozanov & Sokolov 1984). A ferromanganese-enriched discontinuity at the top of R. insolutum beds in England (Brasier & Hewitt 1979) and comparable strata in southeastern

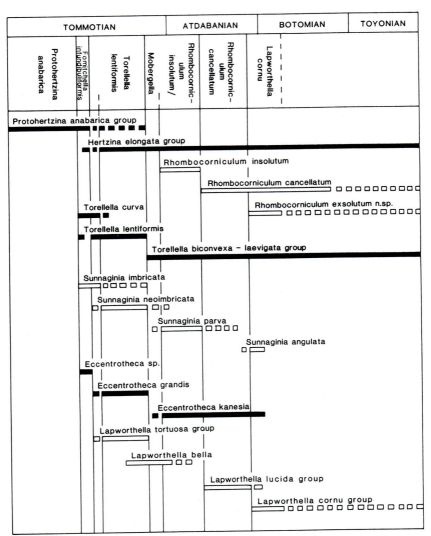

Figure 1. Examples of series of conoidal phosphatic microfossils fro
Precambrian-Cambrian boundary beds.

Newfoundland may compare with that below the R. cancellatum Zone in
Kazakhstan (e.g. Missarzhevsky & Mambetov 1981) and on top of Zone II
in China (e.g. Luo et al. 1984). Thus the iridium and carbon-isotope
anomaly reported by Hsu et al. (1985) from China may be close to the
insolutum-cancellatum boundary. This broad correlation is still tenuous
       Figures 1 and 2 show suggested composite vertical ranges of series
of conoidal phosphatic microfossils useful for 'backbone' biostratigrap
(e.g. Rhombocorniculum) and ranges of other potentially useful marker

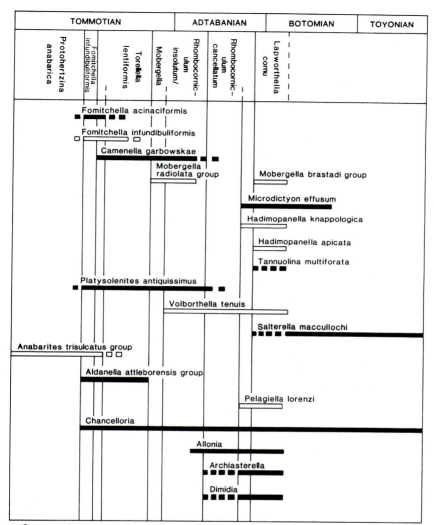

Figure 2. Suggested ranges of some other potentially important marker microfossils and small shelly fossils.

microfossils and small shelly fossils (e.g. <u>Aldanella</u>). The ranges are provisional and the interval names are informal ones relating to the correlation of Avalon-Baltic and Siberian sequences, for discussion only. The term Tommotian is herein used in a loose sense; Tommotian <u>sensu stricto</u> begins above the <u>Protohertzina anabarica</u> interval.

Bioevents

The explosive evolution of invertebrates, including those with skeletons shows distinct phases which may be provisionally determined as follows.

1. The disappearance of the large, soft-bodies <u>Charnia</u> fauna. This may
have been contemporaneous with volcanic and igneous events in the Anglo-
Avalon region (Brasier 1985).

2. The widespread appearance of sapropel-bearing clastics, with large
spherical <u>Chuaria</u> and ribbon-like <u>Vendotaenia</u>. In places, these are
known to have coexisted with organic worm tubes of <u>Sabellidites</u>, phosphatic tubes of <u>Hyolithellus</u> and traces of arthropods (<u>Monomorphichnus</u> plus
complex deposit feeders in transitional Zone 0.

3. Extensive dolomitic carbonate habitats were invaded by a Zone I
assemblage of low diversity but varied skeletal plan, including ? aragonitic tubes (<u>Anabarites</u>, hyoliths), phosphatic tubes <u>hyolithellus</u>)
and phosphatic spines (e.g. <u>Protohertzina</u>). Calcareous cyanobacteria
also permineralised at this time (Riding & Voronova 1984) suggesting
some change in pCO_2 and/or Mg: Ca ratio.

4. A major episode of phosphogenesis was initiated in Zone I of the
Tethyan belt, perhaps in response to a transgressive pulse and oceanic
overturn of stratified waters (Cook & Shergold 1984). Other areas show
transgressive conditions at about this time. The positive $\delta 13C$ spike in
the Yudomian of Siberia (Magaritz et al. 1985) may relate to deposition
of organic carbon during this episode.

5. Coiled, sculptured molluscs of the <u>Latouchella</u> <u>korobkovi</u> and <u>Aldanel</u>
<u>attleborensis</u> groups appeared suddenly and widely, without known pre-
cursors, suggesting invasion from an unknown origin during a transgressive pulse (Zone II). They occur in clastics and carbonates of varying
ecological and palaeoclimatic settings, and may indicate a 'bloom' of
opportunistic forms. Archaeocyathans, allathecid hyoliths and paterinid
brachiopods also appeared here. The negative $\delta 13C$ trend recorded throu
the early Tommotian of Siberia (Magaritz et al. 1985) may relate to increased carbonate deposition, including that of invertebrate skeletons,
rather than to declining organic productivity.

6. Returning carbonates locally culminate in Fe-Mn or limonitic crusts,
suggesting a period of stillstand and emergence, though evidence for
synchroneity at the top of this cycle is needed. Calcite biomineralization of olenellid trilobites first took place in the <u>Rhombocorniculum</u>
<u>insolutum</u> interval.

7. A new depositional cycle was initiated during Zone III in the Tethya
belt, where metaliferous black shales were deposited, and in <u>Rhombocor-</u>
<u>niculum</u> <u>cancellatum</u> times in the Avalon-Baltic region. Calcitic trilobi
skeletons continued to appear widely during this interval, though diachronously and in such a way that ecological controls of chitin biomineralisation are implicated.

8. Pelagic eodiscoid trilobites became significant in Zone V times in China and in late Atdabanian to early Botomian times in Siberia and the Avalon-Baltic region. Archaeocyathan reefs and echinoderm remains also reached their widest extent during this transgressive maximum of the early Cambrian.

Sequential biomineralisation?

This sequence of events, with its strong chemical signals spanning the disappearance of the megascopic Charnia fauna and the phased appearance of diminutive skeletal fossils, suggests that some 'event' took place in the late Precambrian, during a period of relatively low sea level. Plate changes, volcanicity and outgassing, changes in sea level, nutrient upwelling, anoxia and toxicity, major increases of fluctuations in trophic supply and the migration of 'biomineral fronts' all need to be examined for their role in this unique event.

Phases of the Cambrian transgressions seem to have enhanced the sequential evolution of trace fossils -- small shelly fossils-- trilobites (Brasier 1979) and it may be that sequential palaeocenaographic events have left their imprint on biomineralisation (e.g. Brasier, in press).For example, Vendian rifting of a supercontinental assembly (Piper 1982) could have led to the formation of brine-rich, anoxic, marginal and small oceanic basins, where massive removal of light sulphur isotopes by anaerobic bacteria took place (Yudomski 34S event of Holser 1977). Sulphide bacteria around hydrothermal vents could also have provided a major food source for suspension feeding invertebrates. Worm tubes from Carboniferous to Recent vents (e.g. Banks 1985) notably resemble the earliest tubular skeletons and their first remains should perhaps be sought in hydrothermal settings. Such a removal of seawater sulphate by bacteria allows the local formation of dolomite (c.f. Kastner et al. 1983) which is a widespread deposit in latest Precambrian times (e.g. Tucker 1982). Although high sulphate and Mg^{++} concentrations generally inhibit biomineralization, organic or aggulutinated skeletons could occur at this stage, as found in the early remains of Sabellidites and Platysolenites. A 'magnesian event' could ultimately have led to a significant reduction in the Mg/Ca ratios (Riding 1985, Kazmierczak et al. 1985) by the time of the Zone I fauna. Local occurrences of aragonitic and high-Mg calcite skeletons would be a theoretical expectation of this model (e.g. Anabarites tubes and early molluscs). Since Mg^{++} ions inhibit the formation of apatite, its local removal can lead to the precipitation of sedimentary phosphorites (Lucas & Prevot 1984, Kastner et al. 1983), especially in upwelling or transgressive water bodies spilling over into lagoonal settings. This may partly account for the ensuing major phosphate event in Zones I and II, associated with the phosphatic skeletons of Protohertzina, Hyolithellus and others. Since phosphorus also inhibits the formation of calcite, its removal in turn allows the formation of low Mg-calcite skeletons (Simkiss 1964) and carbonates, such as those of trilobites.

Evidence for sequential deposition of sulphate-black shale, dolomit phosphate-aragonite and calcite across the Precambrian-Cambrian boundar should be carefully examined. Such a pattern could have brought about a sequence of biomineral events: e.g. organic -- aragonite -- high Mg calcite -- apatite -- low Mg calcite. This is almost the predicted orde of calcium carbonate precipitation in a brine of decreasing supersatura tion, which could have developed diachronously through the course of the Cambrian transgression, with 'biomineralisation fronts' passing across a series of shallow coastal habitats.

REFERENCES

BANKS, D.A. (1985): A fossil hydrothermal worm assemblage from the Tynagh lead-zinc deposit in Ireland.- Nature 313, 138-31.

BENGTSON, S. & FLETCHER, T.P. (1983): The oldest sequence of skeletal fossils in the Lower Cambrian of southeastern Newfoundland.- Can. J. Earth Sci. 20, 525-536.

BHATT, D.K.; MAMGAIN, V.D. & MISRA, R.S. (1985): Small shelly fossils (early Cambrian (Tommotian) age from Chert-Phosphorite Member, Tal Formation, Mussoorie syncline, Lesser Himalaya, India, and their chronostratigraphic evaluation.- J. Pal. Soc. India 30, 92-102.

BRASIER, M.D. (1979): The Cambrian radiation event.- in: HOUSE; M.R. (ed.): The Origin of Major Invertebrate Groups. Syst. Ass. Spec. Vol 12. Acad. Press, London 103-159.

-- (1984): Microfossils and small shelly fossils from the Lower Cambri Hyolithes Limestone at Nuneaton, English Midlands.- Geol. Mag. 121, 229-253.

-- (1985): Evolutionary and geological events across the Precambrian-Cambrian boundary.- Geology Today 1, 141-146.

-- (1986): The succession of small shelly fossils (especially conoidal microfossils) from English Precambrian-Cambrian boundary beds.- Geo Mag. 123, 237-256.

-- (in press): Why do lower plants and animals biomineralise?- Paleo-biology.

-- & HEWITT, R.A. (1979): Environmental setting of fossiliferous rocks from the uppermost Proterozoic-Lower Cambrian of central England.- Palaeogeogr., Palaeoclimatol., Palaeoecol. 27, 35-57.

-- ; PEREJON, A. & DE SAN JOSE, M.A. (1979): Discovery of an important fossiliferous Precambrian-Cambrian sequence in Spain.- Estudios Geo 35, 379-383.

COOK, P.J. & SHERGOLD, J.H. (1984): Phosphorus, phosphorites and skele evolution at the Precambrian-Cambrian boundary.- Nature 308, 231-23

DORE, F. & REID, R.E. (1965): Allonia tripodophora nov. gen., nov. sp. nouvelle Eponge du Cambrian inferieur de Carteret (Manche).- C.R. S Geol. France, 1965, 20-21.

HOLSER, W.T. (1977): Catastrophic chemical events in the history of th ocean.- Nature 267, 403-408.

HSU, K.J.; OBERHANSLI, H.; GAO, J.Y.; SUN SHU, CHEN, H. & KRAHENBUHL,
    U. (1985): 'Strangelove ocean' before the Cambrian explosion.- Nature
    316, 809-811.
KASTNER, M.; KUSHNIR, J. & ANDERSON, G.E. (1983): Diagenetic relation-
    ships of dolostone-phosphorite-chert: a geochemical interpretation.-
    Abstr. Programs Geol. Soc. Amer. 15, 607.
KAZMIERCZAK, J.; ITTEKOT, V. & DEGENS, E.T. (1985): Biocalcification
    through time: environmental challenge and cellular response.- Palae-
    ont. Z. 59, 15-33.
LANDING, E. (1984): Skeleton of lapworthellids and the suprageneric
    classification of tommotiids (early and middle Cambrian phosphatic
    problematica).- J. Paleont. 58, 1380-1398.
LINAN, E. (1978): Bioestratigrafia de la Sierra de Cordoba.- PhD Thesis,
    Univ. Granada.
LUCAS, J. & PREVOT, L. (1984): Synthese de l'apatite par voie bacterienne
    a partir de matiere organique et de divers carbonates de calcium dans
    eaux douce et marine naturelles.- Chem. Geol. 42, 101-118.
LUO HUILIN; JIANG ZHIWEN; WU XICHE; SONG XUELIANG; OUYANG LIN; XING
    YUSHENG; LIU GUIZHI; ZHANG SHISHAN & TAO YONGHE (1984): Sinian-
    Cambrian boundary stratotype section at Meishucun, Jinning, Yunnan,
    China.- People's Publishing House, Yunnan, China, 154 p.
MAGARITZ, M.; HOLSER, W.T. & KIRSHVINK, J.L. (1986): Carbon-isotope
    events across the Precambrian-Cambrian boundary on the Siberian Plat-
    form.- Nature 320, 258-259.
MISSARZHEVSKY, V.V. & MAMBETOV, A.J. (1981): Stratigraphy and fauna of
    Cambrian and Precambrian boundary beds of Maly Karatau (in Russian).-
    Trudy Akad. Nauka SSSR, Moscow, 326.
NOWLAN, G.S.; NARBONNE, G.M. & FRITZ, W.H. (1985): Small shelly fossils
    and trace fossils near the Precambrian-Cambrian boundary in the Yukon
    Territory, Canada.- Lethaia 18, 233-256.
PIPER, J.D. (1982): The Precambrian palaeomagnetic record: the case for
    the Proterozoic supercontinent.- Earth Planet. Sci. Lett. 59, 61-89.
RIDING, R. (1985): Calcareous algae near the Precambrian/Cambrian bound-
    ary.- Nerc News J. 3, 11-12.
-- & VORONOVA, L. (1984): Assemblages of calcareous algae near the Pre-
    cambrian/Cambrian boundary in Siberia and Mongolia.- Geol. Mag. 121,
    205-210.
ROZANOV, A. Yu. & SOKOLOV, B.S. (1984): Lower Cambrian Stage Subdivision.
    Stratigraphy (in Russian).- Iztdat. Nauka Moscow.
RUSHTON, A.W.A. (1966): The Cambrian trilobites from the Purley Shales
    of Warwickshire.- Palaeont. Soc. Monograph., 511.
SDZUY, K. (1969): Unter- und mittelkambrische Porifera (Chancelloriida
    und Hexactinellida).- Paläont. Z. 43, 115-147.
SIMKISS, K. (1964): Phosphates as crystal poisons of calcification.-
    Biol. Rev. 39, 487-505.
SOKOLOV, B.S. & ZHURAVLEVA, I.T. (1983): Lower Cambrian stage subdivision
    of Siberia (in Russian).- Atlas of fossils. Izdat. Nauka Moscow.
TUCKER, M. (1982): Precambrian dolomites: petrographic and isotopic
    evidence that they differ from Phanerozoic dolomites.- Geology 10,
    7-12.
VORONIN, Yu. I.; VORONOVA, L.G.; GRIGORIEVA, N.V.; DROSDOVA, N.A.; SHE-
    GALLO, E.A.; ZHURAVLEV, A. Yu.; RAGOZINA, A.L.; ROZANOV, A. Yu.;
    SAYUTINA, T.A.; SISOEV, V.A. & FONIN, V.D. (1982): The Precambrian/
    Cambrian boundary in the Geosynclinal areas; the reference section of
    Salany-Gol, MPR (in Russian).- Transactions of the Joint Soviet-
    Mongolian palaeontological expedition, 18. Moscow. Izdat. Nauka.
XING YUSHENG; DING QIXIU; LUO HUILIN; HE TINGGUI & WANG YANGENG (1984):
    The Sinian-Cambrian boundary of China and its related problems.- Geol.
    Mag. 121, 155-170.

ORDOVICIAN TO THE BOUNDARY
ORDOVICIAN/SILURIAN

THE FAUNAL EXTINCTION EVENT NEAR THE ORDOVICIAN-SILURIAN BOUNDARY: A CLIMATICALLY INDUCED CRISIS

BARNES, Christopher R. *)

Introduction

In seeking to provide a definition for the base of the Silurian System, the work of the IUGS Ordovician-Silurian Boundary Working Group over the last decade resulted in much new faunal and stratigraphic data through the boundary interval. Very few sections are known in which there is a continuous and fossiliferous sequence. Most have minor or major hiatuses and/or intervals with barren or poorly fossiliferous strata. This has prevented rigorous faunal analysis through the boundary interval. To date, only a few major fossil groups have been studied in detail and, in several, taxonomic problems create additional difficulties. Despite these qualifications, there is no doubt that a profound faunal change, involving dramatic extinctions, occurs at this level and that it represents one of the major faunal turnovers in the Phanerozoic. Its significance has been ignored or underestimated until recently when new data emerged.

The Late Ordovician continental glaciation (Tamadjert glacial epoch) in North Africa (e.g. Beuf et al. 1966, 1971; Hambry 1985) was documented through the 1960's. As its stratigraphic and geographic extent became evident, the glacial event was used to explain major eustatic and faunal changes that had been particularly appreciated by specialists of shelly fossils from the low paleolatitude, carbonate platforms (e.g. Berry & Boucot 1973; Sheehan 1973, 1975; Lenz 1976; Brenchley 1984). In the clastic facies of the mid to high paleolatitude, the curious development of the distinctive brachiopods and trilobites of the Hirnantia faunas in latest Ordovician time was likewise attributed by some specialists to extensive cold water conditions (e.g. Brenchley & Newall 1980, Rong 1984, Brenchley & Cocks 1984).

As more attention has been focussed on the glacial event, the systemic boundary debate and the faunal extinctions, it has been critical to develop refined biozonations and chronostratigraphy. Unfortunately, graptolite and conodont zonations have few precise tie points and correlations between the regional stages of the type Ashgill and those of the carbonate platforms are difficult. The definition approved by IUGS in 1985 for the base of the Silurian System, at the level of the Parakido-

*) Department of Earth Sciences, Memorial University of Newfoundland, St. John's, Newfoundland, A1B 3X5, Canada.

Lecture Notes in Earth Sciences, Vol. 8
Global Bio-Events. Edited by O. Walliser
© Springer-Verlag Berlin Heidelberg 1986

graptus ? acuminatus Zone, has resulted in the glacial and extinction
events being assigned to the latest Ordovician, within the Ashgillian
Series, rather than straddling the boundary.

Faunal changes

Graptolite taxonomy is still in a state of flux through the boundary
interval. However, late Caradoc-early Ashgill faunas are diverse, typically yielding 15-20 species. A moderate decline occurs in the D. com-
planatus Zone with a severe decline in the Climacograptus extraordinari
Zone where the fauna is reduced to about three species. The Glyptograpt
persculptus Zone shows a modest increase, and a recovery to about half
the diversity levels of the early Ashgill is attained in the P.? acuminatus Zone. A slow and progressive increase continues through most of
the Llandovery. This dramatic diversity decline has been documented by
Koren' and Nikitin (1982) and Rickards (1982) and can be seen in severa
detailed boundary studies (e.g. Williams 1983 for Dob's Linn, Scotland
Koren' et al. 1979 for north-east USSR, and Mu & Lin (1983) for the
Yangtze region, central China). The principal faunal turnover correlate
with the lower Hirnantian Stage (Fig. 1).

Conodonts exhibit a marked faunal provincialism (Midcontinent and
North Atlantic provinces) during the Late Ordovician. Lower Ashgill

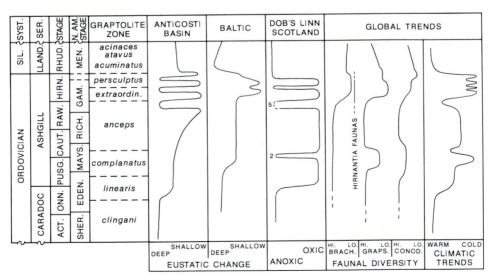

Figure 1.   Interpreted late Ordovician-early Silurian changes in
eustasy, oxygenation, faunal diversity and climate. Changes are considered to be induced by the late Ordovician glacial event.

faunas have yielded about 75 - 100 species. Although several new species appear through the Ashgill, there is a sharp decline in diversity in the upper Ashgill to about 20 - 30 species in the Hirnantian (Gamachian). Only eight genera appear to cross the systemic boundary of which only one, possibly two, have a compound apparatus; most are coniform. The character of Early Silurian faunas is quite different and diversities of platform faunas are typically in the range of 20 - 25 species; Early Silurian basinal to oceanic faunas are poorly known. The faunal change is therefore profound, relatively abrupt, with a decline through the Hirnantian (Gamachian) and the actual turnover occuring at or below the systemic boundary, probably in the upper half of the G. persculptus Zone. This pattern of change has been documented in general by Sweet (1985) and in detail by Barnes & Bergström (in press), and in several specific boundary studies, the most complete being that of McCracken & Barnes (1981) from Anticosti Island, Quebec.

Information on brachiopod and trilobite changes across the boundary interval is less controlled. The faunal turnover at this level and the change from provincial to cosmopolitan faunas was recognized by Berry & Boucot (1973) and Sheehan (1973, 1975). These changes and the reduction in diversity during the Hirnantian Stage has been well documented in studies by Lespérance (1974, 1985), Jaanusson (1979), Amsden (1980), Brenchley & Cocks (1982), Rong (1982), and Owens & Rushton (1984). Although there are still major taxonomic questions to resolve, the general pattern is one of major turnover in these shelly faunas within the Hirnantian. On Anticosti Island, the faunal change is at a similar level to that of the conodonts (Chatterton et al. 1983).

Other fossil groups are too poorly known to provide much firm data, although some seem to be significantly affected at or near the boundary interval (e.g. Duffield & Legault 1981).

Event stratigraphy

Through analysis of key stratigraphic sections from high, mid, and low paleolatitudes it is possible to reconstruct an event stratigraphy for the Late Ordovician and earliest Silurian.

In North Africa, the glacial sequences suggest three or four major glacial phases but their individual dating is still imprecise (late Caradoc through Ashgill). Deep oceanic and basinal sequences, particularly in mid latitudes, show evidence of erosion, oxygenation, and variation in the preservation of carbonates; slope facies exhibit large slumps and biofacies shifts. These and other features can be related to the influence of the glacial event. Low latitude carbonate platforms reveal a distinctive transgressive-regressive pattern and in marginal basins a more subtle eustatic record is preserved suggesting multiple eustatic cycles in the Hirnantian.

Integration of biofacies and biostratigraphic data into the physical stratigraphy allows further refinement of the glacial event stratigraphy on a global scale. This evidence suggests that contintental glacial conditions developed in the late Caradoc or early Ashgill. Two major fluctuations occurred during the Ashgill with maximum glaciation in the late Ashgill (Hirnantian) but with significant fluctuations even within this interval. Deglaciation appears to have been a rapid event in earliest Silurian time (Rhuddanian). The Ashgill was approximately 10 Ma in duration (448 - 438 Ma, Harland et al. 1982) and the Hirnantian may have been approximately 2 Ma.

Faunal extinctions and changes appear to mirror the physical, and interpreted environmental, events suggesting a cause and effect relationship. The influence of plate motions through this brief time interval appears to have been minor. No evidence has yet been advanced for a bolide impact. The glacial event seems to have had greatest impact on mid to high paleolatitude faunas with Silurian replacement faunas originating primarily from low latitude marginal basin and slope environment.

To date only limited isotopic data has been published to suggest trends in oceanic chemistry or temperature through this interval. Largely on stratigraphic, sedimentologic, and faunal evidence it can be argued that the partial, and then virtually complete, draining of warm, somewhat hypersaline epeiric seas of the low latitude platforms, combined with global cooling trends, destroyed their endemic faunas; the main turnover occurred during the Hirnantian (Gamachian). The more cosmopolitan faunas of the slope and oceanic environments are significantl modified during the Asghill with their main turnover also in the Hirnantian. Fundamental temperature changes may alone explain their demise, but the increased circulation and oxygenation in these environments may have been equally disruptive to stable communities that had evolved earlier in the Ordovician. The Ashgill glaciation, of dimensions only slightly smaller than Quaternary ice sheets (Hambrey 1985) and developi and oscillating through a similar time interval, was the first major glacial event to affect the Paleozoic biotas. The actual pattern of extinction for the best known faunal groups is not dissimilar to the Plio-Pleistocene mass extinction described by Stanley (1986) from the Western Atlantic. Sepkoski (1982) calculated that 22 % of all families became extinct; if correct, this extinction event is second only to the Permo-Triassic extinctions in severity.

REFERENCES

BARNES, C.R. & BERGSTRÖM, S.M. (in press): Conodont biostratigraphy of
   the uppermost Ordovician and lowermost Silurian.- in: COCKS, L.R.M:
   & RICKARDS, R.B. (eds.): A global analysis of the Ordovician-Siluri-
   an boundary.- Natl. Mus. Wales.
BERRY, W.B.N. & BOUCOT, A.J. (1973): Glacio-eustatic control of Late
   Ordovician-Early Silurian platform sedimentation and faunal changes.-
   Geol. Soc. Amer. Bull. 84, 275-284.
BEUF, S.; BIJU-DUVAL, B.; STEVAUX, J. & KRILBICKI, G. (1966): Ampleur
   des glaciations "siluriennes" au Sahara: leurs influences et leurs
   consequences sur la sedimentation.- Rev. Inst. fr. Petrole 21, 363-
   381.
-- ; -- ; DE CHARPAL, O.; ROZNON, P.; GABRIEL, O. & BENNACEF, A. (1971):
   Les gres du Paleozoique inferior du Sahara - sedimentation et discon-
   tinuites, evolution, structural d'un craton.- Inst. fr. Petrol.-sci.
   et Techn. du Petrol. 18, 1-464.
BRENCHLEY, P.J. (1984): Late Ordovician extinctions and their relationship to the Gondwana glaciation.- in: BRENCHLEY, P.J. (ed.): Fossils
   and Climate.- J. Wiley & Sons, Chichester, 291-316.
-- & COCKS, L.R.M. (1982): Ecological associations in a regressive
   sequence: the latest Ordovician of the Oslo-Asker district, Norway.-
   Palaeontology 25, 783-815.
-- & NEWALL, G. (1980): A facies analysis of Upper Ordovician regressive
   sequences in the Oslo region, Norway - a record of glacio-eustatic
   changes.- Palaeogeogr., Palaeoclim., Palaeoecol. 31, 1-38.
CHATTERTON, B.D.E.; LESPÉRANCE, P.J. & LUDVIGSEN, R. (1983): Trilobites
   from the Ordovician-Silurian boundary of Anticosti Island, eastern
   Canada.- in: Papers for the symposium on the Cambrian-Ordovician and
   Ordovician-Silurian boundaries, Nanjing, China, October, 1983. Nanjing
   Institute of Geology and Palaeontology, Academia Sinica, 144-145.
DUFFIELD, S.L. & LEGAULT, J.A. (1981): Acritarch biostratigraphy of
   Upper Ordovician-Silurian sequence of Anticosti Island.- in: LESPÉRAN-
   CE, P.J. (ed.): Stratigraphy & Paleontology. Subcomm. Silurian Strati-
   graphy & Ordovician-Silurian Working Group. Dépt. Géol., Univ. Mon-
   tréal, 91-95.
HAMBREY, M.J. (1985): The Late Ordovician-Early Silurian glacial period.-
   Palaeogeogr., Palaeoclim., Palaeoecol. 51, 273-290.
HARLAND, W.B.; COX, A.V.; LLEWELLYN, P.G.; PICKTON, C.A.G.; SMITH, A.G.
   & WALTERS, R. (1982): A geological time scale.- Cambridge Univ. Press,
   Cambridge, 131 p.
JAANUSSON, V. (1979): Ordovician.- in: ROBISON, R.A. & TEICHERT, C.
   (eds.): Treatise du invertebrate paleontology. An Introduction, A136-
   A166.
KOREN', T.N. & NIKITIN, I.F. (1982): Comments on Report No. 45: Grapto-
   lites about the Ordovician-Silurian boundary by R.B. RICKARDS.-
   Ordovician-Silurian Boundary Working Group (I.U.G.S.), unpubl. rep.,
   9 p.
-- ; SOBOLEVSKAJA, R.F.; MIKHAJLOVA, N.F. & TZAI, D.T. (1979): New evidence on graptolite succession across the Ordovician-Silurian boundary in the Asian part of the USSR.- Acta Palaeont. Polonica 24, 123-
   136.
LESPÉRANCE, P.J. (1974): The Hirnantian fauna of the Percé area (Québec)
   and the Ordovician-Silurian boundary.- Amer. J. Sci. 274, 10-30.
-- (1985): Faunal distributions across the Ordovician-Silurian boundary,
   Anticosti Island and Percé, Québec, Canada.- Can. J. Earth Sci. 22,
   838-849.
LENZ, A.C. (1976): Late Ordovician-Early Silurian boundary in the northern Canadian Cordillera.- Geology 4, 313-317.
McCRACKEN, A.D. & BARNES, C.R. (1981): Conodont biostratigraphy and
   paleoecology of the Ellis Bay Formation, Anticosti Island, Quebec,
   with special reference to Late Ordovician-Early Silurian chronostra-

tigraphy and the systemic boundary.- Geol. Surv. Canada, Bull. 329, 51-134.

MU, E. & LIN, Y. (1983): Graptolites from the Ordovician-Silurian boundary sections of Yichang area, W. Hubei.- in: Papers for the symposium on the Cambrian-Ordovician and Ordovician-Silurian boundaries, Nanjing, China, 1983. Nanjing Inst. Geol. Palaeont., Academia Sinica, 107-109.

OWEN, A.W. & RUSHTON, A.W.A. (1984): Trilobites in British stratigraphy. Geol. Soc. London, Spec. Rep. 16.

RICKARDS, R.B. (1982): Graptolites about the Ordovician-Silurian boundary.- Ordovician-Silurian Boundary Working Group (I.U.G.S.), unpubl. rep. 45.

SHEEHAN, P.M. (1973): The relationship of Late Ordovician glaciation to the Ordovician changeover in North American brachiopod faunas.- Lethaia 6, 147-154.

-- (1975): Brachiopod synecology in a time of crisis (Late Ordovician-Early Silurian).- Paleobiology 1, 205-212.

STANLEY, S.M. (1986): Anatomy of a regional mass extinction: Plio-Pleistocene decimation of the Western Atlantic bivalve fauna.- Palaios 1, 17-36.

SWEET, W.C. (1985): Conodonts: those fascinating little whatzits.- J. Paleont. 59, 485-494.

WILLIAMS, S.H. (1983): The Ordovician-Silurian boundary graptolite fauna of Dob's Lin, southern Scotland.- Palaeontology 26, 605-639.

A BIG EVENT OF LATEST ORDOVICIAN IN CHINA

RONG Jia-Yu & CHEN Xu *)

A contribution to Project GLOBAL BIO-EVENTS

The examination of the last appearances of animal genera and species through the Early Palaeozoic showed that there was a mass extinction at the end of the Ordovician in China. This event occurred before the zone of _Glyptograptus persculptus_ **) and affected a wide range of unrelated groups, both benthic and pelagic (Mu et al. 1974, 1984; Rong 1979, 1983, 1984; Mu 1983; Chen & Lenz 1984).

This conclusion is drawn on account of material which has been obtained during the last two decades. A great number of sections across the Ordovician-Silurian boundary have been studied in China. Continuous sequences, representing Ashgillian and an Early Llandoverian interval, are still relatively common in this country (Mu 1983), although the effects of eustatic sea-level fall and regional or local uplift during the Ashgillian caused unconformities between Ordovician and post-Ordovician rocks in North China and other areas, where at least Silurian has not been deposited. The sequences across the O-S boundary in northwest China, northernmost China, and southern South China are mostly continous, with great thickness, but complicated in geologic structure, and rare in fossils (Mu 1983). That is not the case for the central-southwestern China (Yangtze) and Tibet-western Yunnan regions, especially in western Hubei and the border area of Sichuan and Guizhou Provinces, where the Ashgillian and early Llandoverian rocks are unusually well represented by platform deposits, far less in thickness, rich in fossils (both benthic and pelagic) with good exposures. Therefore, all data discussed herein, came from these regions.

The Wufeng Formation (Ashgillian) and the Lungmachi Formation (Llandoverian) are composed mainly of siliceous shales and contain abundant graptolites. Between these two formations, there is a thin bed, the socalled Kuanyinchiao Bed, consisting mainly of mudstone, argillaceous limestone and silty shale, in which a shelly fauna, i.e. _Hirnantia-Dalmanitina_ Fauna, is present. The graptolite successions associated with this shelly fauna are as follows:

**) Early preference by many specialists was to use the _persculptus_ Zone as the Ordovician-Silurian boundary level. This level cannot be defined at Dob's Linn, Moffat, Scotland. The definition that the boundary point coincides with the base of _Parakidograptus acuminatus_ has been recently approved by the International Commission on Stratigraphy.

*) Nanjing Institute of Geology and Palaeontology, Academia Sinica, Nanjing, P.R.C.

Lungmachi Formation (early Llandoverian)	"Eospirigerina" Fauna	Pristiograptus cyphus Zone Cystograptus vesiculosus Zone Parakidograptus acuminatus Zone Glyptograptus persculptus Zone
Wufeng Formation (Ashgillian)	Hirnantia-Dalmanitina Fauna	Diplograptus bohemicus Z. Paraorthograptus uniformis Z Diceratograptus mirus Zone Tangyagraptus typicus Zone Dicellograptus szechuanensis Zone

The significant extinctions occurred not only in shelly organisms, but also in the graptolite fauna. The most striking extinction took place among the brachiopods, rugose corals, trilobites, and nautiloids have obviously been also affected.

The latest Ordovician brachiopod assemblages were named as Hirnantia Fauna (Temple 1965) which is geographically widespread and distinctive (Rong 1979, 1984). The fauna consists of a great number of articulate genera, including Toxorthis, Dalmanella, Trucizetina, Kinnella, Hirnantia, Draborthis, Mirorthis, Dysprosorthis, Aegiromena, Leptaena, Leptaenopoma, Paramalomena, Aphanomena, Eostropheodonta, Fardenia or Coolina, Triplesia, Oxoplecia, Cliftonia, Onychoplecia, Dorytreta, Plectothyrella and Hindella (= Cryptothyrella). It seems that this fauna abundantly appeared at the beginning of the late Ashgillian (i.e. Hirnantian), spread rapidly, and became extinct as a whole before the Glyptograptus persculptus Zone. Among the constituents mentioned-above, there are ten genera (Toxorthis, Trucizetina, Kinnella, Draborthis, Mirorthis, Dysprosorthis, Paromalomena, Leptaenopoma, Dorytreta and Plectothyrella) and many species of such genera as Dalmanella, Hirnantia Leptaena, Aphanomena, Eostropheodonta and Hindella, which disappeared at the latest Ashgillian, i.e. pre-persculptus Zone. The latter itself is overlain by the Parakidograptus acuminatus Zone. Up to now, no irrefutable evidence can prove the Hirnantia Fauna to be in association with the zone of G. persculptus all over the world. G. persculptus forma A and forma B co-occurred with the Hirnantia Fauna in the Durben Bed of the Chu-Illi Mountains in Kazakhstan (Rukavishnikova et al. 1968, Mikhajlova 1970, Nikitin 1976, Koren' et al. 1979) actually belong to the Diplograptus boehmicus or Climacograptus extraordinarius Zone (W_6) of latest Asghill rather than G. persculptus Zone (L_1) (Mu et al. 1984). It is also believed that the Keisley limestone containing the Hirnantia Fauna at Keisley, northern England, is just underlain by the G. persculptus Zone (L_1) (Wright 1985).

The trilobite Dalmanitina fauna (mainly including Dalmanitina and

its close relates, Platycoryphe and Leonaspis), the coral fauna (Bore-
alasma, Brachyelasma, Streptelasma, Gryngwinkia, Lambeophyllum, and
Xinkiangophyllum), and the nautiloid fauna (Pleurothoceras and others)
as a whole, were also severely affected (Zhu & Mu 1984, Ho 1978, Zou
1985). It is apparent that almost all of these taxa had their last
appearance in the Hirnantian Kuanyinchiao Beds, with the exception of a
few genera, such as Dalmanitina and Leonaspis, which rarely survived
into the base of the Lungmachi Formation (Early Llandovery) in Northern
Guizhou (Zhan, Chen et al. 1964).

Concerning graptolites, immediately above the level of the Ordovician-Silurian boundary, the Silurian monograptid fauna begins within
an adaptive radiation, which followed the most intensive phylogenetic
crisis in the history of the Ordovician dicellograptid fauna. There is
a big change in graptolite aspects presented in the duration of the pre-
and post-Hirnantian (Chen & Lenz 1984). The dicellograptid fauna has been
in decline for some time in the late Ordovician. In the middle Wufeng
Formation (W_2 - W_3, middle Ashgillian) of the Yangtze Region, there
occurs a high diversified graptolite fauna, within which Pleurograptus,
Neurograptus, Nymphograptus, Pararetiograptus, Arachniograptus, Plegmato-
graptus, and Phormograptus became extinct at the end of Dicellograptus
szechuaensis Zone (W_2). In addition, Leptograptus and most species of
Dicellograptus, Tangyagraptus, Paraplegmatograptus, aand Yinograptus became extinct by the end of Tangyagraptus typicus Zone (W_3). Dicellograptus made it's last appearance in the latest Ashgillian rocks (W_6, equally
to Climacograptus extraordinarius Zone). It indicates that the crisis of
dicellograptid fauna was manifested in a mass extinction of representatives of almost all phylogenetic lines existing at that time. Only some
conservative stocks, mainly diplograptids (such as Climacograptus, Orthograptus, Glyptograptus, Diplograptus, Amplexograptus, Paraorthograptus,
and Pseudoclimacograptus), survivied into the Silurian. Even in these
conservative genera, the changes of species groups are still obvious.
This big event leads to the replacement of the longispinus group of
climacograptids by the scalaris group, the amplexicaulis (= truncatus)
group of orthograptids by the angustiformis group, the uniformis group
of glyptograptids by the persculptus-sinuatus and tamariscus group, the
bohemicus group of diplograptids by the modestus group, the disjunctus
group of amplexograptids by the elegans group, and the anhuiensis group
of pseudoclimacograptids by the hughesi group. The pacificus group of
paraorthograptids became extinct at the end of Ashgillian, and only a
few species (for example, P. innotatus) survived into the Llandovery.

The monograptid fauna starts from the G. persculptus Zone. As an
ancestor of the family Monograptidae, Atavograptus has been found to

occur in the persculptus Zone, both in China and Great Britain. Although the ancestor of Akidograptus has been known from the Tangyagraptus typicus Zone (W_3), the flourishing of akidograptids was in the persculptus and acuminatus Zones. From the persculptus Zone to the acuminatus Zone, the peak of A. ascensus was followed by a peak of Parakidograptus acuminatus (L_2). Mu et al. (1982) proposed, therefore, that the akidograptid-dimorphograptid subfauna should be considered as the first subfauna of the monograptid fauna. It should be also pointed out that representatives of monograptids have been reported from the persculptus Zone in Great Britain (Richards et al. 1977).

If many important biologically and ecologically unrelated widespread groups disappeared at about the same time, general, rather than particular causes should be sought. These should be of cosmopolitan impact. Many possible causes, such as geomagnetic reversal and extraterrestrial events, climatic and sea-level changes and others have been proposed to account for those mass extinctions. There was an indirect effect in the case of the late Ordovician episode, due to withdrawal of epicontinental seas as a consequence of growth of the ice-sheet, possibly in north-western or western Africa. It was suggested by Sheehan (1973, 1975, 1982) and Brenchley (1985) that shallow water invertebrates suffered a major extinction at the Ordovician-Silurian boundary and the extinction was associated with glacio-eustatic lowering of sea level which destroyed most shallow-water epicontinental sea habitats. This explanation is suitable for interpreting examples from North America, North Europe and other regions, but it is not the case to account for those in South China, where the strata across the Ordovician-Silurian boundary are completely continuous. The disappearing of the shelly fauna (including the Hirnantia-Dalmanitina Fauna) was not due to the main drop in sea-level and withdrawal of epicontinental seas.

The importance of black shales of the mid-Ashgillian Wufeng Formation and its equivalents for an anoxic environment in large pats of the South China region, has been recently discussed by Chen et al. (in press). The ventilation of deep ocean waters would have primarily resulted from the high latitudes during the late Ordovician glaciation. The minimum depth of oxygen would increase at this interval (see Berry & Wilde 1978), when the South China Plate was situated in a low latitude and was surrounded partially by lands. The oxygenated fresh waters from the lands may have emptied into the sea and the superficial waters of the sea may have been ventilated from the intermixing fresh waters. These formed a suitable habitat for rich graptolites, associated with rare inarticulate brachiopods, radiolaria, trilobites (such as Triarthrus) and nautiloids (such as _Pleurothoceras_). It is inferred that water depth of the Yangtze platform, which was restricted and different from open seas, was not deeper than about 150 m. Therefore, there was no deep water ventilated by deeper cold water from the high latitude areas.

In late Ashgillian (= Hirnantian), the _Hirnantia-Dalmanitina_ Fauna (mailly BA 2-3) occurred and flourished in this area (Rong 1979, 1984, 1986). It is postulated that in this case two main controlling factors would be an eustatic sea-level fall and the appearing of ventilation at sea bottom of the Yangtze platform. It seems to the writers that the glacio-eustatic lowering of sea-level did not destroy the shallow-water, epicontinental sea habitats of this area. A large part of the sea bottom came up to the ventilated surface water layer and the minimum depth of oxygen increased when the late Ordovician glaciation reached its climax.

In earliest Llandoverian time, an anoxic condition returned due to an immediately rapid sea-level rise and produced an increasing of the minimum depth of oxygen and a disappearing of ventilation at the sea bottom. It decimated shelly faunas adapted to shallow water and oxic conditions. Thus, the _Hirnantia-Dalmanitina_ Fauna was severely affected and eventually became extinct as a whole at the beginning of the _Glyptograptus persculptus_ Zone (L_1), which occurs at the base of the Lungmachi Formation in South China.

ORDOVICIAN-SILURIAN BOUNDARY EVENT IN BOHEMIA (PRAGUE BASIN-BARRANDIAN AREA)

ŠTORCH, Petr *)

A contribution to Project GLOBAL BIO-EVENTS

Abstract: The event character of the Ordovician-Silurian boundary chan-ges is well recognizable in the Barrandian area (Prague Basin). It could be studied in many sections in detail. There, the late Ordovician and the Ordovician-Silurian boundary sequences are marked by abrupt changes in the facies development and the faunal assemblages, which could well be explained by glacio-eustatic movements of sea-level, with associated environmental changes. The sudden disappearance of the rich, upper Krá-lodvor benthic assemblage is followed, by substitution of muddy sedimen-tation, by a regressive, flysch-like sequence of the lower part of the Ko-sov Series. The Kosov (Hirnantian) regressive sequence culminated with shallow marine sandstones and conglomerates. They were substituted by siltstones and mudstones due to the initial phase of the new transgres-sion in the uppermost part of the Kosov. The rich Hirnantia fauna occurs in the mudstones near the base of dark graptolitic shales, corresponding mostly just to the base of the Silurian.

Late Ordovician sequence

Considerable changes in lithology and faunal assemblages were recorded at the Králodvor/Kosov (Rawtheyan/Hirnantian) boundary in the Barrandian area (Štorch & Mergl, in press). The deep water mudstones of the Králův Dvůr Formation containing deep water faunal assemblages were followed by gravity flow deposition of coarse-grained subgreywackes and by silty sha-les at the base of the Kosov Formation. The high-diversified (more than 70 genera) trilobite-ostracode-brachiopod dominated Proboscisambon Community, containing warm-water elements of Anglo-Scandic origin (Deco-roproetus, Zetaproetus, Leonaspis, Miraspis, Staurocephalus, Trochurus), was replaced by the low-diversified and subsequently quickly disappearing Mucronaspis Community (Štorch & Mergl, in press), of which last fossils (bivalves and trilobite fragments) have been recorded from the shale of the lowermost Kosov Formation.

The basalmost Kosov subgreywackes were succeeded by a flysch-like sequence forming most of the thickness of the Kosov Formation and culmi-nating by deposition of shallow-water sandstones and conglomerates in the upper part of the formation. In heavy-bedded sandstones mass occur-rences of disarticulated, most likely infaunal bivalves give evidence of storm deposition in probably intertidal (Havlíček 1982) environment. In the uppermost part of the Kosov Formation, the quartz sandstones were substituted by siltstones and mudstones. Pale grey, bioturbated mud-stones and claystones contain rich (more than 40 genera), brachiopod dominated Hirnantia sagittifera Community near the top of the formation.

*) Geological Survey, 11821 Praha 1, Č.S.S.R.

Lecture Notes in Earth Sciences, Vol. 8
Global Bio-Events. Edited by O. Walliser
© Springer-Verlag Berlin Heidelberg 1986

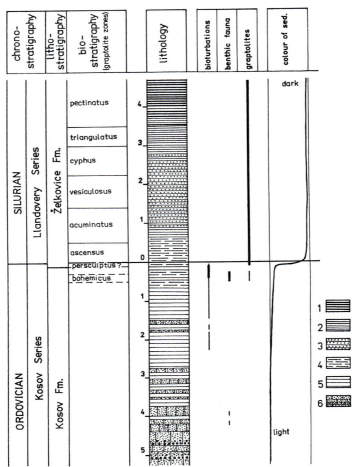

Figure 1. Late Ordovician - basal Silurian history in the Prague Basin (stratigraphy, lithology, faunal distribution, and assumed sea-level changes) - simplified ideal section.
1 - graptolitic shales, 2 - claystones, mudstones and siltstones, 3 - fine and medium-grained sandstones, 4 - coarse-grained sandstones and petromict conglomerates, 5 - coarse-grained subgreywackes, 6 - muddy limestone layer, 7 - limestone nodules and lenses; A - Rafanoglossa leiskowiensis Community, B - Proboscisambon Community, C - Mucronaspis Community, D - bivalves, E - Hirnantia sagittifera Community.

The Hirnantia fauna, referred by Havlíček (1982) to the subtidal environment, was found only in the eastern part of the Prague Basin (Barrandian area) up to now. A gradual deepening of the sea could be assumed in the uppermost Kosov (Hirnantian) of the Prague Basin.

The base of the Bohemian uppermost Ordovician Kosov Series is well comparable to the base of the Hirnantian Stage owing to the presence of widely recognized (Berry & Boucot 1973, Brenchley & Cocks 1982, Brenchle

& Cullen 1984, Brenchley & Newall 1984) considerable changes in lithology and faunal assemblages of which an event character is suggestive. The similarity of the Bohemian Králodvor-Kosov boundary sequences to the Rawtheyan-Hirnantian boundary sequences explained by environmental changes caused by rapid growth of Gondwanaland ice-sheet is remarkable. According to Brenchley (1984) the glacio-eustatic regression culminated in the upper Hirnantian. It well corresponds to the regressive Kosov sequence which culminates with heavy-bedded shallow-water sandstones and petromict conglomerates. The new transgression started in the uppermost Kosov and brought the new faunal assemblage, i.e. the Hirnantia sagittifera Community, after a long period since the Mucronaspis Community disappeared at the base of the series. The graptolite Glyptograptus bohemicus Marek accompanies the world-wide distributed Hirnantia fauna in Bohemia and supports the international biostratigraphic correlation of the sequence (Glyptograptus bohemicus Zone). The layer containing the Hirnantia fauna is separated from the first graptolitic shales by at least 0.2 m thick, mostly heavily bioturbated mudstone.

The Ordovician-Silurian boundary and the early Silurian sequence

In general, the sedimentation is continuous through the Ordovician/Silurian boundary in the Prague Basin, in spite of some differences between the different sections. The most quiet sedimentation in probably the deepest parts of the basin is developed in the sections along the whole south limb of the basin. A complete succession starting with the Akidograptus ascensus Zone has been recorded there in all localities. Clayey shales with climacograptids and rare glyptograptids were recorded even below the first occurrence of Akidograptus ascensus Davies at two localities and could represent the upper part of the Glyptograptus persculptus Zone. The ascensus Zone is developed as clayey shales. Sandy micaceous laminites set on in the course of the following Parakidograptus acuminatus Zone.

The sequence continued without any break through the whole Llandovery in the south limb of the Prague Basin. Laminites are subsequently substituted by siliceous shales which culminate with silty silicites in the Demirastrites convolutus Zone. The most intensive sedimentation with the largest thickness of the basal Silurian graptolite zones was reported by Štorch (1982, 1986) in the Řepy and Běchovice sections.

Comparative studies of the sections from the sedimentological point of view and a detailed intrazonal biostratigraphy (Štorch 1986) revealed that the laminites represent more condensed sedimentation than clayey

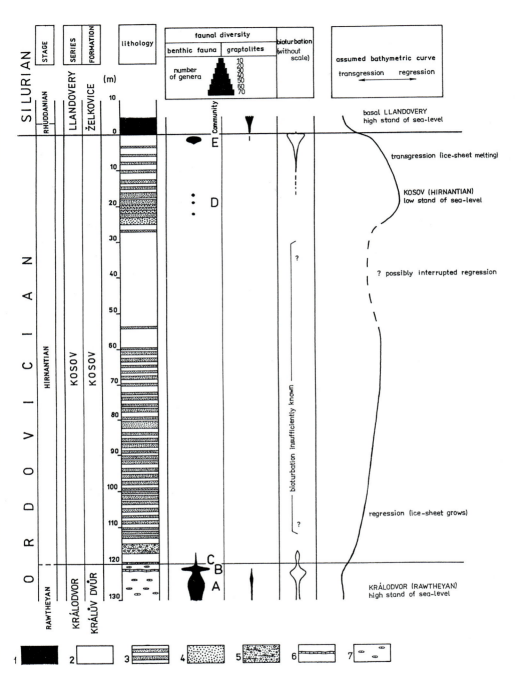

Figure 2. Ordovician-Silurian boundary in the Prague Basin (Barrandia area) - summarized ideal section.
1 - black siliceous shales and silty silicites, 2 - dark grey or black shales, 3 - laminites, 4 - clayey shales, claystones, 5 - light grey mudstones and siltstones, 6 - pale grey quartz sandstones and petromict conglomerates.

and silty shales. The onset of laminites in the acuminatus Zone in the southern limb of the basin and in the Řepy and Běchovice area was assumed to be synchronous with the start of the break in sedimentation in the northern limb of the basin.

A long break in sedimentation is known from the restricted part of the Prague Basin (Malá Chuchle, Pankrác, Nová Ves and Tachlovice sections). The topmost Ordovician mudstones are followed there by graptolitic shales of the Monoclimacis griestoniensis Zone (Litohlavy Formation, upper Llandovery). In this case I presume the possibility of additional washing of previously deposited incoherent clayey and muddy sediments of the basalmost Silurian (the ascensus and the lower part of the acuminatus Zones) and perhaps even topmost Ordovician layers (several tens of centimetres in thickness). In the vicinity of Stodůlky and Řeporyje, the above mentioned break in sedimentation splits into two shorter gaps. The earlier of them starts above the ascensus Zone (and/ or in the course of the acuminatus Zone) and thus supports the present assumption of the time of origin of the long break recorded in the limited part of the basin.

Conclusive remarks

The abrupt extinction of cystoids, highly diversified trilobites and other elements of relatively deep water Proboscisambon Community (Štorch & Mergl, in press) could be correlated with the extinction of trilobite-cystoid-gastropod faunas in the outer shelf conditions near the Rawtheyan -Hirnantian boundary (described by Brenchley 1984). In addition, other important changes as graptolite extinction, decrease of bioturbation and disappearance of limestone nodules could be observed at the top of Králodvor Series and thus support our correlation. When we also consider the following gravity flow sedimentation at the beginning of the Kosov regressive sequence, we have a set of evidences for rapid onset of glaciation which may serve for widespread stratigraphic correlations including the correlation between the Bohemian Králodvor/Kosov boundary and the Rawtheyan/Hirnantian boundary in the Anglo-Scandic province.

The Ordovician/Silurian boundary is marked by abrupt change from pale mudstones and claystones containing the Hirnantia fauna to the euxinic facies of black graptolitic shales. The sudden changeover in colour and structure of sediments mostly corresponds in the Prague Basin to the base of the ascensus Zone. When compared with the British Isles, the base of the ascensus Zone in Bohemia is comparable to the base of the acuminatus Zone at the type section Dob's Linn (described by Williams 1983). A low-diversified climacograptid-glyptograptid assenblage was recorded in the Prague Basin in several localities at the base of graptolitic shales just below the ascensus Zone. It is referred (with a question-mark) to the upper part of the persculptus Zone, in spite of the fact that the true Glyptograptus persculptus (Salter) has not yet been verified there. However, in most of the sections the layer containing the climacograptid-glyptograptid assemblage has not been developed because of the extreme reduction of sedimentary rates during the rapid rise of sea-level at the Ordovician-Silurian boundary.

The speed of sedimentation in the lowermost Silurian (up to the cyphus Zone) was calculated (Štorch 1986) to be about 1 m of shales per 10^6 years (afterwards up to 7.5 m per 10^6 years) in the Prague Basin.

The advancing transgression caused a further deepening of the sea during the acuminatus Zone and probably gave origin to a mildly intensiv bottom current in the deeper central part of the linear depression of the Prague Basin. The current is considered to cause the above mentioned local breaks in sedimentation, somewhere perhaps accompanied with a slight subaquatic erosion (Štorch 1986). In sites with decreased washing ability of the current, condensed sedimentation of laminites appeared.

REFERENCES

BERRY, W.B.N. & BOUCOT, A.J. (1973): Glacio-eustatic control of Late Ordovician - Early Silurian platform sedimentation and faunal changes - Geol. Soc. Amer. Bull. 84, 275-284.
BRENCHLEY, P.J. (1984): Late Ordovician Extinctions and their Relationship to the Gondwana Glaciation.- in: BRENCHLEY, P.J. (ed.): Fossils and Climate. J. Wiley & Sons Ltd., Chichester, 291-315.
-- & COCKS, L.R.M. (1982): Ecological associations in a regressive sequence - the latest Ordovician of the Oslo-Asker District, Norway.- Palaeontology 25, 783-815.
-- & CULLEN, B. (1984): The environmental distribution of associations belonging to the Hirnantia fauna - Evidence from North Wales and Norway.- in: BRUTON, D.L. (ed.): Aspects of the Ordovician System. Pala ont. contr. Univ. Oslo 295, 113-125.
-- & NEWALL, G. (1984): Late Ordovician environmental changes and their effects on faunas.- in: BRUTON, D.L. (ed.): Aspects of the Ordovicia System. Palaeont. contr. Univ. Oslo 295, 65-79.
HAVLIČEK, V. (1982): Ordovician in Bohemia: development of the Prague Basin and its benthic communities.- Sborník geol. Věd., Geologie 37, 103-136.
ŠTORCH, P. (1982): Ordovician-Silurian boundary in the northernmost par of the Prague Basin (Barrandian, Bohemia).- Věstník Ústř. Úst. geol. 57, 231-236.
-- (1986): Ordovician-Silurian boundary in the Prague Basin (Barrandian area, Bohemia).- Sborník geol. Věd., Geologie 41, in press.
-- & MERGL, M. (in press): Králodvor/Kosov boundary and the late Ordovician environmental changes in the Prague Basin (Barrandian area, Boh mia).- Sborník geol. Věd., Geologie.
WILLIAMS, S.H. (1983): The Ordovician-Silurian boundary graptolite faun of Dob's Linn, Southern Scotland.- Palaeontology 26, 605-639.

EARLY ORDOVICIAN EUSTATIC CYCLES AND THEIR BEARING ON PUNCTUATIONS IN EARLY NEMATOPHORID (PLANKTIC) GRAPTOLITE EVOLUTION

A contribution
to Project
GLOBAL
BIO-
EVENTS

ERDTMANN, Bernd-Dietrich *)

Abstract: Early Ordovician nematophorid graptolite morphogenesis and evolution operate on two levels: Gradualistic changes involving species-level and intergeneric clades, and punctualistic (anagenetic) changes operating on suprageneric levels. Anagenetic characteristics, such as loss of bithecae and reduction of first order stipes (progressive atro-phies from quadriradiate, through triradiate to biradiate organizational schemes) are achieved polyphyletically but virtually concurrently at di-stinct stratigraphic levels which coincide with initial phases of global marine transgressions. In depositionally homotaxial deep water facies both modes of morphogenesis are observed, but only anagenetic (major, qualitative) changes appear to involve different cladogenetic lineages simultaneously at such specific stratigraphic positions. These anagene-tic changes represent principal morphological innovations, and their first appearances are consistently related to lithological disruptions in all investigated sections representative of constant deep water litho-facies and may therefore be causally related to global eustatic events. These cyclic events are depositionally verifiable in both outer shelf, shelf marginal, condensed starved basinal, and "oceanic" sequen-ces by sedimentological and/or geochemical signals. The sequence of Ordovician regressive cycles is dated as starting with the Latest Cam-brian Acerocare regressive event, being followed by the early/middle Tremadoc Peltocare, then by the Tremadoc/Hunnebergian Ceratopyge, the late Arenig/Llanvirn "Valhall", then a late Llandeilo (infra-Caradoc) and a terminal Ashgill "Hirnantian" event. The gradualistic introduc-tion of new morphogenetic characters usually involves first peripheral areas of rhabdosomes, then regressing proximally ("gerontomorphic innovation"). Therefore this often escapes the attention of taxonomists because mature and gerontic specimens are rarely preserved. Punctualis-tic changes specifically involve the proximal portion of rhabdosomes and are easily ascertained. Morphogenetic "patterns" or "stipe organi-zational schemes" are introduced at ecologically most productive regions, i.e. along the ocean-facing margins of low latitude continental plates, and then move both shelfward and into stable oceanic areas, although only gerontic forms of trophic generalists apparently inhabited the open Cambro-Ordovician oceans.

Introduction

Recent advances in our knowledge of palaeogeography and palaeooceano-graphy have enabled us to generate models and syntheses which, in turn, have stimulated new ideas and, occasionally, also created charmingly simplistic theories for the explanation of biological evolution. The new sensationalistic science ("cataclysmic geology") calls for physically induced punctuations caused by cosmic impact events (Alvarez et al. 1980), invoking neocatastrophic "calculative" cyclic extinction and radiation patterns (Fischer & Arthur 1977), or reassessing time-speci-

*) Institut und Museum für Geologie und Paläontologie der Universität, D-3400 Göttingen, F.R.G.

Lecture Notes in Earth Sciences, Vol. 8
Global Bio-Events. Edited by O. Walliser
© Springer-Verlag Berlin Heidelberg 1986

fic "opportunistic" accelerations of normal (background) evolutionary rates as responses to differential physical stress phenomena (Walliser 1984) to explain apparent punctuational perturbations of the intricate evolutionary processes. However, regardless of the emphasis which may be placed upon either external "holocaustic" interruptions or internal accelerations of gradualistic evolutionary developments, minor attention has been paid to the fundamentally different "evolutionary behaviour" of benthic as compared to non-benthic organisms with regard to their responses to marine eustatic cycles as well as with respect to the inherently different environmental interactions and spatial distribution strategies employed during adaptive radiations of these ecologically quite distinct biota.

Prior to the brief but profound Latest Cambrian "Acerocare regressive event" (ARE), a notable increase in abundance of pseudo-planktic "rooted dendroids" is observed in homotaxial ocean-facing palaeoslope sequences in western Newfoundland (Erdtmann 1986). The spontaneous rise of holo-planktic nematophorid graptolites together with first euconodonts is causally related to the sharp primary production increase in continent marginal and shelf seas (see Fig. 1) which were affected by the early Tremadocian pandemic "Dictyonema Shale" transgression (Andersson et al. 1985, Thickpenny 1985). The first nematophorids belong to two lineages (horizontal Staurograptus and pendent Rhabdinopora), both displaying a consistent quadriradiate proximal stipe pattern; this group being followed by several triradiate lineages (pendent "Matane-type" Rhabdinopora, horizontal Anisograptus, Radiograptus, and a reclined Psigraptus). A few degenerate triradiate forms survive the early/"middle" Tremadoc regression (PRE) and re-appear in the superjacent beds together with first biradiate Adelograptus tenellus and Bryograptus kjerulfi. The "uneventful" middle to late Tremadoc experienced only minor gradualistic developments, in particular the rise of the dorsally "folded" biradiate and bi- to quadriramous Kiaerograptus (incl. Triograptus) and of sicular-bithecate predecessors of Clonograptus. Upon termination of the global "Ceratopyge regression " (CRE), at the end of the Tremadoc, all bithecae bearing dendroids (Anisograptidae) had disappeared (first major extinction event). Together with the initial Hunneberg transgression "epigonic mimics" of forms which were originally introduced along shelf margins re-appear in open pelagic or cold-water shelf habitats, such as the La2 biradiate pendent Araneograptus (Erdtmann & VandenBerg 1985) as a late "imposter" of the early Tremadoc Rhabdinopora flabelliformis. A probably two-phased radiation of early graptoloids occurs during the subsequent pre-Arenig Hunneberg Stage (La2 to Be1 of Australasia) involving at leas

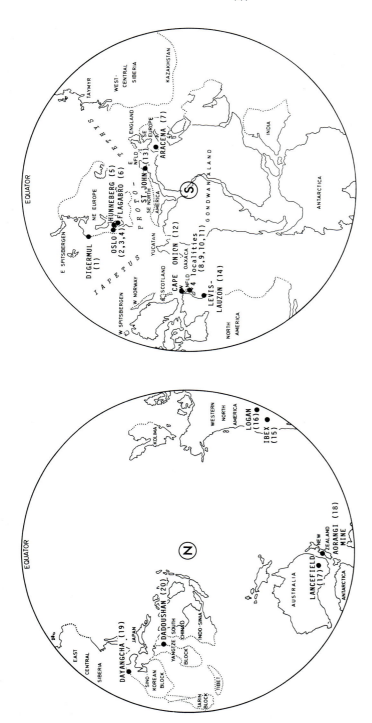

Figure 1. Early Ordovician palinspastic world maps of northern (left) and southern (right) hemispheres, polar projections (from Erdtmann 1982), incl. investigated localities. Numbers (in parenthesis after locality names) correspond to the localities listed in Table 1.

four separate clades (see Fig. 2): 1. a pendent lineage (bryograptids/
pendeograptids), 2. a subhorizontal/deflexed clade (adelograptids/sigma-
graptids and kinnegraptids), 3. a strictly horizontal lineage (adelo-
graptids/clonograptids), and 4. a variably oriented declined/horizontal/
reclined lineage with isolated metathecae and/or dorsal prothecal foldin
(psigraptids and kiaerograptids/sinograptids). The stipe reduction
"events" and the loss of bithecae affected all these lineages virtually
synchroneously during the terminal Tremadoc "Ceratopyge regressive event
(CRE) regardless of deep or shallow, cold or warm biofacies. The Hunne-
berg-Arenig transgressive development fostered the greatest radiation
of planktic graptolites (Graptoloidea) of all times prior to a new crisi
(second major extinction event, see Fig. 3) related to the pre-Whiterock
(late Castlemainian to Yapeenian) Valhallan regression (VRE), which was
followed by the late Arenig to early Llanvirn rise of polyphyletically
derived biserial graptolites (Axonophora). Middle and Late Ordovician
regressions are also succeeded by major radiations of biserial and
pseudo-biserial forms (Fig. 3), such as Orthograptus, Dicranograptus and
Amplexograptus. It is hypothesized that all anagenetic macro-evolutionar
stages of graptolite radiations are causally related to watermass perturbations, which, in turn, resulted from glacially or Pitman-induced
global eustatic cycles.

Investigations in depositionally homotaxial continent margin sequences (e.g. Cow Head Group of western Newfoundland) and correlations acros
different facies have revealed that both conodonts and graptolites experienced mainly innovative phases (without preceding extinction events
of morphogenetic radiations during the Late Cambrian to Early Ordovician
interval. The cladogenetic aspects of these graptolite radiations are
apparently facies (water depth and temperature) controlled and are prop-
agated like "innovative waves" from the ecologically most productive
continental margins (upwelling?, high primary productivity rates?) into
directions of increased physical and/or competitive stress, i.e. both
oceanwards and shorewards (Fig. 4). Anagenetic (first order) innovation
such as morphogenetic changes of familial and suprafamilial rank, are
virtually "spontaneous" and appear to affect globally all graptolite ta
regardless of their phylogenetic relations across different depth and

Figure 2. Diagrammatic illustration of evolution of the dendroid
family Anisograptidae and origin of earliest Hunnebergian and Arenigian
graptoloids. Several of the presumed phyletic lineages (clades) represent families or subfamilies, others are informally designated. Primary
stipe organizational schemes (quadri-, tri-, and biradiate) on right,
stratigraphic subdivisions and sea level curve on left. Note: "sheafed"
stauro-rhabdinoporids are earliest non-dissepimentous, slightly deflexe
"nematophorids" with variable nematic structures which may only represent juveniles of both Staurograptus and Rhabdinopora.

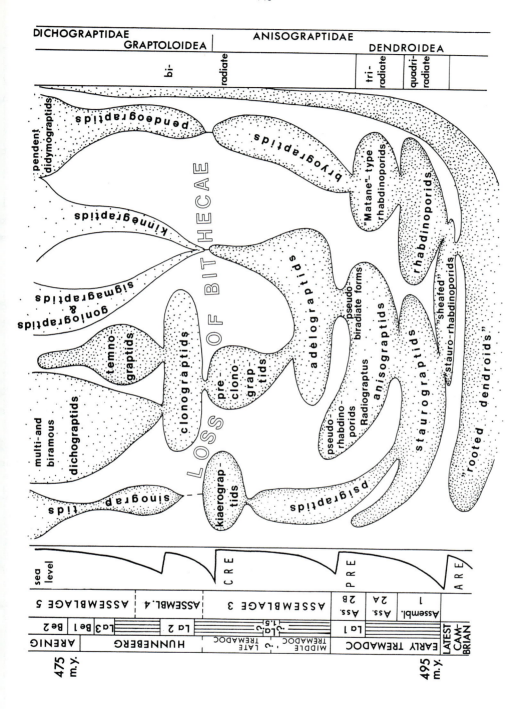

temperature controlled biofacies regimes. The morphogenetic innovations, i.e. the reduction of primary stipe generation from originally four (quadriradiate) to three (triradiate), and then to two (biradiate) withi the Anisograptidae, and eventually the complete loss of bithecae (dendroid/graptoloid transition) appear to be globally interconnected and stratigraphically congruent with worldwide regressive events.

Early Ordovician palaeogeography and sea level oscillations

Graptolite material has been taxonomically analyzed from representative sections of palaeogeographically widely separate locations. These locations include both roughly homotaxial and quite heterotaxial facies regimes, i.e. they belong to pelagic ("geosynclinal", continental slope, outer shelf and inner shelf areas, and palinspastically represent both reconstructed high latitude and low latitude regions (see Fig. 1). The stratigraphic interval here considered extends from the Latest Cambrian (Trempealeauan) to the early Arenig (Late Canadian or Late Ibexian) with preliminary investigations carried forth to the Middle Ordovician (Llanvirn-Llandeilo or Chazyan) strata. Where possible, the graptolites were collected bed-by-bed, i.e. with metric controls, including pertinent observations of the lithologies involved and general analyses of facies-related inventories of these lithologies. Twenty investigated localities and their palaeogeographic and palaeoenvironmental characteristics are listed in Table 1.

1. The Acerocare Regressive Event (ARE)

The global infra-Ordovician "Acerocare regressive event" or ARE (Erdtma 1984) or "Lange Ranch Eustatic Event", LREE of Miller 1984 is based on meteoric or shallow marine erosion phenomena in outer shelf carbonates of the equatorial North American, Siberian, Sino-Korean, and Australian platforms, on faunal and sedimentary (subaqueous and/or subaerial) disconformities on the Baltic Shield (Martinsson 1974), and on meteoric erosional disconformities or shallowing upwards criteria in the high-latitude peri-Gondwana or "Mediterranean" region (South America, NW Africa, "Avalonian-Armorican" Europe, Arabian Peninsula, see Erdtmann & Miller 1981, Cocks & Fortey 1982, Fortey 1984). The potential glacial cause for this profound regression was first mentioned by Erdtmann & Miller 1981 and affirmed by Erdtmann (1982), Fortey (1984) and Miller

Figure 3. Graptolite evolutionary pulses during the Ordovician in relation to eustacy and relative black shale frequency on the Balto-Scandinavian platform.

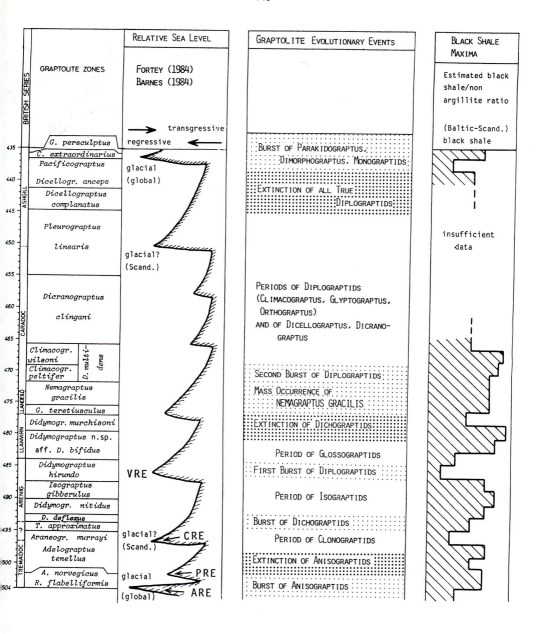

TABLE 1: Investigated graptolitic sections and their palaeogeographic-environmental parameters.

LOCATION OF SECTION	PALAEOGEOGR. SETTING	BATHYMETRY	LITHO. ENVIRONMENT	STRATIGR. INTERVAL
01. Digermul Penins., N. Norway	middle latitude	outer shelf	bl.shales, qtzt. (temp.)	U.€. - M.Tr.
02. Sofienberggte., Oslo, Norway	middle latitude	marg. basin	bl.shales, lst.nod.	L.Tr. - U.Tr.
03. Töyen tunnelbane, Oslo, Norway	middle latitude	marg. basin	bl.shales, lst.nod.	U.Tr. - Ar. - Llv. - L
04. Naersnes, S. Oslo, Norway	middle latitude	marg. basin	bl.shales, lst.nod.	U.€. - L.Tr.
05. Hunneberg, Västergtld., Sweden	middle latitude	outer shelf	glauc.lst., bl.shales	U.€. / U.Tr. - L.Ar.
06. Flagabro, Scania, S. Sweden	middle latitude	marg. basin	bl.shales, lst.nod.	U.€. - Ar. - Llv.
07. N. Aracena, S.ra Morena, S. Spain	high latitude	inner shelf	siltshales, drop st.	U.Tr. - L.Ar.
08. St. Paul's, W. Newfld., Canada	low latitude	mid slope	lst.brecc., lime-mudst.	U.€. - Tr. - Ar. - Llv
09. Broom Point N, W. Newfld., Canada	low latitude	mid slope	lst.brecc., lime-mudst.	U.€. - M.Tr.
10. Martin Point, W. Newfld., Canada	low latitude	distal slope	lime-mud laminarites	U.€. - Tr. - Ar.
11. Green Point, W. Newfld., Canada	low latitude	distal slope	lime-mud laminarites	U.€. - Tr. - Ar.
12. Cape Onion, N. Newfld., Canada	?low latitude	ocean floor	bl.shales, volcanics	U.€. - L.Tr.
13. St. John, New Brunsw., Canada	?low latitude	ocean floor	bl.shales, volcanics	U.€. - M.Tr.
14. Levis-Lauzon, Quebec, Canada	low latitude	slope	lst.brecc., lime mudst.	U.€. - Tr. - Ar. - Llv
15. Ibex area, W. Utah, USA	low latitude	outer shelf	lst. - mudshales	U.€. - Tr. - Ar. - Llv
16. Logan, NE. Utah, USA	low latitude	mid shelf	lst. - mudshales	U.€. - Tr. - Ar.
17. Lancefield, Vict., Australia	?low latitude	starved basin	bl.shales, sandst.	L.Tr. - Ar. - Llv.
18. Aorangi Mine, NW Nelson, N.Z.	?low latitude	starved basin	bl.shales, sandst.	U.Tr. - Ar.
19. Dayangcha, Jilin, P.R. China	low latitude	inner shelf	lst., brecc., mudst.	U.€. - Tr. - Ar.
20. Dadoushan, W. Zhejiang, P.R. China	low latitude	outer sh./slope	lst., brecc., shales	U.€. - M.Tr.

(1984). The glacial cause for the Ordovician regressions is generally hypothesized (Miller 1984, Fortey 1984), although reasonably datable Tremadoc and Arenig diamictites of potential glaciomarine origin are known from Salta in NW Argentina (Saladillo Formation: Keidel 1943, Harrington & Leanza 1957) and from southern Bolivia (Lohmann 1970, Castaños & Rodrigo 1978). In the Cow Head Group palaeoslope sequences limestone breccia intercalations between the Tuckers Cove and Broom Point Members of the proximal Shallow Bay and distal Green Point Formations (Williams, James & Stevens 1985) probably manifested the terminal Cambrian (ARE) regressive cycle boundary (James, Stevens & Fortey 1979)

2. The Peltocare Regressive Event (PRE)

Subsequent to the ARE the transgressive Dictyonema Shales expanded, staged by varieties of Rhabdinopora, from west to east onto an Upper Cambrian disconformity of the Baltic Shield (Lindström & Dworatzek 1979 Erdtmann 1984) or even directly onto a Precambrian basement in Mauritan (southern margin of Tindouf Basin: Destombes, Sougy & Willefert 1969

Figure 4.   Conceptual biofacies diagram across Cambrian-Ordovician boundary for selected significant trilobites, conodonts and graptolites Trophic generalists (pelagic forms) are placed in centre (cross-hachure Note their relative long stratigraphic ranges and shelfward migratory behaviour. A few representative warm-water trophic specialists are on left (obliquely hachured) and cold-water specialists on right (square hachure).

LOW LATITUDE PLATFORMS
(WARM CLIMATES)

OFF-SHELF AND OCEANIC AREAS
INCL. ARC SYSTEMS AND SUSPCT. TERRANES
REGARDLESS OF LATITUDE

HIGH LATITUDE PLATFORMS
(COOL CLIMATES)

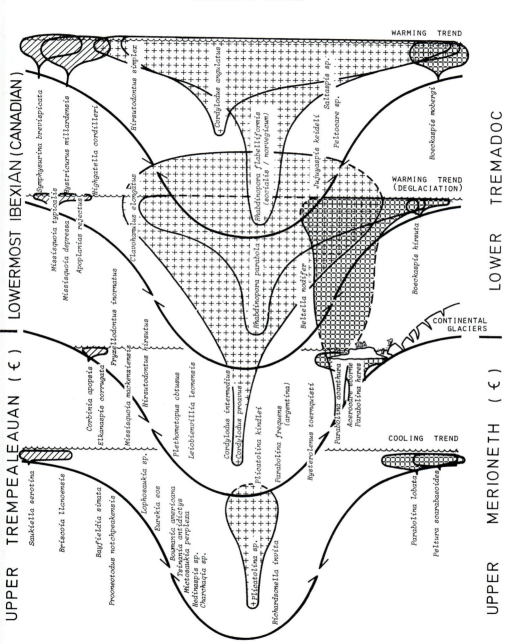

Deynoux, Sougy & Trompette 1985). Another less pronounced regressive event is noted particularly in the carbonate shelf sequences of North America, in the Georgina Basin of NW Queensland, Australia, in North China, possibly in Siberia, and in other parts of the world (shallowing indicated by black _Peltocare norvegicum_ bearing limestone bands in the Oslo area), a marine offlap, coeval with Miller's (1984) "Black Mountai Eustatic Event" (BMEE) and which, in keeping with Scandinavian conventions, is here called "_Peltocare norvegicum_ regression" or PRE. This regressive event (see Fig. 2) coincides with the first occurrences of the conodont _Cordylodus angulatus_ (Miller 1984), which also roughly correlates with the early to "middle" Tremadoc boundary as defined by Bulman (1954), or with the first appearance of biradiate anisograptids (_Adelograptus tenellus_). In the Green Point section (distal slope facie of western Newfoundland this event is documented by a ca. 7 m thick intercalation of slumped calcarenites containing the first _Cordylodus angulatus_ into a finely laminated succession of graptolitic lime-mud-stones below and above.

Middle to late Tremadoc sedimentation on slope sequences (i.e. Gree Point) are characterized by uninterrupted laminarites of green to black mudshales reflecting vacillating interfaces of denitrification and anoxic water layers with the seafloor (Berner 1981, Berry, Wilde & Hunt 1985). These beds are almost devoid of graptolites except for thin blac condensation horizons which contain a scarce low-diversity fauna of adelograptids and bryograptids. The terminal Tremadoc is indicated by brick-red to purple mud-laminarites and glauconitic beds completely devoid of graptolites.

3. The _Ceratopyge_ Regressive Event (CRE)

At Green Point the more than 100 m thick middle to late Tremadoc suite of laminarites is succeeded by only a few meters thick "package" of alternating ribbon limestone with brown to black mud-shale interbeds containing a La2-type graptolite fauna, including relief-preserved spec mens, which are bithecae-bearing in the lower beds and unbithecate in higher strata of this "package". The inception of ribbon limestones in the Cow Head Group is regarded as an indication of renewed transgressi conditions (James, Stevens & Fortey 1979). Upslope, on the carbonate platform of the correlative St. George Group (autochthonous and shelf facies counterpart of the Cow Head Group in western Newfoundland) a "pebble bed" in the upper Boat Harbour Formation is regarded as a mark horizon of the same regressive event (Boyce 1979, Stouge 1982). Else-where this significant regression may be traced to Scandinavia (diagenetic Ceratopyge Limestone intercalation into a virtually continuous black shale succession), to the Lake District of England (Molineux & Rushton 1984), and to many other coeval "breaks", which have been traced across different litho- and biofacies around the world (Fortey 1984, Fig. 5). Apparently, this "Ceratopyge regressive event" (CRE) was more profound than the ARE and the PRE because even in the deeper water Oslo-Scania-Lysogory Depression it is documented by several decimeter thick glauconitic nodular Ceratopyge limestones and a disconformity at its top, whereas the Cambro-Ordovician Acerocare interval is represented by continuous black shales and (diagenetic) anthraconitic black limestone nodules. Graptolite documentation is quite sparse in the lower "Hunne-bergian" beds (this Scandinavian stage name is here used to designate deposits and fossil zones spanning the Tremadoc/Arenig hiatus in Wales) of the marginal Baltic platform depression (Erdtmann 1965), but recent studies of the Hunneberg Mountain and drill core sections in SW Sweden (Lindholm 1981, Erdtmann et al. 1987) have revealed corresponding "transitional" La2-type faunas, similar to those observed at Green Point and Martin Point in Newfoundland. Although reminiscent of the morpho-logical development stages reached during the late Tremadoc with regard to stipe organizational and autothecal schemes, these Hunnebergian graptolites (younger than Kiaerograptus, but older than Eotetragraptus approximatus) are without bithecae, except that sicular bithecae are occasionally observed in very rare instances. A latest "imposter" of Rhabdinopora, the pendent conical and biradiate Araneograptus (Erdtmann & VandenBerg 1985), also occurs in these transitional dendroid to grapto-loid La2-type faunas in off-shore and deeper water starved basinal facies.

4. The Valhall Regressive Event (VRE)

As stated above, the La2-type graptolites are "Tremadocian" in general morphological character, but "Arenigian" by virtue of their graptoloid nature (no bithecae). A full radiation of "new" morphogenetic schemes, replete with reclined and scandent "quadriserial" rhabdosomes (Phyllo-graptus, Pseudophyllograptus), appears first during and after the first appearance of the pandemic horizontal H-shaped Eotetragraptus approxima-tus, when the climax of the Arenig transgression and concomitant shore-ward migrations of chemoclines had been reached (during the Pseudophyl-lograptus densus Zone and concurrent shale development over limestone terranes in central Sweden). During the ensuing middle to late Arenig interval (Bendigo to late Castlemain of Australasia) graptolite diversity reached its all-time high and strongest concurrent biofacies differentiation (provincialism). A postglacial low-latitude "Pacific" fauna (Australasia, North China, North America) is characterized by pendent didymograptids, reclined (late aberrant) isograptids, sigmagraptids and sinograptids (the latter two groups "reiterating" all preglacial dichograptoid stipe patterns: Hsu & Chao 1976). The temperate water Baltic fauna contains a less diversified assemblage than the "Pacific" fauna including the same morphogenetic types -- but stratigraphically delayed. The cold-water Bohemian-British-Iberian-North African (peri-Gondwana) terranes comprise an even more depauperate fauna ("trophic generalists") but it is interesting to observe that morphogenetic patterns of "Pacific" and "Baltic" derivation appear with an even greater time-lag and a more "abstracted" lower diversity in the high-latitude peri-Gondwana graptolite associations. The typical warm water Chewtonian (middle Arenig) tuning-fork Didymograptellus bifidus "re-appears" as Didymograptus artus (British D. 'bifidus'), a nearly indistinguishable "mimic" of the former, in the early Llanvirn (Darriwil) of Britain, Bohemia, Spain and North Africa. Similar morphogenetic "delay strategies" are repeated for various Bendigonian multiramous sigmagraptids which are "mimicried" by Llanvirnian pendent multiramous Pterograptus and Pseudobryograptus in the post-regression (Valhall glacial regression of Fortey 1984) Luarca, tristani, Šarka and coeval transgressive black shales of the Mediterranean region. Similar postglacial "mimics" are the Llanvirn "phyllograptids" which reappear for the early to middle Arenig quadriserial pseudophyllograptids (Cooper & Fortey 1982), and probably also the "glyptograptids", "climacograptids" and other "pseudo-biserials" which appear as "living fossils" during the early Llandovery postglacial transgressive phase.

It is generally accepted (Scheltema 1977, Johnson 1982, Fortey 1985) that organisms possessing planktic or nektic larval and/or adult stages should be naturally less affected by allopatry and thus by major punctuations in their evolutionary record. Some of the investigated sections (i.e. the Green Point, Newfoundland section and certain transects along the strike of the Oslo-Scania-Lysogory Depression) are sufficiently continuous to permit tracing gradual cladogenetic changes of graptolite morphologies. The operation of the principle of "global sympatry" for the holo-pelagic graptolites is, within limits, verifiable for times of "normal" watermass movements during transgressive periods, but perturbations of these watermass patterns inevitably occur in connection with major global regressions. These perturbations, then, probably induced major disjunctions of graptolite migrational patterns, which, in turn, may have produced the punctuations observed and the concomitant qualitative (anagenetic) changes in graptolite morphogenetics.

REFERENCES

ALVAREZ, L.W.; ALVAREZ, W.; ASARO, F. & MICHEL, H.V. (1980): Extraterrestrial cause for the Cretaceous-Tertiary extinction.- Science 208, 1095-1108.
ANDERSSON, A.; DAHLMAN, B.; GEE, D.G. & SNALL, S. (1985): The Scandinavian Alum Shales.- Sveriges Geol. Undersök., Ser. Ca (56), 50 p.
BERNER, R.A. (1981): A new geochemical classification of sedimentary environments.- J. Sed. Petrol. 51, 359-365.
BERRY, W.B.N.; WILDE, P. & HUNT, M.Q. (1985): The graptolite habitat: an oceanic non-sulfide low oxygen zone?- Graptolite Working Group of the Internat. Palaeont. Assoc., 3rd Internat. Conf. 27-30 Aug. 1985, Copenhagen, Denmark, Abstr. for Meetings, 5-6.
BOYCE, W.D. (1979): Further developments in western Newfoundland Cambro-Ordovician biostratigraphy.- Min. Dev. Div. Newfld., Rept. Activ. 79-1, 7-10.
BULMAN, O.M.B. (1954): The graptolite fauna of the Dictyonema Shales of the Oslo Region.- Norsk Geol. T. 33, 1-40.
CASTAÑOS, A. & RODRIGO, L.A. (1978): Sinopsis estratigrafia de Bolivia. I. Parte de Paleozoico.- Acad. Nac. Cien. Bolivia, 146 p.
COCKS, L.R.M. & FORTEY, R.A. (1982): Faunal evidence for oceanic separations in the Palaeozoic of Britain.- J. Geol. Soc. London 139, 465-478.
COOPER, R.A. & FORTEY, R.A. (1982): The Ordovician graptolites of Spitsbergen.- Bull. British Mus. (Nat. Hist.), Geol. Ser. 36, 157-302.
DESTOMBES, J.; SOUGY, J. & WILLEFERT, S. (1969): Revisions et decouvertes paleontologiques (brachiopodes, trilobites et graptolites) dans le Cambro-Ordovicien du Zemmour (Mauritanie septentrionale).- Bull. Soc. Geol. France, Ser. 7, 11, 185-206.
DEYNOUX, M.; SOUGY, J. & TROMPETTE, R. (1985): Lower Palaeozoic rocks of West Africa and the western part of Central Africa.- in: HOLLAND, C. H. (ed.): Lower Palaeozoic Rocks of the World, Volume 4: Lower Palaeozoic of north-western and west-central Africa. J. Wiley & Sons, London, 337-495.
ERDTMANN, B.-D. (1965): Eine spät-tremadocische Graptolithenfauna von Töyen in Oslo.- Norsk Geol. T. 45, 97-112.
-- (1984): Outline ecostratigraphic analysis of the Ordovician graptolite zones in Scandinavia in relation to the paleogeographic disposition of the Iapethus.- Geologica et Palaeontologica 18, 9-15.
-- (1986): Graptolite-based correlation of Earliest Ordovician in eastern North American marginal sequences with coeval successions in northern China and Oslo, Norway.- Bull. New York State Mus., 462 (in print).
-- ; MALETZ, J. & GUTIERREZ MARCO, J.C. (1987): The new Early Ordovician (Hunneberg Stage) graptolite genus Paradelograptus (Fam. Kinnegraptidae), its phylogeny and biostratigraphy.- Paläont. Z. (in print).
-- & MILLER, J.F. (1981): Eustatic control of lithofacies and biofacies changes near the base of the Tremadocian.- US Geol. Surv., Open-file Rep. 81-743 (Short Papers for the 2nd Internat. Symp. on the Cambrian System, 1981), 78-81.
-- & VANDENBERG, A.H.M. (1985): Araneograptus gen. nov. and its two species from the late Tremadocian (Lancefieldian, La2) of Victoria.- Alcheringa 9, 49-63.
FISCHER, A.G. & ARTHUR, M.A. (1977): Secular variations in the pelagic realm.- in: COOK, H.E. & ENOS, P. (eds.): Deep Water Carbonate Environments. Soc. Econ. Paleont. Miner., Spec. Publ. 25, 19-50.
FORTEY, R.A. (1984): Global earlier Ordovician transgressions and regressions and their biological implications.- in: BRUTON, D.L. (ed.): Aspects of the Ordovician System. Palaeont. Contr. Univ. Oslo 295, 37-50.
-- (1985): Gradualism and punctuated equilibria as competing and complementary theories.- in: COPE, J.C.W. & SKELTON, P.W. (eds.): Evolu-

tionary Case Histories from the Fossil Record. Spec. Pap. Palaeont. 33, 17-28.

HARRINGTON, H.J. & LEANZA, A.F. (1957): Ordovician trilobites of Argentina.- Dept. Geol. Univ. Kansas, Spec. Publ. 1, 276 p.

HSU, Chieh & CHAO, Yuting (1976): The evolution and systematics of the family Sinograptidae.- Acta Geologica Sinica 2, 121-140.

JAMES, N.P.; STEVENS, R.K. & FORTEY, R.A. (1979): Correlation and timing of platform-margin megabreccia deposition, Cow Head and related groups, western Newfoundland.- Bull. Amer. Assoc. Petrol. Geol. 63, 474.

JOHNSON, J.G. (1982): Occurrence of phyletic gradualism and punctuated equilibria through geologic time.- J. Paleont. 56, 1329-1331.

KEIDEL, J. (1943): El Ordovicico Inferior de los Andes del Norte Argentino y sus depositos marino-glaciales.- Bol. Acad. Nac. Cienc. Cordoba 36, 140-229.

LINDHOLM, K. (1981): A preliminary report on the uppermost Tremadocian - lower middle Arenigian stratigraphy of the Krapperup 1 drilling core, southern Sweden.- Unpubl. Undergrad. Proj., Dept. Hist. Geol. Palaeont., Univ. Lund, 44 p.

LINDSTRÖM, M. & DWORATZEK, M. (1979): Exkursion A3 und B2: Das Altpaläozoikum von Südschweden.- 131st Ann. Meeting German Palaeont. Soc., Marburg, Univ. Marburg, 60 p.

LOHMANN, H.H. (1970): Outline of tectonic history of Bolivian Andes.- Bull. Amer. Assoc. Petrol. Geol. 54, 735-757.

MARTINSSON, A. (1974): The Cambrian of Norden.- in: HOLLAND, C.H. (ed.) Lower Palaeozoic Rocks of the World. Volume 2. Cambrian of the British Isles, Norden and Spitsbergen. J. Wiley & Sons, London, 185-283.

MILLER, J.F. (1984): Cambrian and earliest Ordovician conodont evolution biofacies,and provincialism.- Geol. Soc. Amer., Spec. Pap. 196, 43-68.

MOLINEUX, S.G. & RUSHTON, A.W.A. (1984): Discovery of Tremadoc rocks in the Lake District.- Proc. Yorkshire Geol. Soc. 45, 123-127.

SCHELTEMA, R.S. (1977): Dispersal of marine invertebrate organisms: paleobiogeographic and biostratigraphic implications.- in: KAUFFMAN, E.G. & HAZEL, J.G. (eds.): Concepts and Methods of Biostratigraphy. Dowden, Hutchinson & Ross, Stroudsburg/Pennsylvania, 73-108.

STOUGE, S. (1982): Preliminary conodont biostratigraphy and correlation of Lower and Middle Ordovician carbonates of the St. George Group, Great Northern Peninsula, Newfoundland.- Mineral Dev. Div., Dept. Mines & Energy, Gvt. Newfoundland & Labrador, Rept. 82-3, 59 p.

THICKPENNY, A. (1985): "Black Shales" and Early Palaeozoic palaeo-oceanography.- Terra Cognita 5, 109.

WALLISER, O.H. (1984): Global events and evolution.- Proc. 27th Internat. Geol. Congr., Palaeont. 2, 183-192.

WILLIAMS, H.; JAMES, N.P. & STEVENS, R.K. (1985): Humber Arm Allochthon and nearby groups between Bonne Bay and Portland Creek, western Newfoundland.- Geol. Surv. Canada, Current Res., Pt. A, 85-1A, 399-406.

GLOBAL BIO-EVENTS IN THE ORDOVICIAN ?

LINDSTRÖM, Maurits *)

A contribution
to Project
GLOBAL
BIO-
EVENTS

Abstract: Search for possible global events within the Ordovician is hampered by the poor time resolution provided by biostratigraphic dating (not better than about 3.5 m.y. average on the global scale). Attention is directed primarily to the leading pelagic and cosmopolitan groups, graptolites and conodonts, as well as to faunas evolving in a relatively stable environment (Baltoscandia). A possible global extinction event is indicated at the end of the Tremadoc, affecting, among others, taxa of conodonts and dendroid graptolites. Another possible global event occurred at the end of the Llanvirn, involving the extinction of most dichograptids, several North American conodont taxa, and non-conodont taxa in Baltoscandia. Neither of these two possible events approaches the extinction event at the end of the Ordovician in importance. The Ordovician global bio-events appear to have been synchronous with major eustatic fluctuations of the sea-level. However, not all such fluctuations led to noteworthy extinctions.

Introduction

Event correlation has opened up new and interesting possibilities in the younger geological Systems, and it appears worthwhile to ask, to what extent the same approach can be useful in an old Phanerozoic System like the Ordovician, as well.

An e v e n t in this context can be a process of sufficiently short duration to be used for exceptionally precise correlation of those beds in which it is documented. On the other hand, it can be a major process that is proven to be contemporaneous and of short duration over a very large area owing to the evidence by which it can be dated.

A synoptic view of the paleontological record fails to indicate that there was any outsize global bio-event within the Ordovician, although the Period ended with one of the greatest mass-extinctions of the Paleozoic Era (Raup & Sepkoski 1982, Sepkoski & Sheehan 1983). Instead, the picture obtained is one of steady and rapid growth of faunal diversity throughout the Period.

Precision of dating in the Ordovician

The possibility to distinguish global events depends very much on the precision of correlation. So far, the only means of long-distance correlation in the Ordovician is provided by paleontology. International standard index fossils are graptolites and conodonts. About 20 grapto-

---

*) Department of Geology, University of Stockholm, S-10691 Stockholm, Sweden.

Lecture Notes in Earth Sciences, Vol. 8
Global Bio-Events. Edited by O. Walliser
© Springer-Verlag Berlin Heidelberg 1986

lite zones can be distinguished (Skevington 1976a). However, this number is no measure of the precision that can be reached in interregional correlation, because there are questions about the identity of graptolite zones between provinces.

A composite conodont zonation for the Ordovician may contain as many as 37 units (zones and subzones; Miller 1984, Lindström 1971a, Löfgren 1978, Bergström 1971, Sweet & Bergström 1984). However, several units are known only from one region, and the number of units that can be used for intercontinental correlation could at the present be 20-23.

At the moment we cannot date signals from different continents with better precision than to a time span corresponding to one of these units If the Ordovician Period lasted for about 70-80 m.y. (Ross & Naeser 1984), the number of interregionally useful zones and subzones means that the average precision of any biostratigraphic dating is about 3.5 m.y. Clearly, arguments about global events must be highly speculative in view of such imprecision.

Graptolite and conodont evidence of events

The paleontological evidence places a mass extinction as one of the important global events at the end of the Ordovician. Within the Period there is no evidence of any event of comparable dimensions.

In a quest for datable events it is natural to begin by taking a look at the two leading groups of index fossils. In the history of Ordovician graptolite faunas there are two possible global events of relatively high rank. The first of these is the transition from the dendrograptid-anisograptid fauna of the Tremadoc to the dichograptid dominated fauna of the early Arenig. The second is the demise of the dichograptids at the end of the Llanvirn.

In most areas there are either stratigraphic breaks, or contrasts of facies, or simply deficient exposure at the Tremadoc-Arenig boundary, so that the succession of graptolite faunas is poorly known. Probably, the best known graptolite succession at this level is the one of the Oslo area (Erdtmann 1965a & b). The uppermost Tremadoc of this area has yielded a fauna of mainly anisograptids but with a possible dichograptid as one component. Above this level there follows the poorly fossiliferous Hagastrand mudstone Member that contains but a sparse anisograptid-dichograptid fauna in its middle portion; further, altogether dichograptid-dominated beds follow above this member.

Because it appears at present unknowable, how much time is represented by the Hagastrand Member (Erdtmann 1965b, assumes this time span to be relatively short), it must remain an open question, if there was a major event that affected the graptolite faunas at the end of the Tremadoc, or if there was a prolonged and gradual faunal change. Certainly, the transitional beds are worth a close scrutiny with these possibilities in mind.

The disappearance of the *Didymograptus* fauna at the end of the Llanvirn appears to have been relatively abrupt (Skevington 1976b). Even though there is a Llanvirn-Llandeilo boundary problem, with a non-classified interval belonging to neither Series (Bergström et al., in press), an apparently continuous succession of Llanvirnian-Llandeilian graptolitic shales occurs in southernmost Sweden (Hede 1951). In the south Swedish section the *Glyptograptus teretiusculus* Zone (that is regarded as following next above the Llanvirn) begins with a 1.8 m thick transition zone that is poor in graptolites. This transition zone, or its basal portion, might be the local expression of a global event that involved the demise of the dichograptids, but at present this is pure speculation.

In the case of the conodont faunas the situation broadly resembles that outlined for the graptolites. One can discern abrupt faunal changes approximately at the Tremadoc-Arenig boundary, and there was another apparent disruption of continuity in the faunal evolution in the early middle Ordovician.

The Tremadoc is characterized by a cosmopolitan lineage of *Cordylodus* species that began in the latest Cambrian but that cannot be followed above the top of the Tremadoc Series. In addition to the practically omnipresent specimens of *Cordylodus*, upper Tremadocian conodont faunas from shallow-water seas near the Cambrian paleoequator contain specimens of *Loxognathus*, *Clavohamulus* and a few other taxa that are not known to occur above the Tremadoc. Differences between Tremadocian conodont faunas and faunas of younger age could be explained by the circumstance that regressive-transgressive cycles occurred during Tremadoc time (Miller 1984), but this explanation hardly negates the possibility that there was a biological event at the end of the Tremadoc.

Sweet & Bergström (1976) noted a sharp separation in conodont faunas between on one hand Whiterock and older deposits, and on the other Chazyan and younger beds. This appears to be the most distinctive break in conodont faunal evolution in the middle and later Ordovician on the North American continent. The authors equate the stratigraphic level with the top of the *Didymograptus murchisoni* Zone, which according to traditional usage is the top of the Llanvirn. However, they do not claim any great precision for this correlation. In the north European succession there are apparently no correspondingly dramatic changes (Dzik 1983).

Events in Baltoscandia

In a search for possible events there is the option of examining sedimentary sequences for any signals that are out of the ordinary and that
appear in more than a few, restricted areas. Such sequences may have
formed in shallow water, in which case the effects of any major change
tend to be conspicuous, unless they are erased by a regressive-transgressive cycle. Because shallow-water sequences can contain stratigraphic breaks at critical intervals and are notoriously difficult to corre
late interregionally, they do not present optimal conditions for the
quest.Faunal shifts in shallow seas can be due to ecologic response to
relatively slight physical changes in the global environment, such as
minor regressions, or to purely local effects. Sequences formed under
more protected conditions hold greater promise of selectively revealing
the effects of global biological events. Baltoscandia is one region that
had protected conditions of sedimentation during much of the Ordovician
(Lindström 1971b, Jaanusson 1982).

   Like other parts of the world Baltoscandia apparently was affected
by regression in the Tremadoc (Jaanusson 1979). After this regression
there followed an interval of relative tranquility during much of the
Arenig (Lindström 1963), with considerable but essentially continuous
evolution of the regional faunas.This development was broken by regression in the late Arenig to early Llanvirn (Lindström & Vortisch 1983,
Bruton & al. 1985). The regression was accompanied by great diversification of the fauna and increase in importance of sessile suspension
feeders compared with the otherwise dominant trilobite fauna. However,
there was no extermination of important lineages, and later in the Llanvirn conditions reverted to the dominance of the same trilobite groups
that had held the field before the temporary regression.

   The next major deviation from the "normal" occurs in the upper Llanvirn (Aseri and Lasnamägi Stages). In some areas there is a major hiatus
of this age. In others, there is condensation combined with incrusta-
tions with hematite, hematitic pisolites, either hematitic or goethitic
ooids, and laminated crusts (Männil 1966, Larsson 1973, Jaanusson 1982).
In North America deposition of the St. Peter Sandstone may have begun
roughly at the same time, marking the final stages of a major regressive
transgressive cycle. As noted above, this shift is coupled with major
changes in the North American fauna, changes that are not matched in
Baltoscanida. However, other faunal elements are strongly affected. Trilobites, ostracodes, brachiopods, and bryozoans experienced great reduction in diversity in the Aseri and Lasnamägi Ages, and many taxa were
exterminated (Männil 1966, Ulst & al. 1982). Nevertheless, basic featu-

res of the limestone facies were not changed. Arthropods and echinoderms remained dominant among the skeletal components of the sand fraction, as they had been before, and cephalopods and asaphid trilobites kept their leading role among the megafossils. In order to obtain clues to the ecologic situation it could be worth-while to analyse the trace element composition across this level.

Above the Lasnamägi Stage the facies variation is much greater than in underlying beds. The Caradoc and Ashgill contain mound facies, and the Ashgill terminates upwards with siliciclastic and carbonatic, regressive beds. The situation in this respect resembles the one found in other parts of the world. Apparently, there are not any event-like faunal changes within the Caradoc to Ashgill in Baltoscandia.

Sedimentological evidence of events

Unfortunately, very few deep ocean sediments are known from the Ordovician. The Southern Uplands of Scotland contain lower to middle Ordovician red chert-shales that are followed upwards by black graptolite shale in a sequence that is regarded as oceanic (Legget 1980). According to conodont and graptolite evidence the transition took place rather abruptly, in the Llandeilo. Leggett (1980) explains the switch to anoxia on the deep-ocean sea-bed by increased organic productivity owing to raised nutrient content of the ocean. This situation is believed to have been caused by late Llandeilian to early Caradocian eustatic transgression. The existence of this transgression has been known for a long time, and is well documented (McKerrow 1979), but its effect on the fauna appears to lie mainly in a marked increase in diversity of several groups, like anthozoa, ostracoda, stenolaemate bryozoa, articulate brachiopods, and crinoids (Sepkoski & Sheehan 1983).

In central and southern Europe the Tremadoc is commonly missing ; this circumstance is likely to be an effect of the widespread regression referred to in a preceding section. The principal deviating feature of the younger, siliciclastic succession appears to be the presence of widely distributed beds of ferruginous oolite near the Arenig-Llanvirn boundary (Babin & al. 1976, Havlíček 1976). On the global stage this level appears to coincide with a major regression (see above). However, in central and southern Europe the same level is characterized by the transition from a predominantly sandy and presumably shallow-water facies to graptolitic mudstones that would normally signal deeper water. One might venture to challenge either of these interpretations as too sweeping, but the paleogeographic evolution at the time of formation of the iron deposits nevertheless remains unclear. These iron ores could serve to underline the importance of geological events at the Arenig-Llanvirn transition, but these events appear to have been of minor significance for the evolution of biota.

Summary

The time resolution of stratigraphic dating in the Ordovician probably is not much better than 3.5 m.y. on the average. Therefore, one should beware of too much assurance in postulating global bio-events within the Period. Furthermore, there are a great deal more indications of gradual increase in diversity in the biota than of sudden, great exterminations followed by the evolution of new, high-level taxa. At the end of the Tremadoc there was a relatively important, global change in faunal composition, but it is uncertain, how abrupt it was. Another significant change happened at the end of the Llanvirn. This change is documented for the graptolite faunas, and it might be of some importance in North American conodont faunas, as well. There are also indications that the same event, if it existed, is documented in Baltoscandian non-conodont faunas.

In Baltoscandia there are also sedimentary indications of an event at the end of the Llanvirn or not much later than this. Other drastic changes, such as for instance the switch from well oxygenated conditions to less oxic conditions, taking place in deeper levels of the Iapetus Ocean in the Llandeilo, apparently cannot as yet be tied up with global bio-events.

REFERENCES

BABIN, C.; ARNAUD, A.; BLAISE, J.; CAVET, P.; CHAUVEL, J.J.; DEUNFF, J.; HENRY, J.-L.; LARDEUX, H.; MÉLOU, M.; NION, J.; PARIS, F.; PLAINE, J. QUÉTÉ; Y. & ROBARDET, M. (1976): The Ordovician of the Armorican Massif (France).- in: BASSETT, M.G. (ed.): The Ordovician System. University of Wales Press & National Museum of Wales, 359-385.

BERGSTRÖM, S.M. (1971): Conodont biostratigraphy of the Middle and Upper Ordovician of Europe and eastern North America.- in: SWEET, W.C. & BERGSTRÖM, S.M. (eds.): Symposium on Conodont Biostratigraphy. Mem. geol. Soc. America 127, 83-162.

-- ; RHODES, F.H.T. & LINDSTRÖM, M. (in press): Conodont biostratigraphy of the Llanvirn - Llandeilo and Llandeilo - Caradoc Series boundaries in the Ordovician of Wales and the Welsh Borderland.

BRUTON, D.L.; LINDSTRÖM, M. & OWEN, A.W. (1985): The Ordovician of Scandinavia.- in: GEE, D.G. & STURT, B.A. (eds.): The Caledonide orogen - Scandinavia and related areas. John Wiley & Sons, 273-282.

DZIK, J. (1983): Relationships between Ordovician Baltic and North American Midcontinent conodont faunas.- Fossils and Strata 15, 59-85.

ERDTMANN, B.-D. (1965a): Eine spät-tremadocische Graptolithenfauna von Tøyen in Oslo.- Norsk geol. T. 45, 97-112.

-- (1965b): Outline stratigraphy of graptolite-bearing 3b (Lower Ordovician) strata in the Oslo Region, Norway. Investigations of a section at Tøyen, Oslo.- Norsk geol. T. 45, 481-547.

HAVLÍČEK, V. (1976): Evolution of Ordovician brachiopod communities in the Mediterranean Province.- in: BASSETT, M.G. (ed.): The Ordovician System. University of Wales Press & National Museum of Wales, 349-358.

HEDE, J.E. (1951): Boring through Middle Ordovician - Upper Cambrian strata in the Fågelsång district, Scania. 1. Succession encountered in the boring.- Lunds Univ. Arsskr., N.F., Avd. 2, 46 (7), 80 p.

JAANUSSON, V. (1979): Ordovician.- in: ROBISON, R.A. & TEICHERT, C. (eds.): Treatise on Invertebrate Paleontology. A. Introduction. Geol. Soc. America & Univ. Kansas Press, A 136-166.

-- (1982): Introduction to the Ordovician of Sweden. - in: BRUTON, D.L. & WILLIAMS, S.H. (eds.): Field Excursion Guide. IV Internat. Symp. Ordovician System. Paleont. Contr. Univ. Oslo 279, 1-10.

LARSSON, K. (1973): The Lower Viruan in the autochthonous sequence of Jämtland.- Sver. geol. Unders. Ser. C 683, 82 p.

LEGGETT, J.K. (1980): British Lower Palaeozoic black shales and their palaeo-oceanographic significance.- J. geol. Soc. London 137, 139-156.

LINDSTRÖM, M. (1963): Sedimentary folds and the development of limestone in an Early Ordovician sea.- Sedimentology 2, 243-292.

-- (1971a): Lower Ordovician conodonts in Europe.- in: SWEET, W.C. & BERGSTRÖM, S.M. (eds.): Symposium on Conodont Biostratigraphy. Mem. geol. Soc. America 127, 21-61.

-- (1971b): Vom Anfang, Hochstand und Ende eines Epikontinentalmeeres.- Geol. Rdsch. 60, 419-438.

-- & VORTISCH, W. (1983): Indications of upwelling in the Lower Ordovician of Scandinavia.- in: THIEDE, J. & SUESS, E. (eds.): Coastal Upwelling. B. Plenum Publishing Co., 535-551.

LÖFGREN, A. (1978): Arenigian and Llanvirnian conodonts from Jämtland, northern Sweden. - Fossils and Strata 13, 129 p.

MÄNNIL, R. (1966): Istoriya razvitiya Baltiyskogo basseyna v ordovike.- Eesti NSV Tead. Akad. geol. Inst., 201 p.

McKERROW, W.S. (1979): Ordovician and Silurian changes in sea level.- J. geol. Soc. London 136, 137-145.

MILLER, J.F. (1984): Cambrian and earliest Ordovician conodont evolution, biofacies, and provincialism.- in: CLARK, D.L. (ed.): Conodont Biofacies and Provincialism. Geol. Soc. America Spec. Pap. 196, 43-68.

RAUP, D.M. & SEPKOSKI, J.J., Jr. (1982): Mass extinctions in the marine fossil record.- Sci. 215, 1501-1503.

ROSS, R.J., Jr. & NAESER, C.W. (1984): The Ordovician time scale - New refinements.- in: BRUTON, D.L. (ed.): Aspects of the Ordovician System. Palaeont. Contr. Univ. Oslo 295, 5-22.

SEPKOSKI, J.J., Jr. & SHEEHAN, P.M. (1983): Diversification, faunal change, and community replacement during the Ordovician radiations.- in: TEVESZ, M.J.S. & McCALL, P.L. (eds.): Biotic interactions in Recent and fossil benthic communities. Plenum Press, 673-717.

SKEVINGTON, D. (1976a): British Ordovician graptolite Zones and inter-regional correlation.- in: KALJO, D.L. & KOREN, T.N. (eds.): Grapto-lites and Stratigraphy. Acad. Sci. Estonian SSR Inst. Geol., 171-179.

-- (1976b): A discussion of the factors responsible for the provincia-lism displayed by graptolite faunas during the Early Ordovician.- in: KALJO, D.L. & KOREN, T.N. (eds.): Graptolites and Stratigraphy. Acad. Sci. Estonian SSR Inst. Geol., 180-201.

SWEET, W.C. & BERGSTRÖM, S.M. (1976): Conodont biostratigraphy of the Middle and Upper Ordovician of the United States Midcontinent.- in: BASSETT, M.G. (ed.): The Ordovician System. University of Wales Press & National Museum of Wales, 121-151.

-- & BERGSTRÖM, S.M. (1984): Conodont provinces and biofacies of the Late Ordovician.- in: CLARK, D.L. (ed.): Conodont Biofacies and Provincialism. Geol. Soc. America Spec. Pap. 196, 69-87.

ULST, R.Zh.; GAJLITE, L.K. & YAKOVLEVA, V.I. (1982): Ordovik Latvii.- Ministerstvo Razov. Promyshlennost. SSR Vses. Nauchn.-Issled. Inst. Morsk. Geol. Geofiz., 294 p.

SILURIAN TO PERMIAN

SIGNIFICANT GEOLOGICAL EVENTS IN THE PALEOZOIC RECORD OF THE SOUTHERN ALPS (AUSTRIAN PART)

SCHÖNLAUB, Hans P. *)

A contribution to Project GLOBAL BIO-EVENTS

The geological history of the Austrian part of the Southern Alps, i.e. the Carnic Alps and the Karawanken Alps ranges from the Upper Ordovician to the middle Triassic. During this long time various fossiliferous rocks were deposited in a mobile sedimentary basin including nearshore rocks and shallow water carbonates as well as deep water basinal deposits such as graptolitic shales or other pelitic rocks.

The total up to 5000 m thick sequence is divided into two major sedimentary cycles separated by the Variscan unconformity in the Upper Carboniferous. Beside this primary geologic event several other distinct events have been recognized in the Paleozoic record in recent years. The majority of these are biological events connected with short-term abiotic processes. They caused, however, only minor biotic changes. The difficulties to recognize true biotic events within a uniformly developed sequence are numerous and well known. In our case we face the following problems:

1. Sudden and distinct changes of the lithofacies mainly occur in open marine pelagic sequences. This realm is dominated by a pelagic fauna consisting of various planktonic and nektonic animal groups such as conodonts, dacryoconarids, graptolites and cephalopods. As a response to short-term tectonic movements its composition never changed significantly. Hence, the bio-event is less distinct than the litho-event.

2. Scarcity of fossil-data. In comparison with other Paleozoic sequences in the world the pelagic sequences of the Southern Alps are generally poor in fossils. Except graptolites and conodonts which occur in great abundance and variety no other fossil group can be used for statistical purposes. Representatives of both groups fit well into the currently used zonal schemes. In comparison with coeval faunas elsewhere differences of their composition merely reflect different geographic and ecologic habitats.

3. Statements about appearances of new taxa (or extinctions) can only be made for few groups, for example graptolites, conodonts and dacryoconarids. Acritarchs from the Silurian offer some promise for the future too. Other groups like trilobites, orthoconic nautiloids, goniatites, bivalves and brachiopods occur in only few layers or make their appearance step by step as a response to long-term geologic processes like a transgression. Exceptions are Upper Devonian trilobites and clymenids.

4. The equivalents of the Silurian and Devonian reach a thickness between 60 and 100 m each. Obviously, these pelagic sequences are more or less condensed and even stratigraphic gaps must be considered. In particular, this concerns the Silurian with its sharp bounded tripartite lithologies. Hence, appearance and last occurrences of fossils may be misleading and are mostly not realistic. In many cases this assumption is supported by

---

*) Geologische Bundesanstalt, 1031 Vienna, Austria.

Lecture Notes in Earth Sciences, Vol. 8
Global Bio-Events. Edited by O. Walliser
© Springer-Verlag Berlin Heidelberg 1986

TRIAS
- Ladin — 500-700 m Schlerndolomit
- Anis — ~30m Muschelkalk / ~40m Tuffe, Tuffite, Andesite, Dazite
- Skyth — 0-30m Muschelkalk-Konglomerat / ~100 m Werfen-Formation

PERM
- Tatar — 250 m Bellerophon-Formation
- Kungur-Murgab — 30-40 m Gröden-Formation
- Artinsk — Tarvis-Breccie / Tressdorf-Kalk / Trogkofel-Konglomerat
- Sakmar — ~400 m Trogkofel-Kalk
- Assel — 175 m Ob. Pseudoschwagerina-Formation / ~125 m Grenzland-Formation / 160 m Unt. Pseudoschwagerina-Formation

KARBON
- Gzhel — 600-800 m Auernig-Formation
- Kasimov
- Moskau — ~40 m Waidegg-Formation)
- Bashkir — ~10 m Goniatitenkalk
- Serpukhov — Dimon-Plenge-Formation / Hochwipfel-Formation
- Visé — Spilit, Pillow-Laven, Rhyolith, Agglomerat, bas. Tuff. / >1000 m Hoch = wipfel-Fm. / Dazit-Keratophyr
- Tournai — Kronhof-K. ~10m / -0.2 m Kronhof-Schiefer / Kirchbach-Kalk

DEVON
- Famenne — ~40 Pal-Kalk / >100 m
- Frasne — 150 m Korallen-Stromatoporen-Schuttkalk / ~50 m Philipsostrea-Brachiopoden-Kalk / Zollner-Formation
- Givet — 200-300 m Elskar-Kalk / ~400 m Cellon-Kalk / >100 m Freikofel-Kalk / ~15 m Valentin-Kalk / 30-40 m Hohe Trieb-Formation
- Eifel — 220 m Amphipora-K. / 150 m Kor./Crin.-Kalk / ~500 m Kunzkopf-Kalk / Grünschiefer SW Feistritz
- Ems — ~150m Consuolo-Lamit-Kalk / ~80 m dkl. Plattenk. / 40-60 m Findenig-Kalk
- Prag — ~40 m Seewarte-Kalk / ~150 m Kellerwand-Kalk / 350 m Conjugula- und Riffkalk

230 Mio Jahre — 245 — 290 — 360

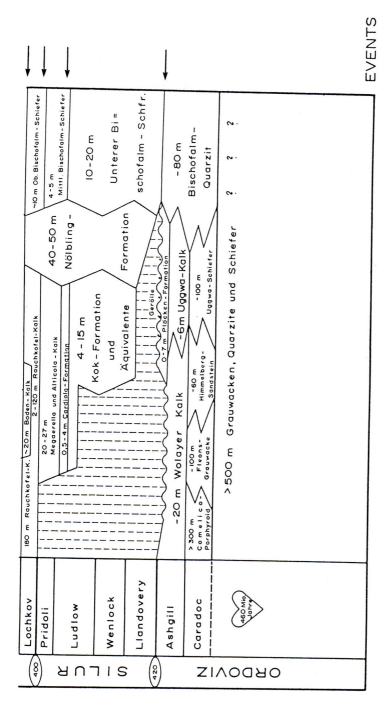

Figure 1. Events between 460 and 230 Mio years in the sequence of the Carnic Alps.

the missing ancestor-descendant relationship of new taxa.

5. In contrast to the pelagic realm the shallow water deposits are very fossiliferous. Direct comparison with the former, however, has not been proved in detail. Hence, correlation between synchronous events in both facies and its effect on the distribution of the shallow water fauna is not fully understood yet.

In spite of all these limitations within the Paleozoic sequences of the Southern Alps 12 events have been recognized which have a local and in some cases regional importance. Our knowledge about these events is based on detailed studies of several sections and its faunal content, and a careful mapping of the whole area under consideration. Due to the fact that megafossils are very rare we first recognized litho-events and then worked out its possible connection with a bio-event. In stratigraphic order the following events were discovered:

1. Ordovician/Silurian Boundary Event: Culmination of a regression at the end of the Ordovician followed by a transgression at the very beginning of the Llandovery (acuminatus Zone). Process which caused stratigraphic gaps, the first appearance of graptolites, a highly diversified conodont fauna, trilobites and molluscs compared with the late Ordovician new environments were established, climate became moderate and water temperatures probably increased.

2. Cardiola Event in the late Ludlovian: Abiotic event which caused in both the shelly facies and the graptolitic facies ("Mittlere Bischofalm-Schiefer") distinct lithologies. This event maybe attributed to a short-term regression the termination of which was at the end of the Silurian. The event caused an income of new taxa, e.g., conodonts graptolites and bivalves and its abundance increased. In the graptolitic environment the black alum shales were replaced by green shales.

3. Silurian/Devonian Boundary Event: Abiotic event with effect on the fauna. New environments were established followed by the income of new faunas and/or replacement of old ones (conodonts, echinoderms, graptolites, dacryoconarids, trilobites).

4. Lochkovian/Pragian Boundary Event: At or near this level significant biological changes took place, i.e., distinct changes of the conodont fauna, disappearance of graptolites and appearance of new dacryoconarids. The event is also indicated by changes of the lithology (different coloured flaser limestones).

5. The otomari Event: Occurrence of a black shale and chert unit within a limestone sequence. This event caused only minor faunal changes but led to favourable conditions for the microfauna (conodonts and styliolinids). Useful event for correlation purposes as it occurred within a short time span in the ensensis conodont Zone of the latest Eifelian or early Givetian.

6. Middle/Upper Devonian Boundary Event: Abiotic short-term geologic event connected with tectonic movements which caused stratigraphic gaps, reworking, and mixing of faunas, and different lithologies.

7. Kellwasser Event: In contrast to other places this event -- if any -- is only poorly represented in the Southern Alps leading to different limestone types in both the pelagic and shallow water realm. Traditionally it coincides with the end of the reef and Amphipora limestone development.

8. Devonian/Carboniferous Boundary Event: Geologic event with some bearing on the fauna, e.g. conodonts and goniatites. Less distinct are changes in the lithology.

9. Late Tournaisian Event: Short-term abiotic event which caused a black chert unit widely distributed between the top of the Variscan limestone sequence and the clastic Kulm deposits.

10. Variscan orogeny event: Geologic (tectonic) event between the Serpukhovian and the Moscovian Stages causing folding, overthrust and nappe structures of the older strata. Also, this event produced new environments followed by the income of before unknown lithologies and faunal and floral constituents, e.g., fusulinids, algae, different plants, sponges, and others.

11. Saalic phase event: Geologic (tectonic) event which caused block faulting but only minor biologic changes.

12. Permian/Triassic Boundary Event: This event caused most dramatic biotic changes. This fact was outlined by several authors in the past based on a worldwide data source. Due to facies reasons these biotic changes are less expressed in the continuous rock sequence of the Austrian part of the Southern Alps. However, first results of a geo-chemical study clearly demonstrated a C-isotope anomaly (and a S-isotope anomaly elsewhere). A more detailed interdisciplinary research program has recently been proposed and approved by both the Austrian FWF and the US-NSF. In the center of this program is an approx. 400 m long core across the P/T boundary, i.e. through the Werfen and Bellerophon Forma-tion to study changes in paleontology, paleobotany, sedimentation, mineralogy, marine chemistry, paleomagnetism, and geophysics, and in particular to analyse the content of iridium and related meteoritic elements and of stable isotopes of carbon, oxygen, sulfur and strontium.

In conclusion, event stratigraphy in the Austrian sector of the Southern Alps is well represented and documented. Most events are ex-pressed as short-term tectonic movements of the crust within a highly mobile sedimentary trough. In any case these geologic processes had consequences on the distribution of faunas and floras. Whether or not these changes were linked with simultaneously occurring chemical alter-ations of the sea-water can not yet be decided. True biotic events can be assumed at the Ordovician/Silurian boundary (probably a climatic triggered biologic event), at the Lochkovian/Pragian boundary, the Kellwasser Event (?), at the Devonian/Carboniferous boundary, and at the Permian/Triassic boundary. These 5 events will be studies in more details in the coming years.

REFLECTION OF POSSIBLE GLOBAL DEVONIAN EVENTS IN THE BARRANDIAN AREA, C.S.S.R.

CHLUPÁČ, Ivo & KUKAL, Zdeněk *)

Abstract: The Lower and Middle Devonian of the Barrandian area of Czechoslovakia shows reflections of several events of possibly global importance. The most important ones are the following: the Lochkovian-Pragian Boundary Event (regressive), the Dalejan Event (transgressive and gradual), the Basal Choteč Event (transgressive, rather quick), and the Kačák Event (transgressive, abrupt). These events, clearly expressed in lithology and fossil content, are probably caused by eustatic sea level changes. They are demonstrated in a similar manner in many other areas, even distant and of different tectonic regimes.

The Lower and Middle Devonian of the classical Barrandian area of Central Bohemia allows to study the stratigraphical turning points of the event character in great details. The following events seem to be of global importance within the preserved Devonian sequence:

The Lochkovian-Pragian Boundary Event

This Lower Devonian event is expressed in l i t h o l o g y by a change from darker and obviously shale-rich Lochkovian to markedly lighter Pragian carbonates (Fig. 1): In the SE limb of the basin, the platy biomicritic and fine-grained biosparitic Radotín and Kosoř Limestones with dark shale interbeds (locally with graptolites) are overlain by light grey micritic Dvorce-Prokop or biosparitic crinoidal rose-coloured Slivenec Limestones of Pragian age. In the other parts of the basin, the upper Lochkovian consists of grey biosparitic Kotýs Limestone and the transition into the Pragian Koněprusy Limestone is marked by a lighter colour, increase of grain-size and thicker bedding. Although the change in colour is usually not quite abrupt and some sections show two or up to three phases of colour changes, the boundary is distinct and distinguished by an onset of sediments of a greater energy and shallow-water origin, the area occupied by coarser bioclastic limestones becoming larger.

The f a u n a l c h a n g e   at or near the Lochkovian-Pragian boundary is characterized by a decline of evolutionary relationships traceable from the Silurian up to the late Lochkovian e.g. among graptolites, trilobites, brachiopods, chitinozoans etc., which caused that the Lochkovian was regarded for a long time as the topmost Silurian. The influx of new elements in benthic (trilobites, brachiopods, bivalves etc.) and plank-

*) Ústřední ústav geologický, 11821 Praha 1, Č.S.S.R.

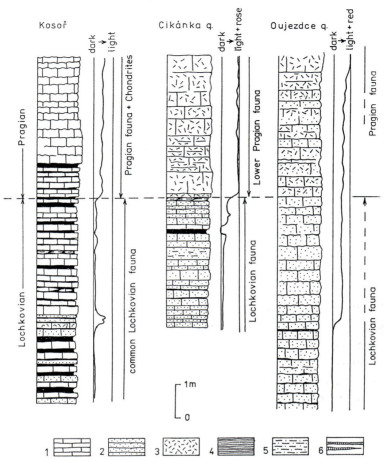

Figure 1. Examples of sections showing the Lochkovian-Pragian Boundary Event in different facies developments: Kosoř - basinal development in the SE limb, Cikánka quarry - transition into the NW limb near Prague, Oujezdce quarry - shallow-water facies in the Koněprusy area.
    Lithology (also for Figs. 2 to 4): 1 - micritic limestones, 2 - fine-grained sparitic limestones, 3 - coarse biodetrital limestones, 4 - calcareous shales, 5 - siltstones and flyschoid sediments, 6 - cherts.

tonic faunas (conodonts: the lineage of <u>Latericriodus</u> <u>steinachensis</u>, dacryoconarid tentaculites: appearance of the typical <u>Nowakia</u> <u>acuaria</u>, chitinozoans: the base of the <u>Angochitina</u> <u>comosa</u> Zone, Chlupáč et al. 1985) is demonstrable both in deeper-water benthic assemblages 4 to 5 and shallow-water assemblages 3 in Boucot's (1975) classification. This is exemplified also by changes in trilobite assemblages: The Lochkovian <u>Lochkovella</u>-<u>Lepidoproetus</u> and <u>Coniproetus</u>-<u>Decoroscutellum</u> Assemblages are replaced by markedly more diversified Pragian assemblages (Chlupáč

1983).

The change at the Lochkovian-Pragian boundary is synchronous in many sections and seems to be relatively quick but non-drastic for most biota. It is commonly manifested within an interval of several cm up to a few metres. A possible global lowering of the sea level, i.e. a regressive event, may be the fitting explanation .

Analogous changes in corresponding stratigraphical levels are known e.g. from Saxonian-Thuringian and Frankenwald regions (Alberti 1981, 1983a), Carnic Alps (Schönlaub 1980, 1985), Sardinia (Jaeger 1976, Alberti 1983a), NW Africa (Alberti 1981, 1983a), Massif Armoricain (Paris 1981), Western and Eastern North America (Johnson & Murphy 1984, Johnson et al. 1985).

The Daleje Event

This late Lower Devonian event named by House (1985) is marked in the Barrandian by a gradual onset of calcareous shale or deeper-water lime- stone facies (Fig. 2): In the greatest part of the basin, the uppermost Zlíchov Formation shows a gradually increasing amount of grey, locally even black calcareous shales up to the predominance of shales at the base of the Daleje Shale. In the NW limb, the topmost Zlíchov Formation is developed as the red crinoidal Chýnice Limestone which is rather sharply overlain by greenish or reddish calcareous Daleje Shale or by the red micritic nodular lowest part of the Třebotov Limestone. The Koněprusy reef area shows a special development: The Pragian reef of the Koněprusy Limestone was drowned after a period of erosion and karsti- fication, and the Suchomasty Limestone, coeval with the Daleje Shale and the Třebotov Limestone, sharply overlies the reef, filling also fissures (neptunian dikes) penetrating the reef bodies.

The change falls biostratigraphically into an interval of the tenta- culite zones Nowakia barrandei - elegans - cancellata corresponding to the Polygnathus gronbergi (upper) to P. laticostatus conodont Zones, cul- minating within the N. elegans - base of the N. cancellata Zones (=bound- ary interval between the regional Zlíchovian and Dalejan Stages). Faunal changes are apparent especially in trilobites (the Zlíchovian Phacops- Crotalocephalus or Orbitoproetus-Scabriscutellum Assemblages grade into the Dalejan Phacops-Cyrtosymboloides Assemblage, Chlupáč 1983) but ana- logous changes are demonstrable in other groups (Chlupáč et al. 1979, Chlupáč & Turek 1983) and the level shows some significance even in the representation of palynomorphs (McGregor 1979).

The lithologic and faunal changes are in most cases - except the Koněprusy reef area - expressively g r a d u a l , occupying an interval

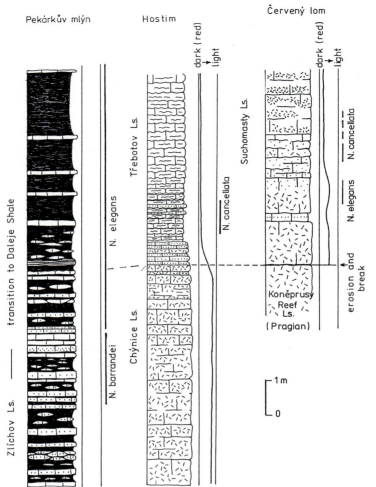

Figure 2. Examples of sections showing the Daleje Event: Pekárkův mlýn - basinal development in the SE limb, Hostim - NW limb (the Daleje Shale is replaced by the lowest Třebotov Limestone), Červený lom near Suchomasty - the Suchomasty Ls. overlies, after a break, the Pragian Koněprusy reef.

of several metres in sections dominated by low-energy sediments. The character of the change points to a gradual, possibly world-wide rise of the sea level, i.e. a transgressive event.

Changes in a correlatable stratigraphical level are reported from many areas, e.g. N-Spain (lower part of the Arauz Form., Henn 1985), Massif Armoricain (Paris 1981), Asia Minor (onset of Bohemian faunas near the Dede-Gebze Formations boundary, Haas 1968), NW-Africa (onset of the Rahal Shale, Alberti 1969 a.o.), Central Asia - Tian Shan (Kimovsk-

Dzhaus Form., Kim et al. 1978), S-China (Yang et al. 1981), Novaya
Zemlya (base of the Pakhtusov Form., Čerkesova et al. 1982), Western
North America and New York (deepening within the Cycle 1b, Johnson et
al. 1985).

House (1985) stressed the importance of this level for the development of goniatites which is clearly demonstrated in the Barrandian sections (Chlupáč & Turek 1983). The difference of Upper Emsian (= Dalejan)
faunas from those of the Lower Emsian (= Zlíchovian) is demonstrated
even in the clastic Rhenish development (Solle 1972), and the decrease
of provinciality in the Dalejan = late Emsian faunas may be influenced
by this transgressive event.

The Basal Choteč Event

This event, designated also as the "jugleri Event" (Walliser 1985),falls
in the Barrandian at the base of the Choteč Limestone and its equivalents, close above the newly defined Lower-Middle Devonian boundary
drawn according to the base of the Polygnathus costatus partitus conodont
Zone.

This level is represented in lithology by a change from light to
dark colour in the micritic limestone development and by an onset of
darker grey and sorted biosparitic layers and dark shale interbeds which
characterize the base of the Choteč Limestone (Fig. 3 , description of
sections in Chlupáč 1959, Chlupáč et al. 1979). In the reefal area of
Koněprusy, the reddish biosparitic and biomicritic Suchomasty Limestone
is rather sharply overlain by the light grey sparitic and sorted biodetrital Acanthopyge Limestone, locally with breccious layers at the base.

A marked faunal change is demonstrable in almost all groups: in
trilobites the change affects the shallow- and deep-water assemblages
(Chlupáč 1983) and analogous changes are expressed in other groups, e.g.
brachiopods, dacryoconarid tentaculites (the base of the Nowakia sulcata
sulcata Zone), conodonts (the base of the P. costatus costatus Zone,
Klapper et al. 1978), nautiloids, ostracods etc. Among goniatites, the
starting occurrence of Pinacites jugleri (Roem.), representatives of
Agoniatites (Fidelites) and a decline of Anarcestes are very marked
(Chlupáč & Turek 1983).

Although the change may appear as abrupt in some sections, its
rather gradual character cannot be overlooked in detailed study: changes in lithology and colour are gradually developing within an interval
of several tens of cm up to about 2 m, and new faunal elements also
appear gradually within this interval, although their accumulations in
some biodetrital beds accentuate the limit.

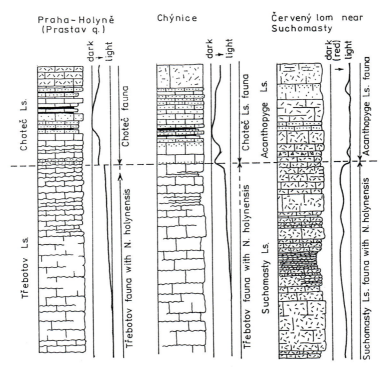

Figure 3. Three selected sections demonstrating the Basal Choteč Event: Praha-Holyně, Prastav quarry - parastratotype of the Lower-Middle Devonian boundary, Chýnice - outcrop near Ve mlýnci farm, Červený lom near Suchomasty - shallow-water development in the Koněprusy area.

The event is in the deeper-water facies clearly anoxic, possibly best explainable by a r a t h e r   q u i c k rise of the sea level.

Coinciding changes are known from Thuringia (Schwärz Shale, Alberti 1980, 1985), Rheinisches Schiefergebirge and Harz (Requadt & Weddige 1978 a.o.), N-Spain (the base of the La Loma Mbr, Henn 1985), the Ural Mts. (the upper limit of the Zdimir pseudobaschkiricus Zone, Sapelnikov & Mizens 1980), NW-Africa (Alberti 1980 a.o.). The transgressive event character of this level might have a positive influence on the wide recognition of the base of the Polygnathus costatus costatus conodont Zone and its equivalents (shown e.g. by Klapper & Johnson 1980).

The Kačák Event

This significant Middle Devonian event named by House (1985) and known also under the designation "otomari Event" or "rouvillei Event" (Walliser 1984, 1985) is the most expressive event within the Devonian of the Barrandian.

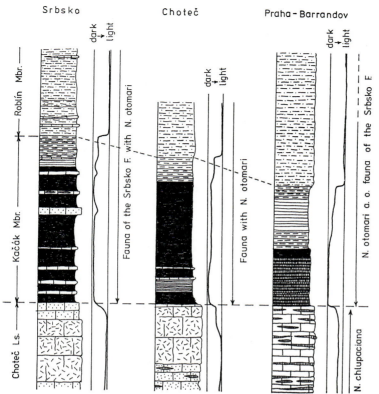

Figue 4. Examples of sections reflecting the Kačák Event: Srbsko - SW part of the basin, Choteč - central part of the basin, Praha - Barrandov - NE part of the basin.

It is characterized by an abrupt onset of dark calcareous shales of the Kačák Member which sharply overlies the Choteč Limestone (Fig. 4). Although the uppermost Choteč Limestone shows local facies differences - thick bedded and light biodetrital limestones in the SW, dark grey micritic laminated limestones with cherts in the NE part- the sharp contact with the Kačák Shale, differing only locally in proportions of impure limestone intercalations and dark radiolarian cherts, is always characteristic (description of sections in Chlupáč 1960).

This level represents a distinct biostratigraphic boundary - the base of the Nowakia otomari tentaculite Zone corresponding to the base of the Cabrieroceras crispiforme = rouvillei goniatite Zone. The limit is accentuated by a drastic impoverishment of all benthic forms - e.g. none of the almost 50 trilobite species known from the Choteč Limestone and its equivalents continues into the Kačák Shale where only 2, but differing, species are known, and similar relations are known in brachio-

pods, crinoids, corals etc. Planktonic and nektonic groups were evidentl
less affected and the level of the Kačák Event falls still within the
upper part of the <u>Tortodus</u> <u>kockelianus</u> <u>kockelianus</u> conodont Zone, close
to the base of the <u>P. ensensis</u> Zone. The fauna of the Kačák Shale (about
60 species described) is of a markedly lesser diversity than that of
the Choteč and Acanthopyge Limestones (about 250 species described). The
Kačák Shale commonly contains only dacryoconarid tentaculites (<u>Nowakia</u>
<u>otomari</u>, <u>Styliolina</u>), nautiloids, small posidoiid bivalves, inarticulate
brachiopods, sponge spicules, rare phyllocarids etc., accompanied by
fragmented land plants. Goniatites exhibit a new, Givetian character
stressed by frequent <u>Agoniatites</u> of the <u>costulatus</u> and <u>vanuxemi</u> groups,
tornoceratids etc.

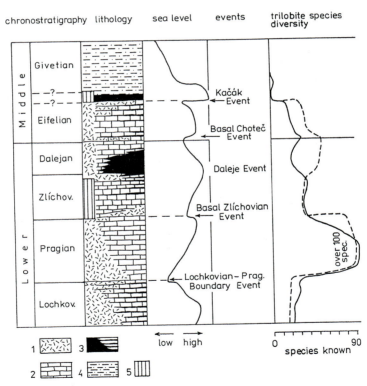

Figure 5. Stratigraphic position of main events, sea-level curve
and trilobite diversity in the Lower and Middle Devonian of the Barrandian. Full line in the trilobite diversity curve indicates the number
of species in the deeper-water micritic and shale facies, interrupted
line the diversity in the shallow-water biodetrital and reefal facies.
1 - biodetrital and reef facies, 2 - other limestones, 3 - shale seque-
ces and intercalations, 4 - flyschoid sediments, 5 - breaks.

The Kačák Event probably reflects a q u i c k rise of the sea level causing deepening of the basin and an abrupt predominance of anoxic conditions.

Correlative changes are known e.g. from Rheinisches Schiefergebirge (the Odershausen Limestone, Walliser 1985), Thuringia and Carnic Alps (Alberti 1983b, 1985; Schönlaub 1985), NW-Africa (Alberti 1980), S-China (Yang et al. 1981). In the classic New York section, the onset of the dark Union Spring Shale with the Cherry Valley Limestone correlates with the Kačák Event, and in Western North America the transgression near the base of the P. ensensis Zone falls in the approximately same level (Johnson & Murphy 1984, Johnson et al. 1985).

House (1985) stressed the importance of this event for the evolution of goniatites which would be in the ammonoid biostratigraphy the "natural" base of the late Middle Devonian "Maenioceras Stage". The global character of the Kačák Event is very probable.

Other events in the Devonian of the Barrandian

Within the Lower and Middle Devonian of the Barrandian area also other events may be recognized, from which especially the B a s a l Z l í c h o - v i a n E v e n t in the Lower Devonian and the B a s a l R o b l í n E v e n t (onset of the flysch-like sedimentation in the Middle Devonian) are markedly expressed. These events, however, seem to have a rather local importance and tectonic control.

The S i l u r i a n - D e v o n i a n b o u n d a r y, which may in its onset of new forms demonstrate a discrete bio-event of global scale, is not connected in the Barrandian with any marked lithologic boundary and also generally uninterrupted lineages within different faunal groups do not point to an expressive and drastic event in strict sense (comp. description of sections in Chlupáč et al. 1972 and general discussion in McLaren 1971).

REFERENCES

ALBERTI, G.K.B. (1969): Trilobiten des jüngeren Siluriums sowie des Unter- und Mitteldevons. I.- Abh. Senckenb. naturforsch. Ges. 520, 1-692.
-- (1980): Neue Daten zur Grenze Unter-/Mittel-Devon, vornehmlich aufgrund der Tentaculiten und Trilobiten im Tafilalt (SE-Marokko).- N. Jb. Geol. Paläont. Mh. 10, 581-594.
-- (1981): Daten zur stratigraphischen Verbreitung der Nowakiidae (Dacryoconarida) im Devon von NW-Afrika (Marokko, Algerien).- Senckenbergiana leth. 62, 205-216.
-- (1983a): Trilobiten des jüngeren Siluriums sowie des Unter- und

Mittel-Devons IV.- Senckenbergiana lethaea 64, 1-87.
-- (1983b): Zur Biostratigraphie und Paläontologie des kalkig entwik-
kelten Unterdevons im Frankenwald.- Geol. Bl. NO-Bayers 33, 91-133.
-- (1985): Zur Tentaculitenführung im Unter- und Mittel-Devon der Zentralen Karnischen Alpen (Österreich).- Cour. Forsch.-Inst. Senckenberg 75, 375-378.
BOUCOT, A.J. (1975): Evolution and extinction rate controls.- Elsevier,
Amsterdam.
ČERKESOVA, S.V.; PATRUNOV, D.K.; SOBOLEV, N.N.; SMIRNOVA, M.A. & EGORO-
VA, A.A. (1982): Ranica nižnego i srednego devona na Novoj Zemle.-
in: Biostratigrafia prograničnyh otloženij nižnego i srednego devona,
158.
CHLUPÁČ, I. (1959): Facial development and biostratigraphy of the Daleje
Shales and Hlubočepy Limestones (Eifelian) in the Devonian of Central
Bohemia.- Sbor. Ústř. Úst. geol. 24, 446-511.
-- (1960): Stratigraphical investigation of the Srbsko Beds (Givetian)
in the Devonian of Central Bohemia.- Sbor. Ústř. Úst. geol. 26, 143-
182.
-- (1983): Trilobite assemblages in the Devonian of the Barrandian area
and their relations to palaeoenvironments.- Geologica et Palaeonto-
logica 17, 45-73.
-- ; JAEGER, H. & ZIKMUNDOVÁ, J. (1972): The Silurian-Devonian boundary
in the Barrandian.- Bull. Can. Petrol. Geol. 20, 104-174.
-- ; LUKEŠ, P. & ZIKMUNDOVÁ, J. (1979): The Lower/Middle Devonian
boundary beds in the Barrandian area, Czechoslovakia.- Geologica et
Palaeontologica 13, 125-156.
-- ; LUKEŠ, P.; PARIS, F. & SCHÖNLAUB, H.P. (1985): The Lochkovian-
Pragian boundary in the Lower Devonian of the Barrandian area,
Czechoslovakia.- Jb. Geol. B.-A. 128, 9-41.
-- & TUREK, V. (1983): Devonian goniatites from the Barrandian area,
Czechoslovakia.- Rozpr. Ústř. Úst. geol. 46, 1-159.
HAAS, W. (1968): Trilobiten aus dem Silur und Devon von Bithynien (NW-
Türkei).- Palaeontographica 130A, 60-207.
HENN, A.H. (1985): Biostratigraphie und Fazies des hohen Unter-Devon
bis tiefen Ober-Devon der Provinz Palencia, Kantabrisches Gebirge,
N-Spanien.- Göttinger Arb. Geol. Paläont. 26, 100 p.
HOUSE, M.R. (1985): Correlation of mid-Palaeozoic ammonoid evolutionary
events with global sedimentary perturbations.- Nature 313, 17-22.
JAEGER, H. (1976): Das Silur und Unterdevon vom thüringischen Typ in Sar
dinien und seine regionalgeologische Bedeutung.- Nova Acta leopold.
45, 263-299.
JOHNSON, J.G.; KLAPPER, G. & SANDBERG, C.A. (1985): Devonian eustatic
fluctuations in Euramerica.- Geol. Soc. Amer. Bull. 96, 567-587.
-- & MURPHY, M.A. (1984): Time-rock model for Siluro-Devonian continenta
shelf, western United States.- Geol. Soc. Amer. Bull. 95, 1349-1359.
KIM, A.T.; YOLKIN, E.A.; ERINA, M.V. & GRATSIANOVA, R.T. (1978): Type
section of the Lower and Middle Devonian boundary beds in Middle Asia
- Field Session Internat. Subcommiss. Devonian System, Guide to fiel
excursions, 1-54.
KLAPPER, G. & JOHNSON, J.G. (1980): Endemism and dispersal of Devonian
conodonts.- J. Paleont. 54, 400-455.
-- ; ZIEGLER, W. & MASHKOVA, T. (1978): Conodonts and correlation of
Lower-Middle Devonian boundary beds in the Barrandian area of Czecho
slovakia.- Geologica et Palaeontologica 12, 103-116.
McGREGOR, D.C. (1979): Devonian spores from the Barrandian region of
Czechoslovakia and their significance for interfacies correlation.-
Current Research, Geol. Surv. Canada, B, 79-1B, 189-197.
McLAREN, D.J. (1970): Presidential address: Time, life, and boundaries.
J. Paleont. 44, 805-815.
PARIS, F. (1981): Les Chitinozoaires dans le Paleozoique de sud-ouest
de l'Europe.- Mém. Soc. géol. et minér. Bretagne 26, 1-412.

REQUADT, H. & WEDDIGE, K. (1978): Lithostratigraphie und Conodonten-
   faunen der Wissenbacher Fazies und ihrer Äquivalente in der südwest-
   lichen Lahnmulde (Rheinisches Schiefergebirge).- Mainzer geowiss.
   Mitt. 7, 183-237.
SAPELNIKOV, V.P. & MIZENS, L.I. (1980): Novoe v probleme granicy nižnego
   i srednego devona na Urale.- Paleont. biostr. srednego paleozoja Urala
   Urala, 23-38.
SCHÖNLAUB, H.P. (1980): Field Trip A: Carnic Alps. Second European Cono-
   dont Symposium (Ecos II), Guidebook.- Abh. Geol. B.-A. 35, 5-57.
-- (1985): Devonian conodonts from section Oberbubach II in the Carnic
   Alps (Austria).- Cour. Forsch.-Inst. Senckenberg 75, 353-360.
SOLLE, G. (1972): Abgrenzung und Untergliederung der Oberems-Stufe, mit
   Bemerkungen zur Unterdevon-/Mitteldevon-Grenze.- Notizbl. hess. L.-
   Amt Bodenforsch. 100, 60-91.
WALLISER, O.H. (1984): Geologic processes and global events.- Terra cog-
   nita 4, 17-20.
-- (1985): Natural boundaries and Commission boundaries in the Devonian.
   - Cour. Forsch.-Inst. Senckenberg 75, 401-408.
YANG SHIH-PU; PAN KIANG & HOU HUNG-FEI (1981): The Devonian System in
   China.- Geol. Mag. 118, 113-224.

AMMONOID EVOLUTION BEFORE, DURING AND AFTER THE "KELLWASSER-EVENT" - REVIEW AND PRELIMINARY NEW RESULTS

BECKER, R. Thomas *)

A contribution to Project
GLOBAL BIO-EVENTS
IUGS UNESCO

New work on ammonoid evolution

A new detailed cephalopod biostratigraphy is now possible before, during and after the so-called "Kellwasser-event". This follows from a critical restudy of the Devonian ammonoid literature, unpublished work (especially Ph. D. theses Price and Gatley at Hull) and recent investigations in Europe and North Africa. This will replace the old zonation of Wedekind and Schindewolf. 18 - 19 faunal levels have been distinguished between the base of the old <u>Manticoceras</u> <u>cordatum</u> - zone (m. <u>asymmetricus</u>-zone in conodont terms) and the <u>Prolobites</u> <u>delphinus</u>-zone (u. <u>marginifera</u>-zone in latest sense).

Most of the old <u>lunulicosta</u>-zone of Wedekind (1913) now belongs to the Givetian and the <u>Pharciceras</u>-Stufe (House 1985). Levels with <u>Koenenites</u> and <u>Timanites</u>, which Wedekind (1913) referred to the <u>lunulicosta</u>-zone too, are now known to be Lower <u>asymmetricus</u>-zone in age and to lie between the base of the Frasnian and the <u>cordatum</u>-zone (see House et al. 1985) and should perhaps be called "Lower Frasnian". The <u>nodulosum</u>-zone (do I β) of Wedekind (1913) contains <u>Beloceras</u> s. str., a genus not occurring before the <u>An. triangularis</u>-zone (see Glenister & Klapper 1966 etc.). It was called "lower part of <u>cordatum</u>-zone" by House & Ziegler (1977) in their restudy of the Martenberg section.

Extinction and innovation

80 genera, subgenera and species groups, which require future distinction on an equivalent taxnomic level, were considered. Their reviewed stratigraphic ranges formed the basic datas for the diversity analysis in Figure 1. The most important results are:

A. Diversity distribution of Devonian ammonoids ususally shows alternating phases of small extinctions with radiations. But in the late Frasnian (obvious already from work of Wedekind and Matern) there was a major biocrisis.

The Gephuroceratina are now known to become extinct in the late <u>holzapfeli</u>-zone. Famennian records of manticoceratids are misidentifications (the "<u>M</u>." <u>superstes</u> group are tornoceratids) or depend on erroneous correlations (the Famennian records of Helms 1959 and Ziegler 1971). Whether there is an abrupt or gradual decrease of these forms in the end of the Frasnian is still an open question.

---

*) Geologisches Institut der Ruhr-Universität Bochum, D-4630 Bochum, F.R.G.

Lecture Notes in Earth Sciences, Vol. 8
Global Bio-Events. Edited by O. Walliser
© Springer-Verlag Berlin Heidelberg 1986

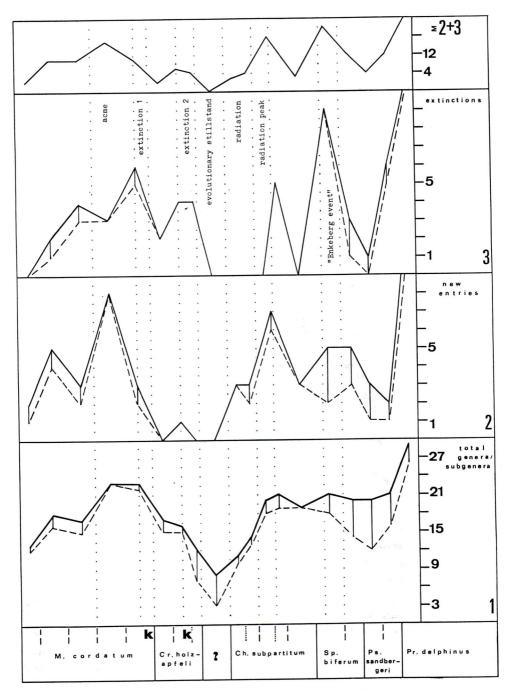

Figure 1. Diversity distribution of ammonoids in the middle Frasnian middle Famennian, showing alternation of minor extinctions and radiations, interrupted by the different phases of the "Kellwasser-event". K = Kellwasser-levels.

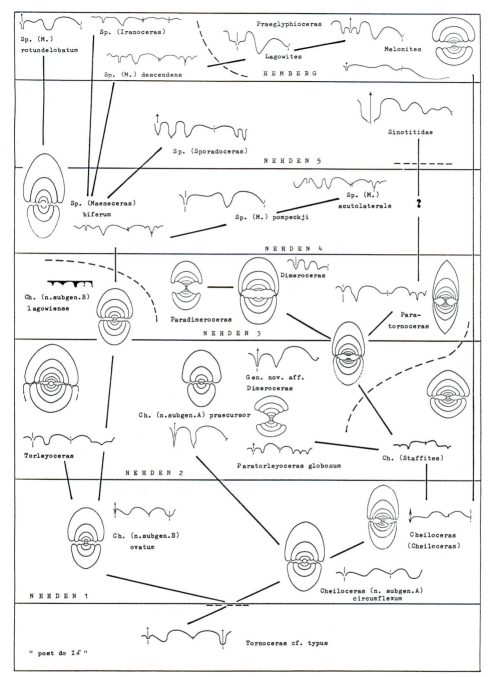

Figure 2. Evolution within Cheiloceratidae, Dimeroceratidae, Sporadoceratidae und Praeglyphioceratidae in the lower-middle Famennian (only first entries of taxa illustrated), showing wide-spread Proterogenesis of conch form development. Interrupted lines = boundaries between families.

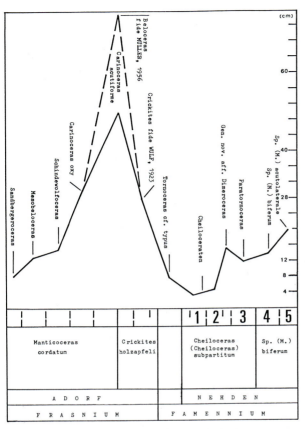

Figure 3.
Influence of the
late Frasnian bio-
crisis on the ma-
ximal size (dia-
meter) of middle
Adorfian-Nehdenian
ammonoids.

B. Beside bactritids only primitive anarcestids (<u>Archoceras</u>) and tornoceratids survive the crisis. Both groups dominate in special facies
situations.

C. There is a gap in the generic record in the "post I δ" (middle to
lower part of upper <u>triangularis</u>-zone) except for <u>Tornoceras</u> and <u>Pol</u>
(<u>Aulatornoceras</u>).

Stages in extinction and evolution

The biocrisis occupied certainly several 100,000 years and a sequence
of phases can be recognized:

A. The acme of diversity in the upper part of <u>cordatum</u>-zone.

B. This is followed by a first extinction (1) which is probably synchronous with the "Lower Kellwasser-level".

C. A further extinction (2) occured during the "Upper Kellwasser-level"
(higher <u>gigas</u> to lower <u>P</u>. <u>triangularis</u>-zone). At this time the diver
sity is 50 % below of the average for the period <u>cordatum</u>-to <u>sand-
bergeri</u>-zone.

D. The low diversity in the "post I δ" represents an evolutionary still-stand with no extinction and no new entry of genera/subgenera. Poor ecological conditions may explain why no immediate radiation occured. Perhaps it is significant that all aulatornoceratids of this time are of very small size.

E. The radiation in the lowermost Cheiloceras-Stufe is not dramatic but it is noticeable that there is an absence of extinctions.

F. The radiation peak lies in the middle part of the subpartitum-zone (m. - u. crepida-zone) and is followed by a new onset of extinctions and perhaps a return to "normal" evolutionary conditions.

G. The "Enkeberg Event" (House 1985) shows a marked faunal change at a high diversity level within the upper Nehdenian (not at the end of it). It is characterized by the loss of many cheiloceratids and the entry of Sporadoceras (Sp.) muensteri-group before the first pseudoclymenids (data taken from a restudy of the Enkeberg section, compare also Fig. 2).

H. The rapid evolution of clymenids in the delphinus-zone produced the first major diversity maximum for Devonian ammonoids.

Evolutionary processes during the biocrisis

During the period considered there are several phases of extinctions and diversifications. Following generalisations seem possible:

A. The taxa surviving the crisis belong,according to conch form and suture line,to more primitive groups than the Gephuroceratina. Bactrites, Lobobactrites and Tornoceras are bradytelic genera which remained alive during several extinction phases in the evolution of ammonoids (see House 1985). Of special interest is the fact that later all survivors seem to die out as a result of a replacement by more advanced forms (indicated by gradual dissappearance in times when other families dominate). Archoceras vanishes in the middle part of the lower Nehdenian, Lobobactrites and Tornoceras in the Hembergian, Bactrites during the Permian. This reflects the existance of two macroevolutionary regimes (compare Jablonski 1986 and Lewin 1986):

regime 1) higher adapted forms cause diminishing of unspecialized taxa; background extinction rate or alternation of minor extinctions and radiations typical

regime 2) higher adapted forms die out, while primitive taxa (which may have had special refuges) survive; mass extinction rate or extinctions without new groups arising typical, followed by a radiation without loss of any genera

B. The return to better living conditions and the absence of many feading competitors (large number of free ecological niches) lowered the selection pressure drastically. As a result of this the following evolutionary phenomena can be observed in the lower Cheiloceras-Stufe

extreme intraspecific variability
Frech (1902) was the first who recognized the enormous variability of cheiloceratids and tornoceratids in the lower Nehdenian faunas. The discrepancy compared with many "phylogenetic stabilized" goniatites of Adorfian or Hembergian age is indeed striking. Obviously mutations were much more less suppressed by selection than in "normal" times.

proterogenesis
This is much more widespread in the Nehdenian than in the Adorfian or Hembergian. It is illustrated in Figure 2 by ontogenetic development of conch forms in several parallel phylogenetic lines within Cheiloceratidae and Dimeroceratidae. Examples:

Ch. (n.subgen.A.) - Ch. (Cheiloceras) - Melonites
                 - Ch. (Staffites) - Paratorleyoceras
                 - Gen. nov. aff. Dimeroceras
                 - Dimeroceras - Paradimeroceras
                 - Paratornoceras - (?)-Sinotitidae

This is an astonishing parallel to the evolution of Prionoceratidae near the Devonian-Carboniferous boundary:

Prionoceras - Acutimitoceras - Gattendorfia - Pseudarietites
            - Zadelsdorfia, Costimitoceras, Gattenpleura

Within tornoceratids proterogenesis is typical for Falcitornoceras (appearance of falcate ribs in juvenile stages, loss of that feature in the adult, see House & Price 1985)

paedomorphosis (sensu McNamara 1986)
Persistance of larval, convex type of growth lines into mature stages (origin of cheiloceratids deriving from Tornoceras, less typical within tornoceratids: Gen. nov. aff. Epitornoceras, Linguatornoceras

All three processes allowed a strong acceleration of evolution!

C. The late Frasnian biocrisis is connected with a drastic size reduction of ammonoids which is tentatively illustrated in Figure 4. While species of Gephuroceratina reached diameter up to 50 (? 75) cm, adult cheiloceratids (with 7 or 8 whorls) are smaller than 5 cm. Gen. nov. aff. Dimeroceras from the Enkeberg and possibly also from the Aachen area (Wulf 1923) is the first genus crossing again the 10 cm mark. The type of Sp. (M.) acutolaterale had a diameter of 20 cm or more.

Causes and mechanisms of the "Kellwasser-event"

Ammonoids can contribute two points to these open problems:

A. The main extinction phases are related to the occurrence of anoxic lithofacies in areas with usually well oxygenated water.

B. Tornoceras, the least affected genus, is the only Devonian ammonoid from South America. Following paleomagnetic reconstructions this continent had a polar position in Upper Devonian times, which implies cold water conditions. All other faunas come from tropical or sub-tropical regions. In the Montagne Noire rich Tornoceras assemblages of the lower Famennian are combined with an unique ferromanganese-bearing lithofacies (described by Tucker 1973).

Both aspects plead for the explanation-model prefered by Geldsetzer et al. (1985): overflooding of shelf areas by cold, anoxic deep-sea-water caused extinction of many benthonic or to warmer climate adapted organisms as well as a transgressive pulse. Its triggering mechanism remains uncertain but may be linked with plate tectonic activities like continental break-ups.

REFERENCES

BECKER, R.T. (1984): Stratigraphie, Fazies und Ammonoideenfauna im oberen Frasnium - unteren Famennium bei Nehden und südlich von Bleiwäsche (östliches Rheinisches Schiefergebirge).- Dipl.-Arb. Ruhr-Univ. Bochum (unpublished), 128 p.

FRECH, F. (1902): Über devonische Ammoneen.- Beitr. Paläont. Geol. Österreich-Ungarns u. d. Orients 14, 27-111.

GATLEY, S.S. (1983): Frasnian (Upper Devonian) Goniatites from South Belgium.- Ph. Thesis, Univ. Hull (unpublished).

GELDSETZER, H.J.; GOODFELLOW, W.D.; ORCHARD, M.J. & McLAREN, D.J. (1985): The Frasnian-Famennian Boundary near Jasper, Alberta, Canada.- 98th Ann. Geol. Soc. Amer. et al. Mtg., Abstract No. 66549, Abstr. with Progr. 17, 589.

GLENISTER, B.F. & KLAPPER, G. (1966): Upper Devonian Conodonts from the Canning Basin, Western Australia.- J. Paleont. 40, 777-842.

HELMS, J. (1959): Conodonten aus dem Saalfelder Oberdevon (Thüringen).- Geologie 8, 634-677.

HOUSE, M.R. (1985): Correlation of Mid-Paleozoic ammonoid evolutionary events with global sedimentary perturbations.- Nature 313, 17-22.

-- & BECKER, R.T. (1986): Comments on Devonian Ammonoid Biostratigraphy in the Sub-Sahara resulting from brief visits.- Internat. Dev. Biostr., Rep. 12, Univ. Hull (unpublished), 49 p.

-- ; KIRCHGASSER, W.T., PRICE, J.D. & WADE, G. (1985): Goniatites from Frasnian (Upper Devonian) and adjacent strata of the Montagne Noire.- Hercynica, I, 1-21.

-- & PRICE, J.D. (1985): New late Devonian genera and species of tornoceratid goniatites.- Paleont. 28, 159-188.

-- & ZIEGLER, W. (1977): The Goniatite and Conodont sequences in the early Upper Devonian at Adorf, Germany.- Geol. Paläont. 11, 69-108.

JABLONSKI, D. (1986): Background and Mass Extinctions: Alternation of Macroevolutionary Regimes.- Science 231, 129-133.

LEWIN, R. (1986): Mass Extinctions select different victims.- Science 231, 219-220.

MATERN, H. (1931a): Die Goniatiten-Fauna der Schistes de Matagne von Belgien.- Bull. Mus. roy. hist. nat. Belg. 7, 1-15.

-- (1931b): Das Oberdevon der Dill-Mulde.- Abh. preuß. geol. L.-Anst. N.F. 134, 139 p.

McNAMARA, K.J. (1986): A Guide to the nomenclature of Heterochrony.- J. Paleont. 60, 4-13.

PRICE, J.D. (1982): Some Famennian (Upper Devonian) ammonoids from north-western Europe.- Ph. D. Thesis, Univ. Hull (unpublished), 555 p

SCHINDEWOLF, O.H. (1923): Beiträge zur Kenntnis des Palaeozoikums in Oberfranken, Ostthüringen und dem Sächsischen Vogtlande. 1. Stratigraphie und Ammoneenfauna des Oberdevons von Hof a.d. Saale.- N. Jb. Min. Geol. Paläont., Beil.-Bd. 49, 250-357, 393-509.

TUCKER, M.E. (1973): Ferromanganese nodules from the Devonian of the Montagne Noire (S. France) and West Germany.- Geol. Rdsch. 62, 138-153.

WEDEKIND, R. (1908): Die Cephalopodenfauna des höheren Oberdevon am Enkenberg.- N. Jb. Min. Geol. Paläont., Beil.-Bd. 26, 565-635.

-- (1913): Die Goniatitenkalke des unteren Oberdevon vom Martenberg bei Adorf.- Sber. Ges. naturforsch. Freunde Berlin 1913, 23-77.

WULF, R. (1923): Das Famennium der Aachener Gegend.- Jb. pr. geol. L.-A. 43, 1-70.

FRASNIAN MASS EXTINCTION - A SINGLE CATASTROPHIC EVENT OR CUMULATIVE?

(Faunistic investigations at the Frasnian - Famennian Boundary in South - Central Asia)

FARSAN, Noor M. *)

A contribution to Project GLOBAL BIO-EVENTS

Introduction

Devonian rocks are widespread over many places in South-Central Asia (Fig. 1). In Iran Devonian rocks are exposed mainly in the Alborz Ranges (Assereto 1963; Gaetani 1965, 1967; Brice et al. 1978), in the Shotori-Shirgesht area and in the Osbak-Kuh area (Stöcklin et al. 1965, Ruttner et al. 1968, Stöcklin 1972, Weddige 1984), in the Kirman area (Huckriede et al. 1962) and in the Zagros Ranges. In Afghanistan the Devonian outcrops have large areal extent in different parts of the western and central Paleozoic synclines, in the North Afghan geosyncline and in the Logar syncline (Brice 1970, Dürkoop et al. 1967, Wolfart & Wittekindt 1980). In Tadžikistan the main Devonian sections are exposed in the eastern and western Serawshan Ranges (Kim et al. 1985, Wolfart & Wittekindt 1980). Devonian rocks are exposed only in the north of Pakistan (Desio 1966, Gaetani 1967, Klootwijk & Conghan 1979, Stauffer 1967, Talent & Mawson 1979, Wolfart & Wittekindt 1980) and India (Chaterji et al. 1967, Gupta & Erben 1983). The main outcrops are exposed in Chitral, in the Khyber Pass-Peshawar-Nowshera-Saidu areas, in the Naubug-Valley and in the Muth area.

All of these sections, especially in the Upper Devonian, produce abundant fauna of many different groups of fossils, but in most cases, because of lack of key fossils, the localisation of the Frasnian-Famennian boundary is not obvious.

The Frasnian-Famennian boundary

There are only 4 localities in South-Central Asia in which an uninterrupted transition from Frasnian sediments to Famennian strata can be verified by means of fossils. These are: Rebat-e-Pai in Western Afghanistan, Ghuk in West-Central Afghanistan, Iraq in Central Afghanistan and Kalate Hadji Abdul Hossein in East Alborz, Iran. All these 4 sections, consisting mainly of marly limestones (Fig. 2), limestones and argillites (Brice 1970, Brice et al. 1978, Brice & Lang 1968, Brice & Lapparent 1969, Farsan 1981a, b and 1986) yield abundant brachiopod assemblages.

*) Leisberg 30, D-6900 Heidelberg, F.R.G.

Lecture Notes in Earth Sciences, Vol. 8
Global Bio-Events. Edited by O. Walliser
© Springer-Verlag Berlin Heidelberg 1986

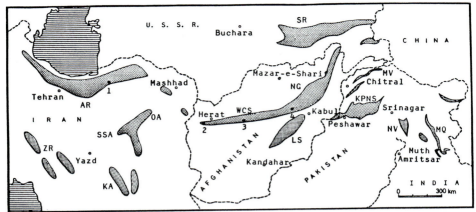

Figure 1. Distribution of main Devonian outcrops in the southern part of Central Asia. Locations of sections discussed in this study are indicated by numbers 1 to 4. Abbreviations: 1 = Kalate Hadji Abdul Hossein, 2 = Rebat-e-Pai, 3 = Ghuk, 4 = Iraq, AR = Alborz Ranges, KA = Kirman Area, KPNS = Khyber Pass - Peshawar - Nowshera - Saidu Areas, LG = Logar Syncline, MQ = Muth Quartzite, MV = Mastuj Valley, NG = North Afghan Geosyncline, NV = Naubug Valley, OA = Ozbak-Kuh Area, SR = Serawshan Range, SSA = Shotori-Shirgesht Area, WCS = West- and Central Afghan Synclines, ZR = Zagros Range. (Data from Brice 1970, Dürkoop et al. 1967, Wolfart & Wittekindt 1980.)

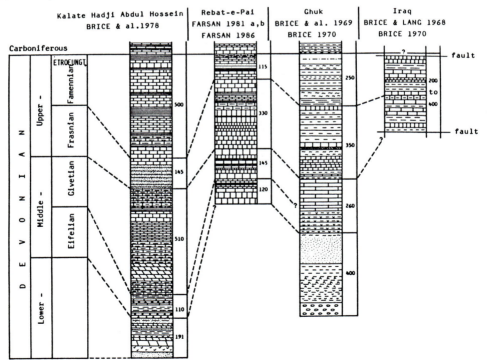

Figure 2. Schematic stratigraphic sections in the early to late Devonian of the East-Alborz (Iran), Rebat-e-Pai (Western Afghanistan), Ghuk (West-Central Afghanistan) and Iraq (Central Afghanistan).

	Kalate Hadji Abdul Hossein	Rebat-e-Pai	Ghuk	Iraq		
	32	71	58	17	total number of Devonian brachiopods	
		41	41		number of brachiopod species common between studied sections	
	21	21				
	18		18			
		13		13		
			11	11		
	5			5		
	4	4	4	4	number of common species	
FAMENNIAN	+	+	+	+	*Cyrtospirifer*	brachiopod genera persisting into Famennian
	+	+	+	+	*Productella*	
	-	+	+	+	*Ripidiorhynchus*	
FRASNIAN	+	+	+	+	*Cyrtospirifer*	
	+	+	+	+	*Productella*	
	+	+	+	+	*Ripidiorhynchus*	

Figure 3. Numerical comparison of brachiopod species occurring in section used in this study. Brachiopod genera Productella, Ripidiorhynchus and Cyrtospirifer cross the Frasnian-Famennian boundary unchanged.

The section of Rebat-e-Pai (Western Afghanistan) with 71 brachiopod species is the richest (Fig. 2), followed by Ghuk with 58 species, Kalate Hadji Abdul Hossein with 32 species and Iraq with 17 brachiopod species. The Upper Devonian sediments of Rebat-e-Pai and Ghuk with 41 brachiopod species in common show the closest faunistic relationship to each other, followed by Rebat-e-Pai and Kalate Hadji Abdul Hossein with 21 species in common, Kalate Hadji Abdul Hossein and Ghuk with 18 shared species, Rebat-e-Pai and Iraq with 13, Ghuk and Iraq with 11 and Kalate Hadji Abdul Hossein and Iraq with 5 species. There are only 4 species which are common to all 4 sections. The brachiopod genera Productella, Ripidiorhynchus and Cyrtospirifer cross the Frasnian-Famennian boundary in South-Central Asia and provide evidence for continuous uninterrupted sedimentation at the Frasnian-Famennian boundary in the region under discussion.

Section of Rebat-e-Pai (Western Afghanistan)

The process of faunal turnover at this boundary can be demonstrated only in Rebat-e-Pai as the result of its rich and diverse shallow epicontinental biota. This section can be subdivided by means of brachiopods into Middle Devonian, Frasnian, Famennian, Tournaisian and Visean. On

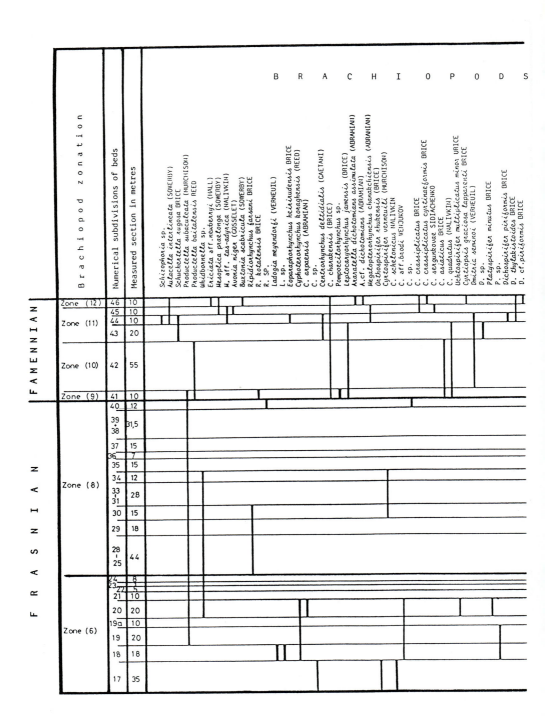

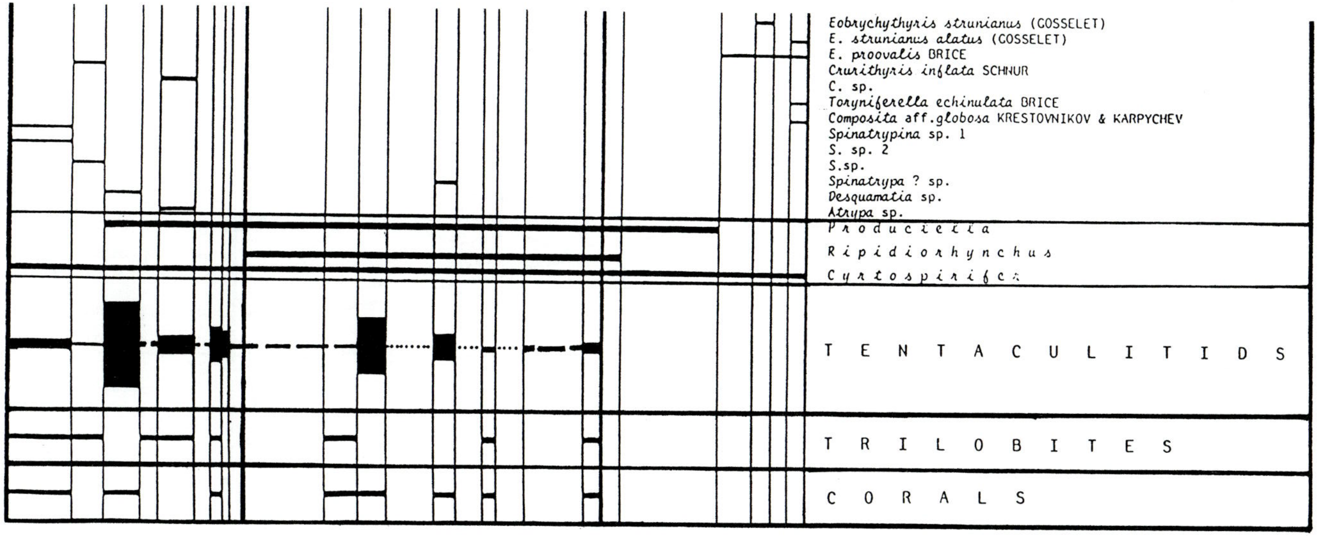

193

Figure 4. Detailed scheme of faunal turnover in the Upper Devonian of Rebat-e-Pai (Western Afghanistan). The evolution of brachiopods, trilobites, tentaculites and corals is punctuated by many episodes
of extinction and origination during the Frasnian. The evolution of brachiopods in the Famennian shows
the opposite phenomenon, i.e. a gradual increase in diversity.

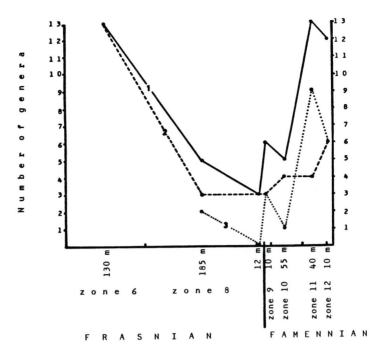

Figure 5. Pattern and magnitude of faunal change of brachiopod genera in Rebat-e-Pai (Western Afghanistan) during the Upper Devonian indicated by the Curves: 1 = standing diversity at zone level; 2 = survivors persisting into subsequent zone; 3 = new genera appearing in each zone (1 = 2 + 3).

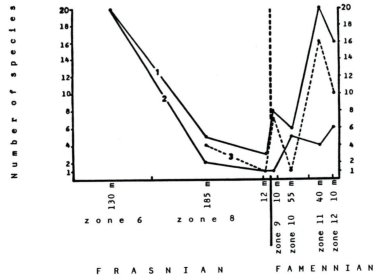

Figure 6. Pattern and magnitude of faunal turnover of brachiopod species in Rebat-e-Pai (Western Afghanistan) during the Upper Devonian indicated by the curves: 1 = standing diversity at the zone level; 2 = survivors persisting into subsequent zone; 3 = new species appearing in each zone (1 = 2 + 3).

the basis of brachiopod zonations we distinguish 7 brachiopod zones in the Devonian of Rebat-e-Pai (Farsan 1986).

The process of faunal turnover happens in the following manner (Fig. 4): In the Lower Frasnian there are many short term extinctions which replace each other. At the end of the early Frasnian, 7 out of 9 brachiopod species do not survive into the Middle and Upper Frasnian. The pattern of extinction continued during the Middle and Upper Frasnian. At the end of the Frasnian there remain 3 brachiopod species , only one of which survives beyond the Frasnian-Famennian boundary.

The evolution, extinction and appearances of the associated fauna (trilobites, tentaculitids and corals) follows a similar pattern. There are few tentaculitids, trilobites and corals in the late Upper Frasnian all of which go extinct at the end of Frasnian. This pattern of gradual extinction continues but in the Famennian there were more appearances than extinctions resulting in a radiation especially in the late part of the early Famennian.

The pattern and magnitude of the faunal turnover of brachiopod genera (Fig. 5) during the Upper Devonian is demonstrated as follows: a high diversity at the zone level in the Lower Frasnian is followed by a decrease in diversity in the Upper Frasnian. After a limited recovery in the early Famennian and a small decrease in zone 10 the diversity increases until the Upper Famennian.

The pattern and magnitude of the faunal turnover of brachiopod species is similar to that of genera (Fig. 6). In the Lower Frasnian, origination exceeds extinction resulting in the high diversity during the Lower Frasnian. The drop in diversity at the end of the Lower Frasnian is marked and continues into the Upper Frasnian. The recovery starts in the early Famennian, after a small decrease in zone 10 the diversity occurs in the Upper Famennian with high origination of brachiopod species.

Details of faunal change at the Frasnian-Famennian boundary in Rebat-e-Pai show the disappearance of the last representatives of the trilobites, tentaculitids and corals at the end of the Frasnian and the continuation of the three surviving brachiopod genera (Productella, Ripidiorhynchus, and Cyrtospirifer) into the early Famennian.

All Frasnian brachiopod superfamilies, with the exception of Atrypacea, are represented in the Famennian. All taxonomic categories of Frasnian brachiopod (superfamilies, genera and species) exhibit distinct increase in number during the Famennian. The percentage increase of genera and species of the main brachiopod superfamilies (Productacea, Spiriferacea and Rhynchonellacea) in the Famennian is very high.

Conclusions

1. There are at least 4 localities in the Southern part of Central Asia in which continuous sedimentation at the Frasnian-Famennian boundary can be verified by means of fossils.

2. The biologic evolution is characterized by successive decreases in diversity during the Frasnian followed by successive increases in diversity during the Famennian. The result was replacement of the Frasnian brachiopod-bryozoan-tentaculitid -trilobite -crinoidal fauna association by the Famennian fauna in which brachiopods and bryozoans were dominant.

3. The Frasnian "extinction" was not a single event. It was an accumulation of many successive episodes of extinctions which changed the character of the fauna.

4. There exists no evidence for a catastrophic change at the Frasnian-Famennian boundary. Instead, the documented changes occurred over a time period of approximately 6 to 8 million years.

REFERENCES

ASSERETO, R. (1963): The Paleozoic formations in Central Elburz (Iran).- Riv. Ital. Paleont. 69, 503-543.

BRICE, D. (1970): Étude paléontologique et stratigraphique de Dévonien de l'Afghanistan. Contribution à la connaissance des Brachiopodes et des Polypiers rugueux.- Notes Mém. Moyen-Orient 11, 1-364.

-- ; JENNY, J. et al. (1978): Le Dévonien de l'Elbourz oriental: stratigraphie, paléontologie (Brachiopodes et Bryozoaires), paléogeographie - Riv. Ital. Paleont. 84, 1-56.

-- & LANG, J. (1968): Sur un nouveau gisement de Dévonien supérieur à Iraq (Bamian, Afghanistan).- C.R. Soc. géol. France 168; 120-121.

-- & DE LAPPARENT, A.F. (1969): Stratigraphie du Dévonien de Ghouk (province du Ghor, Afghanistan).- C.R. Acad. Sci. 269 D, 17, 1595-1598.

CHATERJI, G.C. et al. (1967): Devonian of India.- in: OSWALD, D.H. (ed.) Int. Symp. Devonian Syst., Calgary, 1967, 1, 557-563.

DESIO, A. (1966): The Devonian sequence in Mastuj Valley (Chitral, NW Pakistan).- Riv. Ital. Paleont. Strat. 72, 293-320.

DÜRKOOP, A. et al. (1967): Devonian of Central and Western Afghanistan and Southern Iran.- in: OSWALD, D.H. (ed.): Intern. Symp. Devonian Syst., Calgary, 1967, 1, 529-543.

FARSAN, N.M. (1981a): Middle Devonian (Givetian) Tentaculites from West-Afghanistan.- Palaeontographica A 175, 89-105.

-- (1981b): New Asteropyginae (Trilobita) from the Devonian of Afghanistan.- Palaeontographica A 176, 158-171.

-- (1986): Faunenwandel oder Faunenkrise? Faunistische Untersuchung der Grenze Frasnium/Famennium im mittleren Südasien.- Newsl. Stratigr. (in press).

GAETANI, M. (1965): Brachiopods and Molluscs from Geirud Formation, Member A (Upper Devonian and Tournaisian).- Riv. Ital. Paleont. 71, 679-770.

-- (1967): Devonian of Northern and Eastern Iran, Northern Afghanistan and Northern Pakistan.- in: OSWALD, D.H. (ed.): Intern. Symp. Devonian Syst., Calgary, 1967, 1, 519-527.

GUPTA, V.J. & ERBEN, H.K. (1983): A late Devonian ammonoid faunula from
   Himachal Pradesh, India.- Paläont. Z. 57, 93-102.
HUCKRIEDE, R. et al. (1962): Zur Geologie des Gebietes zwischen Kerman
   und Sagand (Iran).- Beih. Geol. Jb. 51, 197 p.
KIM, A.I.; ERINA, M.V. et al. (1985): Subdivision and correlation of the
   Devonian of the Southern Tien Shan and the South of Western Siberia.-
   Cour. Forsch.-Inst. Senckenberg 75, 79-82.
KLOOTWIJK, C.T. & CONAGHAN, P.J. (1979): The extent of greater India, I.
   Preliminary palaeomagnetic data from the Upper Devonian of the
   eastern Hindukush, Chitral (Pakistan).- Earth and Planetary Science
   Letters 42, 167-182 (Elsevier Sci. Publ. Co.).
RUTTNER, A. et al. (1968): Geology of the Shirgesht area (Tabas Area,
   East Iran).- Geol. Surv. Iran, Rep. 4, 133 p.
STAUFFER, K.W. (1967): Devonian of India and Pakistan.- in: OSWALD, D.H.
   (ed.): Int. Symp. Devonian Syst., Calgary, 1967, 1, 545-556.
STÖCKLIN, J. (1972): Iran - Lexique stratigraphique intern. 3, Asie.-
   9b, 376 pp, Paris.
-- et al. (1965): Geology of the Shotori range Tabas area, East Iran).-
   Geol. Surv. Iran 3, 69 p.
TALENT, J.A. & MAWSON, R. (1979): Paleozoic-Mesozoic biostratigraphy of
   Pakistan in relation to biogeography and the coalescence of Asia.-
   in: Geodynamics of Pakistan. Geol. Surv. Pakistan, Quetta 1979, 81-
   102.
WEDDIGE, K. (1984): Zur Stratigraphie und Paläogeographie des Devons
   und Karbons von NE-Iran.- Senckenbergiana lethaea 65, 179-223.
WOLFART, R. & WITTEKINDT, H. (1980): Geologie von Afghanistan.- Beitr.
   Regional. Geol. Erde, (Borntraeger), Berlin-Stuttgart.

EVOLUTION OF THE LAST TROPIDOCORYPHINAE (TRILOBITA) DURING THE FRASNIAN

A contribution
to Project
GLOBAL
BIO -
EVENTS

FEIST, Raimund *) & CLARKSON, Euan N.K. **)

Abstract: Early Upper Devonian Tropidocoryphinae (Proetida, Trilobita) show unequivocal evidence of eye-reduction leading to blindness and extinction slightly before or during the Kellwasser Event. The obviously gradualistic evolution can be related to environmental change.

The Kellwasser Event, which in time is placed just below the Frasnian-Famennian boundary, had severe effects upon the faunas living within the Variscan realm. Prior to the event itself, however, there were significant evolutionary changes in several groups of Frasnian trilobites, which may be closely correlated with environmental changes during this time. The Tropidocoryphinae, which had been a stable group for some 40 million years exhibit some fundamental transformations of the cephalon, giving rise to regression and then virtual disappearance of the eye within a relatively short space of time (Lower asymmetricus to Lower gigas Zones), before becoming extinct at the Kellwasser boundary.

Current research on evolutionary patterns in the Tropidocoryphinae (R. Feist and E.N.K. Clarkson) is based essentially on material collected in the Montagne Noire (southern France). In this region the entire carbonate succession shows a complete record of conodont zones from the Middle Givetian to the topmost Frasnian, at which point the facies representing the Kellwasser Event is seen (Buggisch 1972, Klapper in Feist 1983, Feist & Klapper 1985, Feist 1985).

This succession shows a transition from distal neritic to pelagic facies during the late Eifelian. Within the pelagic realm, from the beginning of the Frasnian onwards, two distinct palaeogeographical environments are manifested, oxygenated and oxygen-deficient respectively.

All these facies were favorable to trilobites, especially the Tropidocoryphinae, which within the time-scale considered here, show some 10 distinct assemblages, both successional and contemporaneous. It has proved possible not only to trace the evolution of the last representatives of this group, but also to clarify some aspects of their extinction.

The earliest Tropidocoryphinae originated from Decoroproetus in late Silurian times. Common features of the cephalon retained by the early members of the main root-stock are (i) stable, divergent anterior

*)   Laboratoire de Paléontologie, Université des Sciences et Techniques du Languedoc, 34060 Montpellier, France.

**) Grant Institute of Geology, The University of Edinburgh, Edinburgh EH9 3JW, U.K.

Lecture Notes in Earth Sciences, Vol. 8
Global Bio-Events. Edited by O. Walliser
© Springer-Verlag Berlin Heidelberg 1986

branches of the facial suture, (ii) a pronounced tropidia and (iii) medium-sized, high, kidney-shaped eyes with more than 1000 lenses. The main line of descent (A) led from T. (Tropidocoryphe) through Longicoryphe (n.g.) to Erbenicoryphe (n.g.) in which early changes were comparatively minor. By the end of the Middle Devonian, however, the tropidia was disappearing, and there was some shift of habitat of these trilobites from the outer shelf to deeper pelagic environments.

The deposition of red cephalopod-bearing limestones began at around the Middle-Upper Devonian boundary (Lower asymmetricus Zone), a sedimentary event which is coupled with an abrupt change in the lineage. The new genus Erbenicoryphe, appearing for the first time, retains the somewhat divergent anterior branches of the facial suture of the ancestral stock, but the palpebral lobe straightens out and disappears, and the eye is reduced to a flat surface lacking lenses. Although these characters remain stable throughout the Lower asymmetricus Zone, the genus then becomes extinct, and following this episode of eye-reduction, the main lineage ends.

Apart from the main lineage, two other lines of descent can be distinguished. Firstly there is (B) a line originating in the early Devonian and leading with a rapid shift of characters to Astycoryphe, which then survived with little change until the late Givetian, giving rise to no descendants. Astycoryphe is invariably distinguished by its broader glabella and less divergent anterior suture. This genus is restricted to shallow-water and peri-reefal facies. Secondly there is (C) the Pterocoryphe - Pteroparia lineage which originated within the Longicoryphe (n.sg.A) - group in the grey stylioline pelagic limestones of the hermanni-cristatus Zone, where some instability in the anterior branch of the facial suture occurs. In the reduced pelagic-sapropelic environment which first develops in some places at the beginning of the Lower asymmetricus Zone, there is observed a successional change in the course of the anterior branch of the facial suture, in which the B-point migrates from a lateral-anterior, through a lateral-opposite to a lateral-posterior position. The latter stage is reached in the late form of Pterocoryphe (n.sg.A). In spite of the considerable reduction of the free cheek, and especially the height and size of the visual surface, the eye is still present, and is of typical kidney-like form. The adjacent fixed cheek displays a true palpebral lobe. In the early Pteroparia (n.sg.A), however, the palpebral lobe straightens again, but a relict of the visual surface is evident on the free cheek, though it is not known whether this functioned as an eye. The last Pteroparia has generally a smooth visual surface but in some cases small tubercles are found thereon which may or may not be lenses. By this stage the course of the facial suture has definitely stabilized.

The morphology of Pteroparia is relatively stable, and this genus spread from the reducing environment once more into the well-oxygenated, red carbonate facies after the extinction of Erbenicoryphe and all other Tropidocoryphines. The genus then persists until the top of the Upper gigas Zone, and then becomes extinct at the Kellwasser Event. As in the Montagne Noire "blind" forms of Pteroparia are reported from the southeastern Rhenish Schiefergebirge (Sessacker) and the Harz Mountains in the topmost Frasnian but do not seem to cross the Upper Kellwasser horizon.

This evolutionary series is based upon the very complete, and wellzoned successions in the Montagne Noire, but many of the genera are known from elsewhere. Pterocoryphe and Erbenicoryphe however, are not at present recorded outside southern France. Further research may reveal whether this development is global or localized with southern Europe as a centre of dispersal.

The sharp differences in the rate of evolutionary change in the Tropidocoryphinae, which coincides with the Middle/Upper Devonian boundary may be used as a supplementary tool for localizing the boundary itself in the pelagic trilobite-bearing facies.

REFERENCES

BUGGISCH, W. (1972): Zur Geologie und Geochemie der Kellwasserkalke und ihre begleitenden Sedimente (Unteres Oberdevon).- Abh. hess. L.-Amt Bodenforsch. 62, 1-68.
FEIST, R. (1985): Devonian Stratigraphy of the Southeastern Montagne Noire (France).- Cour. Forsch.-Inst. Senckenberg 75, 331-352.
-- & KLAPPER, G. (1985): Stratigraphy and Conodonts in pelagic sequences across the Middle-Upper Devonian boundary, Montagne Noire, France.- Palaeontographica A 188, 1-18.
KLAPPER, G. (1983): The Devonian of the eastern Montagne Noire (France).- in: FEIST, R. (ed.): I.U.G.S. Subcomm. Devonian Stratigraphy, Guidebook, Field Meeting, September 1983, U.S.T.L. Montpellier.

MIDDLE TO UPPER DEVONIAN BOUNDARY BEDS OF THE HOLY CROSS MTS: BRACHIOPOD RESPONSES TO EUSTATIC EVENTS

A contribution
to Project
GLOBAL
BIO-
EVENTS

RACKI, Grzegorz *)

Abstract: The succession of Givetian to Frasnian faunas in the Holy Cross Mts is analysed in context of biotic responses to cyclic depositional events, primarily of eustatic origin. This cyclicity stimulated the succession of brachiopod bottom-level assemblages from stringocephalid to ambocoeliid to atrypid (cyrtospiriferid) and to rhynchonellid faunas. This obviously supraregional pattern reflects sequential benthic habitat replacement within the intermittently drowning carbonate shelves. The faunal studies show rather constant exchange rates, and only a few species extinctions could be identified. Consequently, there are no marked faunal breaks across the Middle/Upper Devonian boundary in the Holy Cross Mts.

Introduction

Large progress in understanding of carbonate depositional systems and conodont biochronology have become a key to explanation of many basic questions of widespread Middle to Upper Devonian carbonate complexes. Facies-chronostratigraphy synthesis permitted J.G. Johnson et al. (1985, 1986) to document the 12 Devonian eustatic cycles in Euroamerica. This scheme gives a solid base to elucidate the significance of lithoevents occurring in particular sections and regions.

The current studies of the Givetian to Frasnian sequences in the Holy Cross Mts deal with the cyclic nature of strata (Racki 1985b) and succession of many faunal groups, particularly with brachiopods (Racki 1982). The present article is an attempt to outline the biotic responses to recognized depositional events, especially well-known for eustatic fluctuations (e.g. Roberts 1981, M.E. Johnson et al. 1985).

Opinion is given, that ecostratigraphic approach to regional geobiological history is a prerequisite toward trustworthy considerations on global bioevents (see McGhee 1982). In the shelf succession studied, it gives the possibility to estimate the probable extinction events discussed for the Middle/Upper Devonian boundary by some workers (e.g. McLaren 1970, House 1975).

Paleogeographic setting

The main object of the present study are the stromatoporoid-coral limestones being an important part of the Eifelian to Famennian transgressive

*) Laboratory of Paleontology and Stratigraphy, Silesian University, 41-200 Sosnowiec, Poland.

Lecture Notes in Earth Sciences, Vol. 8
Global Bio-Events. Edited by O. Walliser
© Springer-Verlag Berlin Heidelberg 1986

204

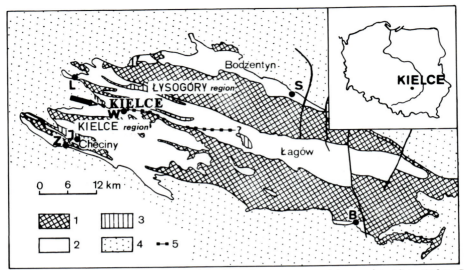

Figure 1. Location of the cited reference sections in the Holy Cross Mts. 1 Lower Paleozoic, 2 Devonian, 3 Upper Paleozoic, 4 Mesozoic and Tertiary, 5 boundary between Kielce and Łysogóry facies regions (arrowed , western part only) near Givetian/Frasnian boundary. B: Jurkowice-Budy Quarry, J: Jaźwica Quarries, L: Laskowa Hill Quarry at Kostomłoty, S: Skały, W: Wietrznia Quarries at Kielce, Z: Zamkowa Hill at Checiny.

carbonate sequence. This facies is typical of the extensive carbonate platform (or biostromal bank sensu Burchette 1981) extended over large areas of aggraded carbonate shelf in southern Poland.

In the Holy Cross Mts (Fig. 1) this shallow-water deposition was gradually limited to centrally situated Frasnian bioherm-fringed bank ("reef" of the Kielce region; see Szulczewski 1971, Szulczewski & Racki 1981). Its northern boundary is sharply marked by the Łysogóry intrashelf basin expanding during the Middle Devonian.

Near the Givetian/Frasnian boundary in the southernmost part of the Holy Cross region developed the Checiny intershoal area. It evolved as a result of the Frasnian transgression into an intrashelf basin. The shrinking parts of the Kielce region "reef" were submerged in the latest Frasnian, and this bioevent is well-recorded in the Holy Cross Mts.

Eustatic events

Four cycles, with more or less distinct shallowing-upward aspect (cyclothemic PAC-sequences, Busch et al. 1985; see also Wilson 1975), were recognized (Racki 1985b) within the Givetian to Frasnian boundary beds in the southern part of the Holy Cross Mts (Fig. 2, 3). The current

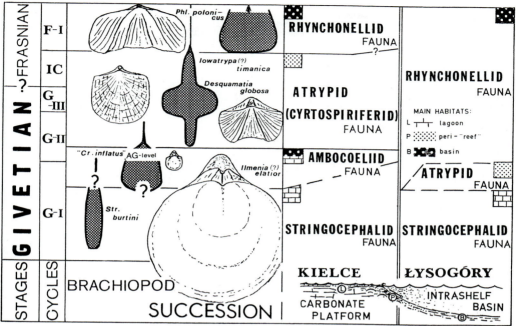

Figure 2. Brachiopod succession across the Givetian/Frasnian boundary in the western part of the Holy Cross Mts (see Fig. 1); G-I, F-I: sedimentary cycles (after Racki 1985b, modified). In the left side column of brachiopod succession there are presented the most typical species and listed are the most common assemblages (Str. burtini - Stringocephalus burtini, "Cr. inflatus" - "Crurithyris inflatus", Phl. polonicus - Phlogoiderhynchus polonicus).

conodont studies allow to correlate this modified event pattern with transgressive-regressive (T-R) cycles of Johnson et al. (1985).

Two most significant deepening pulses are well comparable to incipient transgressional cycles IIb and IIc (Fig. 3). The older event is manifested in the ambocoeliid-gastropod level (= AG-level), a widespread open shelf set within the biostromal sequence, initiating cycle G-II. It is dated recently at most probably basal Lowermost asymmetricus Zone.

The early Frasnian transgression (cycle F-I) begins here in the Middle asymmetricus Zone, according to data of Biernat & Szulczewski (1985, also Szulczewski & Racki 1981) for Phlogoiderhynchus polonicus-bearing marly set (= Phlogoiderhynchus level). On the other hand, the complex nature of the cycle IIb is evident in the area considered, as the two minor deepening episodes (? influenced by local epeirogeny) are present within the hardly divisible Lowermost to Lower asymmetricus Zone. The older event (= atrypid-crinoid level) seems to be limited to the southern periphery of the Kielce facies region. The younger deepening is differ-

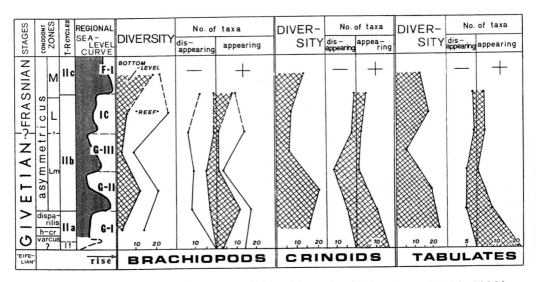

Figure 3. Diversity changes of brachiopods (data from Racki 1982), as well as crinoids (after Głuchowski, in prep.) and tabulate corals (after Nowiński, in prep.) and their relationships to global eustatic (T-R) cycles (after J.G. Johnson et al. 1985) and regional cyclicity in the SW-part of the Kielce region; lists of taxa include data from the better known sites: Jurkowice-Budy (for cycle G-I) and Kielce (Kadzielnia-type bioherms, for IC complex). Number of appearing and disappearing brachiopod taxa is estimated with exclusion of possible speciation cases. Subdivision of conodont zone: Lm: Lowermost, L: Lower, M: Middle.

ently expressed at various localities: it is best indicated by micritic-cherty set H at the Checiny section (see Racki & Baliński 1981, Fig. 2). It is treated herein as a beginning of the basal Frasnian intracyclic (IC) complex containing the Kadzielnia-type bioherms (Szulczewski & Racki 1981) deposited under relatively uniform sea-level conditions. Deepening pulse in the basal Lower _asymmetricus_ Zone is well recorded in more northern sections, e.g. Wietrznia-I Quarry (set C of Szulczewski 1971) and Kostomłoty (goniatite level of Racki et al. 1985).

The oldest unit G-I is not quite clear due to serious secondary dolomitisation. Its base is connected with termination of the dolomitic ("Eifelian"-type) sedimentation, but this facies event is not yet fully understood (Narkiewicz & Olkowicz-Paprocka 1981) and cannot be dated precisely. It is field of speculation, that the significant rebuilding of vast shelf areas (see below) should record the deepening pulse IIa or If. The latter event correlation is more attractive, as the prominent transgression, initiating the development of an intrashelf basin, has just occurred before the Eifelian/Givetian boundary in the eastern Łysogó region (see Malec 1984). It is documented by the very fossiliferous ope shelf Skały Beds above the chiefly dolomitized platform carbonates (Pajchlowa 1957, Adamczak 1976).

Surprisingly, the important onlap IIa is hardly traceable in the Kielce bank, contrary to the Łysogóry basin (Laskowa Hill section, Racki et al. 1985). Scarce conodonts suggest that the open shelf deposition typical for AG-level began locally earlier, maybe even in the *varcus* Zone.

Brachiopod succession

The most striking feature of the studied Givetian to Frasnian brachiopod faunas is strong predominance of a few relatively common and recurrent species. They constitute low-diversity, frequently monospecific assemblages typical for the non-"reef" facies. These bottom-level assemblages contrast with more diversified "reef" associations dwelling in organic buildups. The first category is main object of the present discussion due to the far more abundant material and distinct sequential pattern (Racki 1985a).

Within the sequence of the Kielce region there were distinguished four broadly-defined brachiopod faunas in the following order (Fig. 2):

1. Stringocephalid fauna: exemplified by the widespread giant *Stringocephalus burtini* Defrance (or other akin species), although in the eastern part (Jurowice-Budy section) this species is preceded by *Rensselandia* (Racki, in press);

2. Ambocoeliid fauna: monospecific shell beds built of different species of these small-sized smooth spiriferids; *Ilmenia* (?) *eliator* (Gürich) is succeeded by minute "*Crurithyris inflata* (Schnur), e.g. in Jaźwica section;

3. Atrypid (cyrtospiriferid) fauna: recurring *Desquamatia globosa* assemblage (Racki & Baliński 1981) replaced in the highest part by small *Iowatrypa* (?) *timanica* (Markovsky); association with cyrtospiriferids is notable;

4. Rhynchonellid fauna: *Phlogoiderhynchus polonicus* (Roemer) is a well-established guide fossil typical for the Lower Frasnian marly facies (Biernat & Szulczewski 1975); domination of different rhynchonellids in most of the Upper Devonian bottom-level assemblages is evident (e.g. Biernat & Racki, in press).

Several other assemblages, most frequently productellid, flourished only locally and/or periodically (see Racki, in press, Fig. 1, for ecostratigraphic scheme). "Reef"-associations are frequently linked with bottom-level fauna, e.g. ambocoeliid- or *Desquamatia*-dominated (Baliński

1973, Racki 1981a). Highly diversified and partly endemic association
of the Kadzielnia-type bioherm is the most remarkable exception (see
Szulczewski & Racki 1981).

The sequential replacement of the sessile benthos assemblages is an
evident response to changing biotopes, typically for the Devonian carbonate complexes (see Burchette 1981). It is an explanation for possible
diachronous relationships, especially among the stringocephalid- and
ambocoeliid-faunas. Also the paleogeographic setting strongly influenced
the succession, as the uniform (oldest) assemblage is found only before
submergence of the carbonate platform in the western Łysogóry region
(Fig. 2). It is also notable, that Phlogoiderhynchus polonicus and exter
nally similar forms appeared in the Łysogóry Basin in the late Givetian
already (see Racki et al. 1985), typically in association with diversified faunas, chiefly chonetids, ambocoeliids and orthids.

Closer analysis of the facies control mechanisms revealed variable,
but obvious interrelationships between the sedimentary cycles and the
stages in the regional "brachiopod story".

The "upward shoaling" model of a cycle assumes the successive deterioration of benthic habitats as a result of regressive outbuilding and
continuous restriction of circulation (Wilson 1975). This pattern is
well visible in the bottom-level faunas, which are primarily linked with
thin transgressive (and most open-marine) units at the base of cycles.

Exceptional are the two oldest faunas interpreted as inhabitants
of different biotopes within vast shelf lagoon with more or less restricted circulation (Racki, in press). Only "Crurithyris inflatus
assemblage successfully adapted itself to the novel situation during
the strong deepening pulse G-II; this cycle is marked by final disappear
ance of ambocoeliid assemblages and the domination of atrypids. The acme
of Desquamatia globosa is related to the local persistence of peri-
biostromal habitats with reduced marine biota especially typical of the
cycle G-III at Checiny area. Finally, the early Frasnian transgression
stimulated explosive development of more diversified rhynchonellid-domi-
nated faunas in expansive basinal biotopes.

Discussion

World-wide transgressive events established new areas and changed the
dominant types of habitats for benthic colonizations; they were also
unique occasions to great migrations due to breaking down of geographic
barriers (J.G. Johnson 1979). The size of the faunal influx is estimated
by a number of, in a biogeographic sence, new taxa appearing within a particular cycle (Fig. 3). It is helpful for the analysis, that the evidence

of gradual evolution and well-documented phyletic lineages are very rare
both in the Holy Cross region, and in other areas of coeval brachiopod
faunas (e.g. Roberts 1981, J.G. Johnson 1982).

The main migration waves near the Givetian/Frasnian boundary were
associated with the cycles G-II and F-I in case of bottom-level brachiopod faunas; it is in good agreement with supposed origin and pattern of
the deepening events in the region. Generally similar trends of bottom-
level brachiopods and crinoids, as well as "reef"-dwelling brachiopods
and tabulate corals (Fig. 3) are obscured only by the different pattern
of appearances in the cycle G-I already. It indicates a great range of
ecosystem rebuilding soon after the end of dolomitic deposition, which
occurred in almost abiotic conditions.

On the other hand, disappearances of taxa in a regional scale could
be either a reflection of a short-term local crisis or/and truely global
extinctions. The question is difficult to answer because of insufficient
knowledge of absolute range of most taxa. Nevertheless, some conclusions
seem to be reasonable for the recognized brachiopod succession of the
Kielce carbonate platform:

1. Sudden flourishing of Phlogoiderhynchus polonicus assemblage and
many other species, Desquamatia globosa (Gürich) including, is an obvious
manifestation of their southward expansion from the Łysogóry Basin as
a result of progressive creation of open-shelf habitats, chiefly during
early Frasnian (IIc) onlap.

2. Local disappearance of highly specialized, stenotopic stringocephalids, as well as of some ambocoeliids (e.g. Rhynchospiriferinae) would
coincide with their global extinction (see J.G. Johnson et al. 1980,
Struve 1982). It is related with final termination of their biotopes
typical of bank-phase of the Devonian carbonate complexes only in course
of the latest Givetian (IIb) transgression (J.G. Johnson 1979; Racki,
in press). Also ending of Desquamatia acme seems to be related to its
global extinction (Copper 1966, Fig. 1) due to strong reduction of
suitable low-energy peri-"reef" biotopes after the IIc deepening event.

3. The literature data point to supraregional importance of the
established links between the eustatic cycles resulting in evolution of
dominant benthic habitats and the succession of bottom-level brachiopod
faunas in the Kielce region. The following pattern can be outlined for
many shallow-water (interior) shelf sequences of Euroamerica:

3.a. Cycles If-IIa - cosmopolitan Stringunc Fauna of Struve (1982)
and/or rich ambocoeliid assemblages (see Racki, in press, for review).
The similar faunas have existed since Eifelian (Struve 1982, Fig. 5),
although they displayed different dominant genera (see Rzhonsnitskaja

1952, 1982; Biernat 1953; Jux & Strauch 1965; Caldwell 1967), i.e. Bornhardtina among stringocephalids, broadly-defined Emanuella and Rhynchospiriferinae among ambocoeliids (refinement of taxonomy of this familiy is a growing need).

3.b. Cycle IIb - abundant atrypid (chiefly Desquamatia or akin genera) and/or cyrtospiriferid (Cyrtospirifer, Tenticospirifer, Uchto-spirifer), frequently monospecific assemblages (see Copper 1967, Racki & Baliński 1981, Sartenaer 1982, Godefroid & Jacobs, in press). It corresponds to the cyrtospiriferid fauna of Struve (1982), distinguished from the chronostratigraphic point of view. On the other hand, the last ambocoeliid assemblage at the base of the cycle (AG-level) in the studied region seems to be a rather local variety (but see "Crurithyris ? Hauptlager" in Eifel area, Stuve 1964); in the eastward (Łagów section at this level there occurs the Tenticospirifer assemblage (see also Fig. 2).

3.c. Cycle IIc-large, broadly shaped and weakly-ribbed rhyncho-nellids referred formerly chiefly to Calvinaria (see Sartenaer 1979); highest part of Phlogoiderhynchus Zone of Sartenaer (1980) including (see also Garcia-Alcalde 1985). However, this zone is extended to the cycle IIb in areas displaying the deeper-water (external) shelf sequen-ces (e.g. Marocco; Drot 1982). Consequently, the pattern observed in Łysogóry Basin corresponds to this biofacies setting, known also e.g. from Nevada (J.G. Johnson 1977).

4. In the Holy Cross Mts there is no evident faunal break at the Givetian/Frasnian boundary, as recently defined. The rate of taxa exchange after initial colonization of this part of the shelf is rather uniform (Fig. 3). As for possible extinctions in the faunas, they seem to be mostly linked with the transgression IIb. It is also possible that the global bioevent discussed by some workers as potentially signi ficant, especially among brachiopods, is only artifact of diversity change studies, as evidenced by McLaren (1982), or it has occurred in the other depositional or biogeographic setting.

Acknowledgements

The author is indebted to Professor G. Biernat, Dr. M. Narkiewicz and Dr. T. Wrzołek for the review of the early version of the article.

REFERENCES

ADAMCZAK, F. (1976): Middle Devonian Podocopida (Ostracoda) from Poland their morphology, systematics and occurrence.- Senckenbergiana letha 57, 265-467.

BALIŃSKI, A. (1973): Morphology and paleoecology of Givetian brachio-
    pods from Jurkowice-Budy, Poland.- Acta Palaeont. Polon. 18, 269-
    297.
BIERNAT, G. (1953): On the three new brachiopods from the "Stringocepha-
    lus" limestone of the Holy Cross Mts (in Polish).- Acta Geol. Polon.
    3, 299-324.
-- & RACKI, G. (in press): Late Famennian rhychonellid-dominated assem-
    blages from the Holy Cross Mts, Poland.- Acta Palaeont. Polon.
-- & SZULCZEWSKI, M. (1975): The Devonian brachiopod Phlogoiderhynchus
    polonicus (Roemer, 1866) from the Holy Cross Mountains, Poland.-
    Acta Palaeont. Polon. 20, 199-221.
BURCHETTE, T.P. (1981): European Devonian reefs: a review of current
    concepts and models.- SEPM Spec. Publ. 30, 85-142.
BUSCH, R.M.; WEST, R.R.; BARRETT, F.J. & BARRETT, T.R. (1985): Cyclo-
    thems versus a hierarchy of transgressive-regressive units.- Recent
    Interpretations of Late Paleozoic Cyclothems, Conf. Symp., 141-153.
CALDWELL, W.G.E. (1967): Ambocoeliid brachiopods from the Middle Devonian
    rocks of northern Canada.- in: OSWALD, D.H. (ed.): International
    Symposium on the Devonian System. Alberta Soc. Petrol. Geologists,
    Calgary, Alberta, 1967, 2, 601-616.
COPPER, P. (1966): The Atrypa zonata brachiopod group in the Eifel, Ger-
    many.- Senck. lethaea 47, 1-55.
-- (1967): Frasnian Atrypidae (Bergisches Land, Germany).- Palaeonto-
    graphica A 126, 116-140.
DROT, J. (1982): Brachiopods near the Givetian-Frasnian boundary in
    Tafilalt and Ma'der (south Marocco).- Pap. Fr./Givet. boundary, Geol.
    Surv. Belgium, Spec. Vol., 70-84.
GARCIA-ALCALDE, J.L. (1985): La Extension de la Biozona de Phlogoide-
    rhynchus (Bracquiopodo rinconellido, Givetiense terminal - Frasniense
    inferior).- Trab. Geol. 15, 77 - 86.
GŁUCHOWSKI, E. (in prep.): Givetian and Frasnian crinoids of the Holy
    Cross Mts.
    GODEFROID, J. & JACOBS, L. (in press): Atrypida (Brachiopoda) de la
    Formation de Fromelennes (fin du Givetian) et de la partie inférieure
    de la Formation de Nismes (debut du Frasnian) aux du Synclinorium
    de Dinant (Belgique).- Bull. Inst. r. Sci. Nat. Belg.
HOUSE; M.R. (1975): Faunas and time in the marine Devonian.- Proc.
    Yorks. geol. Soc. 40, 459-490.
JOHNSON, J.G. (1977): Lower and Middle Devonian faunal intervals in
    central Nevada, based on brachiopods.- Calif. Univ. Riverside, Campus
    Mus. Contr. 4, 16-32.
-- (1979): Devonian brachiopod biostratigraphy.- The Devonian System,
    Spec. Pap. Palaeont. 23, 291-306.
-- (1982): Occurrence of phyletic gradualism and punctuated equilibria
    through geologic time.- J. Paleont. 56, 1329-1331.
-- ; KLAPPER; G. & TROJAN, W.T. (1980): Upper range of Stringocephalus
    (Devonian Brachiopoda).- Newsl. Stratigr. 8, 232-235.
-- & SANDBERG, G.A. (1985): Devonian eustatic fluctuations in Euro-
    america.- Geol. Soc. Amer., Bull. 96, 567-587.
-- & -- (1986): Late Devonian eustatic cycles around margin of Old
    Red Continent.- Ann. Soc. geol. Belg. 109, 141-147.
JOHNSON, M.E.; RONG JIA-Yu & YANG XUE-CHANG (1985): Intercontinental
    correlation by sea-level events in the Early Silurian of North Ameri-
    ca and China (Yangtse Platform).- Geol. Soc. Amer., Bull. 96, 1384-
    1397.
JUX, U. & STRAUCH, F. (1965): Die "Hians"-Schille aus dem Mitteldevon
    der Bergisch Gladbach- Paffrather Mulde.- Fortschr. Geol. Rheinl. u.
    Westf. 9, 51-86.
MALEC, J. (1984): New data on the stratigraphy of the Devonian in the
    Grzegorzowice-Skały section (in Polish).- Kwart. Geol. 28, 782-783.
McGHEE, G.R. (1982): The Frasnian-Famennian extinction event: a pre-
    liminary analysis of Appalachian marine ecosystems.- Geol. Soc. Amer.

Spec. Pap. 190, 491-500.
McLAREN, D.J. (1970): Presidential address: time, life, and boundaries.-
    J. Paleont. 44, 801-815.
-- (1982): Frasnian-Famennian extinctions.- Geol. Soc. Amer., Spec. Pap.
    190, 477-484.
NARKIEWICZ, M. & OLKOWICZ-PAPROCKA, I. (1981): Stratigraphy of the De-
    vonian carbonates in the eastern part of the Góry Świetokrzyskie Mts.
    (in Polish).- Kwart. Geol. 27, 225-256.
NOWIŃSKI, A. (in prep.): Tabulata and Chaetetida from the Givetian and
    Frasnian of the Holy Cross Mts.
PAJCHLOWA, M. (1957): The Devonian in the Grzegorzowice-Skały profile
    (Świety Krzyz Mts).- Biul. Inst. Geol. 122, 145-254.
RACKI, G. (1982): Brachiopods and ecostratigraphy of the "Givetian"
    limestones of the Holy Cross Mts (in Polish).- Unpubl. Dr. Thesis,
    Inst. of paleobiol. Pol. Acad. Sci. Warsaw.
-- (1985a): A new atrypid brachiopod, Desquamatia macroumbonata sp.n.,
    from the Middle to Upper Devonian boundary beds.- Acta Geol. Polon.
    35, 61-72.
-- (1985b): Cyclicity of sedimentation versus stratigraphic subdivision
    of Devonian stromatoporoid-coral limestones in the Holy Cross Mts.
    (in Polish).- Przegl. Geol. 1985, 267-270.
-- (in press): Brachiopod ecology of the Devonian carbonate complex, and
    problem of brachiopod hyposalinity.- Proc. First Internat. Congr.
    Brachiopodes.
-- & BALIŃSKI, A. (1981): Environmental interpretation of the atrypid
    shell beds from the Middle to Upper Devonian boundary of the Holy Cross
    Mts and Cracow Upland.- Acta Geol. Polon. 31, 177-197.
-- ; GŁUCHOWSKI, E. & MALEC, J. (1985): The Givetian to Frasnian succes-
    sion at Kostomłoty in the Holy Cross Mts, and its regional signifi-
    cance.- Bull. Polish Acad. Sci. Earth Sci. 33, 159-171.
ROBERTS, J. (1981): Control mechanism of Carboniferous brachiopod zones
    in eastern Australia.- Lethaia 14, 123-134.
RZHONSNITSKAJA, M.A. (1952): Spiriferids of the Devonian rocks of the
    adjacent areas of Kuznetsk basin (in Russian).- Trudy VSEGEI, Moscow.
-- (1982): Significance of brachiopods for zonal subdivisions of the
    Devonian of USRR (in Russian).- in: Sovremiennyje Znatschenje Paleon-
    tologii dla Stratigrafii, Nauka, 63-71.
SARTENAER, P. (1979): Deux nouveaux genres de rhynchonellides Frasniens
    précédemment inclus dans le genre Calvinaria (Brachiopoda).- Geo-
    bios 12, 535-547.
-- (1980): Appartenance de l'espèce Terebratula formosa de l'Eifel au
    genre Phlogoiderhynchus du debut du Frasnien.- Senck. lethaea 61, 17-
    43.
-- (1982): The presence and significance of Spirifer bisinus, S. malais:
    S. supradisjunctus and S. seminoi in Early Frasnian beds of Western
    Europe.- Pap. Fr./Givet. boundary, Geol. Surv. Belgium, Spec. Vol.,
    123-196.
STRUVE, W. (1964): Bericht über die geologischen Exkursionen in der
    Prümer Mulde (20.5.1964) und in der Eifeler Kalkmulden-Zone (21.5.
    1964).- Decheniana 117, 224-244.
-- (1982): The Eifelian within the Devonian frame, history, boundaries,
    definitions.- Cour. Forsch.-Inst. Senckenberg 55, 401-431.
SZULCZEWSKI, M. (1971): Upper Devonian conodonts, stratigraphy and
    facial development in the Holy Cross Mts.- Acta Geol. Polon. 21,
    1-129.
-- & RACKI, G. (1981): Early Frasnian bioherms in the Holy Cross Mts.-
    Acta Geol. Polon. 31, 147-162.
WILSON, J.L. (1975): Carbonate Facies in the Geologic History.- Springe:
    Verl., Berlin-Heidelberg-New York.

THE KELLWASSER EVENT IN MORAVIA

HLADIL, Jindřich, KESSLEROVÁ, Zuzana &
FRIÁKOVÁ, Olga *)

A contribution
to Project
GLOBAL
BIO-
EVENTS

Abstract: The intensive rebuilding of the ecosystems including the gradual extinction of the reef communities which is observed in the late
Frasnian and early Famennian ages corresponds to the large restriction
of the shelf areas as well as to the increase of humidity. While
the nearshore areas were often uplifted and/or destroyed, as concerned
their previous stable ecosystem, the frontal margin of the reefoid
communities progressed deeper basinwards with some ingressions and consequent replacement by the basinal, both pelagic and benthic communities.
The terrigenous phyto-debris, oogonies of Charophyta, fish remains and
other indications of rising climatic humidity are noticed from Moravian
shelfs.
   Additionally to this, the rough evolutionary pattern of these changes
appears to be eustatic changes of the sea level. The raising of the anoxic water level in the depression areas accompanied by the extinction
of some basinal taxa is supposed to be contemporaneous with the particular transgressions on the reefoid shoals. On the other hand the extinctions of the reefoid communities can be placed into the regression
levels. "Black" transgressional (lagoonal, sebkha) horizons occur on the
carbonate shoals. However, the particular crisis in the reefoid communities which were under stress of the great eco-factors can be iniciated
also by other very fine causes. The main crises of the reefoid communities in Moravia are correlated with the three levels:
i)   near the limits of the Lower and Upper Pa. gigas Zone (top of the
      Crassioalveolites domrachevi coral Zone)
ii)   at the top of the Upper Pa. gigas Zone (top of the lower Scolio-
      pora denticulata vassinoensis coral Zone)
iii) in the Middle to Upper Pa. triangularis Zone;
always somewhat higher then the "black" crisis of the basinal facies are
referred.
   The total ceasing of the late Frasnian to early Famennian shallow
water corals and stromatoporoids in Moravia was found much higher on profile, at the Pa. crepida Zone (Křitiny HV 105 borehole, Mokrá - Western
Quarry).

The true Kellwasser Limestone, containing a large amount of organic
matter as well as abundant pelagic fauna, is mentioned in Moravia only
from the middle, so-called basinal part of the Moravo-Silesic depositional region and from the frontal margins of the carbonate platforms.
The distribution of the facies during the Upper Pa. gigas interval is
shown in Fig. 1. The nice example of the Kellwasser Event level was documented in Hranice Quarry (Dvořák et al. 1958): The dish-like bodies
of the Kellwasser Limestone were bounded by the grey nodular limestones
which yielded conodonts of the Upper Pa. gigas Zone. Some examples are
known also from the Horní Benešov belt in the Nízký Jeseník Mts. It is
assumed, that the Kellwasser Limestones are connected in the basinal

*) Ústřední ústav geologický, 60200 Brno, Č.S.S.R.

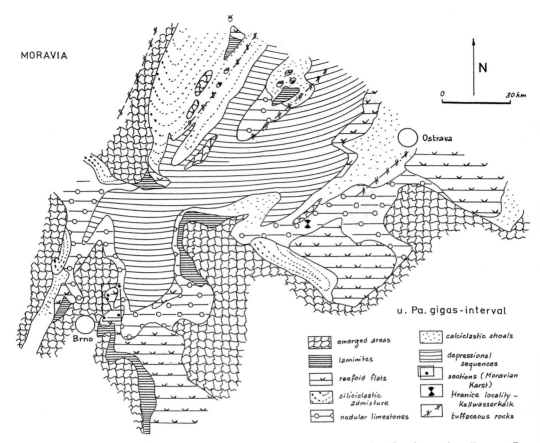

MORAVIA

N

0   30 km

Ostrava

u. Pa. gigas-interval

Brno

	emerged areas		calciclastic shoals
	laminites		depressional sequences
	reefoid flats		sections (Moravian Karst)
	siliciclastic admixture		Hranice locality – Kellwasserkalk
	nodular limestones		tuffaceous rocks

Figure 1.   Distribution of the facies in Moravia during the Upper Pa.
gigas interval.

sequences with 3 levels: 1. About the boundary between Lower and Upper
Pa. gigas Zones; 2. In the Upper Pa. gigas Zone; 3. About the middle
part of the Pa. triangularis Zone. The generation of these horizons can
be explained accordingly to the state-of-the-art which was presented at
the conference as a transgressive overflowing by anoxic water masses.
These dark horizons from the basinal sequences correspond to the reefoid
banks on the platforms. The sedimentation of reefoid limestones was
interrupted by regressions which are marked by gaps and dark lagoonal
sediments (see also the diagram in Fig. 3).

It is remarkable that the rear sides of the platforms were tectonically uplifted during the late Frasnian and early Famennian. It is documented by the sedimentological pattern and condensation accompanied
by the ceasing of some sediments in the nearshore environments (Dvořák
et al. 1984). Also the frontal margin of the carbonate platform, begin-

215

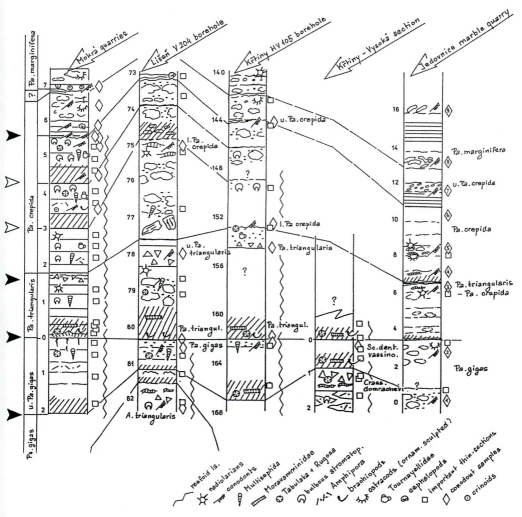

Figure 2.   Diagrammatical columns of some boreholes and outcrops at
the eastern margin of the Moravian Karst.

ning from the Lower/Upper Pa. gigas boundary, was moved predominantly
basinwards (compare Hladil 1983). Despite of the tectonic influences,
the eustatic pattern is visible in the sections (Fig. 2). The sediments
of the carbonate platforms are documented by numerous deep boreholes on
the slopes of the Bohemian Massif as well as by the outcrops in the SE
part of the Moravian Karst.

Namely in the outcrops we can characterize the collapses and rebuildings of the biota in the Kellwasser interval.

The first sharp faunal rebuilding is observed in the Upper Pa. gigas

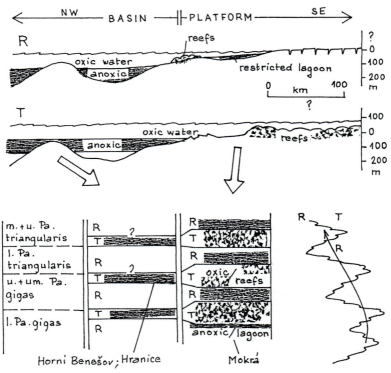

Figure 3. Correlation of the anoxic conditions of the sedimentation in the 3 main levels of the Kellwasser interval.

Zone. Many Tabulata-species extinct: <u>Crassialveolites</u> <u>domrachevi</u> (Sok. <u>Cr</u>. <u>multiperforatus</u> (Salée), <u>Cr</u>. <u>smithi</u> (Lec.), <u>Cr</u>. <u>obtortus</u> (Lec.), <u>Alveolites</u> cf. <u>complanatus</u> Lec., <u>Alv</u>. <u>regularis</u> Sok., <u>Thamnopora</u> <u>boloniensis</u> <u>langi</u> Hl., <u>Gracilopora</u> div. sp. On the other hand <u>Scolio-</u> <u>pora</u> <u>denticulata</u> <u>vassinoensis</u> Dub. and <u>Sc</u>. <u>denticulata</u> <u>rachitiforma</u> Hl. radiate intensively. The change is accompanied by the rebuilding among the endo- and epibionts.

The second level is placed at the top of the <u>Pa</u>. <u>gigas</u> Zone. Many upper Frasnian stromatoporoids and the tabulate coral <u>Sc</u>. <u>kaisini</u> Lec. extinct. Above this level the first occurrence of <u>Syringopora</u> <u>volkensis</u> Chern. and <u>Labechia</u> <u>cumularis</u> Yavorski are observed. The level is of regressional origin. The leached hardground is colonized by coral sheet The Oppenheimer's brachiopod horizon follows it.

The third level can be correlated with the upper part of the <u>Pa</u>. <u>triangularis</u> Zone. Intensive depression among the Actinostromatidae in the favour of genera <u>Atelodictyon</u>, <u>Stromatoporella</u>, and <u>Labechia</u> is documented in the sections. The conodont development, being continuous till this level, shows the pattern of extensive rebuilding (comp. Kesslerová 1985, in other regions, e.g. Buggisch et al. 1983).

It should be noticed that the impoverished stromatoporoid and coral communities (with Amphipora tschussovensis Yav., Natalophyllum perspicuum Hl. etc.) reach the Upper Pa. crepida level. It is described by Friáková et al. (1985) and/or Dvořák et al. (1984). Some "microevents", marked by diversity oscillations, can be distinguished among the 3 levels described above.

Event pattern

The correlation between the Kellwasser Limestones in the basinal sequences with the reefoid banks on the mid- and back-platform areas can be explained by sea-level rise, i.e. as eustatic peaks. These maxima are followed by a subsequent depression, marked on the platforms by gaps and dark lagoonal sediments. The regressive tendency, caused by the teconic uplifting of the rear side of the carbonate platform, seems to be characteristic for the Frasnian/Famennian boundary. The eustatic changes can be considered as a cause for the 3 main Kellwasser levels, but the restriction of the shelf areas and increase of humidity (Hallam 1983) are important for the general pattern of the biotic rebuildings in the late Frasnian and early Famennian as a whole. In addition, small ecological causes could be effective in the communities which were stressed by great and long-persisting factors.

REFERENCES

BUGGISCH, W.; RABIEN, A. & HÜHNER, G. (1983): Stratigraphie und Fazies des kondensierten Oberdevon-Profils "Diana" nördlich Oberscheld.- Geol. Jb. Hessen 111, 93-153.
DVOŘÁK, J.; CHLUPÁČ, I. & SVOBODA, J. (1958): Geologické poměry devonu u Hranic na Moravě.- Sbor. Ústř. Ust. geol. 24, 237-276.
-- ; FRIÁKOVÁ, O.; GALLE, A., HLADIL, J. & SKOČEK, V. (1984): Correlation of the reef and basin facies of the Frasnian age in the Křtiny HV 105 borehole in the Moravian Karst.- Sbor. Geol. Věd.,Ř.G. 39, 73-103.
FRIÁKOVÁ, O.; GALLE, A.; HLADIL, J. & KALVODA, J. (1985): A lower Famennian fauna from the top of the reefoid limestones at Mokrá (Moravia, Czechoslovakia).- Newsl. Stratigr. 15, 43-56.
HALLAM, A. (1983): Interpretacia facii i stratigraficheskaya posledovatelnost (Facies interpretation and the stratigraphic record).- Mir. Moskva.
HLADIL, J. (1983): The biofacies section of Devonian limestones in the central part of the Moravian Karst.- Sbor. Geol. Věd., Ř.G. 38, 71-94.
KESSLEROVÁ, Z. (1985): Biostratigrafie frasnu a famenu okolí jeskyně Balcarky na základě konodontů.- Master's theses, Univ. Brno.

GEOCHEMICAL ANALYSES OF THE LATE DEVONIAN "KELLWASSER EVENT" STRATIGRAPHIC HORIZON AT STEINBRUCH SCHMIDT (F.R.G.)

A contribution to Project GLOBAL BIO-EVENTS

MC GHEE, George R., Jr. *); ORTH, Charles J. **); QUINTANA, Leonard R. **); GILMORE, James S. **) & OLSEN, Edward J. ***)

The Frasnian-Famennian (Late Devonian) extinction event in Europe has been called the "Kellwasser Event" by House (1985), as the event preceded any historical use of the Frasnian-Famennian Stage boundary and as the event corresponds stratigraphically to the geographically widespread unit called the "Kellwasser Limestones". The Kellwasser Limestones usually occur as two black shale and bituminous beds (known as the "Lower" and "Upper" Kellwasser Limestones), which can be traced in the present world from Morocco northeast to the Harz Mountains in Germany, and northwest to the northern Variscan shelf in Belgium (Buggisch 1972).

Refined considerations of conodont zonal data are reported by Ziegler (1984), who stated that a low diversity crisis occurred in the evolution of the conodonts themselves during the span of the Late Devonian. This crisis began at the end of the Uppermost _gigas_ Subzone and was terminated by the end of the Middle _triangularis_ Subzone (Ziegler 1984), and was of two subzones in duration at the most (i.e., Lower-Middle _triangularis_). The new evolutionary radiation of typical Famennian palmatolepid conodonts started at the beginning of the Upper _triangularis_ Subzone (Ziegler 1984), and thus is clearly post-event.

The Lower Kellwasser Limestone falls in the Lower _gigas_ Subzone, and the Upper Kellwasser Limestone falls in the Uppermost _gigas_ Subzone to Lower _triangularis_ Subzone, and is the horizon of the mass extinction event in Europe (Buggisch 1972, House 1985). The purpose of the present paper is to report some of the extensive geochemical analyses of sediments of the Upper Kellwasser Limestone (McGhee et al. 1986), as well as sediments above and below this critical horizon, for evidence of asteroidal impact.

Upper Kellwasser Limestone at Steinbruch Schmidt

The stratigraphic section chosen for study is the exceptionally well preserved exposure in the Steinbruch Schmidt, located just north of the

---

*)   Department of Geological Sciences, Wright Geological Laboratory, Rutgers University, New Brunswick, New Jersey 08903, U.S.A.

**) Los Alamos National Laboratory, Los Alamos, New Mexico 87545, U.S.A.

***) Department of Geology, Field Museum of Natural History, Chicago, Illinois 60605, U.S.A.

Lecture Notes in Earth Sciences, Vol. 8
Global Bio-Events. Edited by O. Walliser
© Springer-Verlag Berlin Heidelberg 1986

village of Braunau an der Ense and near the city of Bad Wildungen, in the Federal Republic of Germany (Fig. 1). The section is one of 56 out-crops of the Kellwasser Limestones analyzed in an earlier geochemical study by Buggisch (1972), and the reader is referred to this excellent paper for additional locality, stratigrpahic, and sedimentological data not given in this paper.

The paleogeography of the sedimentary basin of the Rhenish Slate Mountains where the Steinbruch Schmidt is located is given in schematic fashion in Figure 1, modified from Meischner (1971). The stratigraphic section consists of pelagic limestones and shales deposited on a sub-marine rise (below wave base), flanked to the northwest and southeast by deeper basinal regions.

During periods of normal marine sedimentation, the submarine rises experienced largely carbonate sedimentation, with the clay content of the sediments increasing proportionally with water depth to the north-west and southeast into the deep basin regions. During periods of Kell-wasser Limestone deposition this "normal" pattern of sedimentation was virtually destroyed (Buggisch 1972), and sedimentation patterns were vir-tually independent of bathymetric profile, such that little facies dif-ferentiation existed between shallow and deep water regions.

The Kellwasser Limestones were deposited during periods of basin stagnation and anoxic bottom conditions (Buggisch 1972). Benthic life wa.

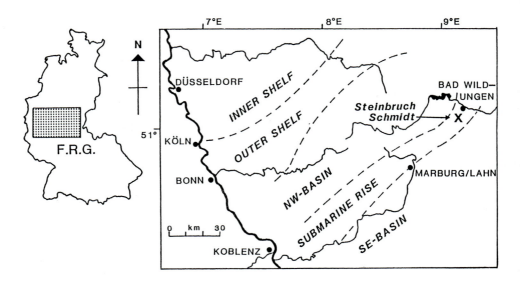

Figure 1.   Location of the field locality.

absent, and endobenthic life was also non-existent, as evidenced by the undisturbed sediments and lack of trace fossils. The upper water column remained oxygenated, however, as witnessed by the presence of planktic, pseudoplanktic, and nektic organisms preserved in the Kellwasser Limestones (Buggisch 1972).

Measurements

A stratigraphic profile of the Steinbruch Schmidt section is given in the left of Figure 2. In Figure 2 A the entire profile is given, from the Lower Kellwasser Limestone to the beginning of the Upper triangularis Subzone, above the Upper Kellwasser Limestone. Figure 2B is an expanded view of the critical Upper Kellwasser Limestone region, up to the Upper triangularis Subzone.

A suite of 165 samples was collected continuously, with no gaps, over the entire 5.00 meters of section. From these samples 100 were selected, with emphasis on clay-rich partings and the two crucial horizons of (1) Uppermost gigas-Lower triangularis, and (2) at the base of

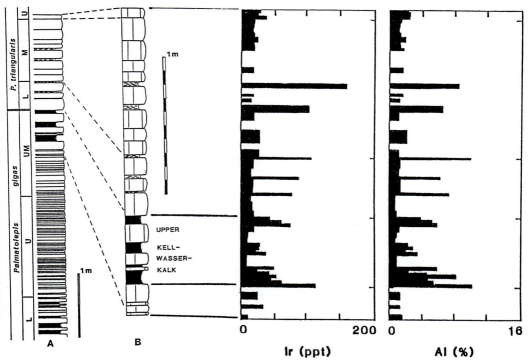

Figure 2.   Stratigraphic profile of the Steinbruch Schmidt section, and Ir and Al data for the crucial Upper Kellwasser Limestone to Upper triangularis Subzone interval.

the Upper <u>triangularis</u> Subzone (where an Ir anomaly has been reported from Australia (Playford et al. 1984)), for instrumental neutron activation analysis. This technique provides abundances for about three dozen elements. A parallel set of 100 samples were irradiated with thermal neu trons for 24 hours and radiochemical separations were performed to achie ve a sensitivity for Ir of one part per trillion (ppt).

In addition to elemental abundances, Oxygen-18 and Carbon-13 isotopic ratios were determined for 24 and 29 samples, respectively, more or less evenly spaced through the section as a check on potential isotopic variation.

The results of part of these analyses are given in Figures 2, 3, and 4 -- where the entire stratigraphic section is given in Figures 3 and 4.

Discussion of results

Ir concentrations vary from 4.5 to 159 ppt. Fifteen peaks of 75 to 159 ppt rise above the background that averages 27 ppt. There is no distinct impact related Ir peak in the distribution, because the higher Ir values are clearly associated with the clay partings in the limestone sequence (Figure 2); Ir correlates strongly with total Al from clay minerals (cor relation coefficient = 0.84). A plot of the Ir/Al ratio is fairly flat throughout the section (Figure 3). Furthermore, Ir on a carbonate-free basis also shows no anomalous peak (Figure 3).

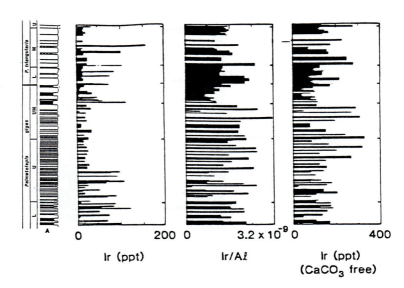

Figure 3. The entire stratigraphic profile, from the Lower Kellwasser Limestone through the Upper Kellwasser Limestone to the base of the Uppe <u>triangularis</u> Subzone, with some elemental abundances of interest.

Any sudden input of excess Ir from an extraterrestrial source (large-body impact) should not only show a spike in the raw Ir plot, but also a spike in the Ir/Al and Ir (on a $CaCO_3$-free basis) plots. Furthermore, elements not enriched in meteorites (Sc, La, Hf, Th, and U) follow essentially the same distribution pattern as the Ir -- higher concentrations in the clay partings (McGhee et al. 1986).

Oxygen isotope ratios (PDB) plot as a flat distribution (Figure 4), showing essentially litte variation with stratigraphic position. A small positive shift in Oxygen-18 occurs at the level of the Upper Kellwasser Limestone, the significance of which is not clear at the present.

Carbon-13, however, decreases upsection from a high in the Upper gigas Subzone to a low of 0.5 per mil (PDB) -- then increases sharply to approximately 3 per mil (PDB) at the biological crisis zone in the Stein-bruch Schmidt section. No indication for a "Strangelove Effect" -- a sharp drop in Carbon-13 values hypothesized to be due to the abrupt

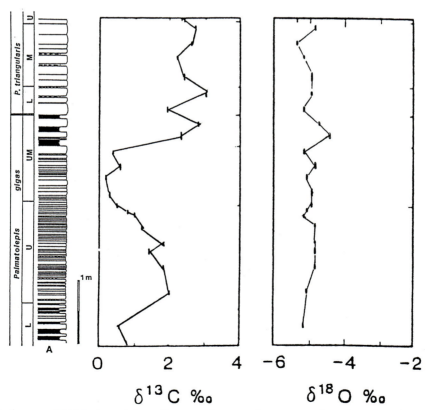

Figure 4.   Isotope ratios for carbon and oxygen (each per mil, PDB reference standard), with respect to stratigraphic position.

cessation of phytoplankton activity -- is thus seen from the Steinbruch
Schmidt strata. On the contrary, the carbon isotopic values reported
would suggest that an abrupt "bloom" in phytoplankton activity occurred
in this interval of time. Such a phytoplankton bloom may have been
regional in nature, however, as no comparable Carbon-13 signatures have
been seen in either western Canada (Geldsetzer et al. 1985) or in NW
Australia (Playford et al. 1984).

Conclusions

We report here the absence of an Ir anomaly in an exceptionally complete
stratigraphic sequence across the Frasnian-Famennian boundary in the
Rhenish   Slate Mountains of the Federal Republic of Germany. There
exists no anomaly either at the "Kellwasser Event" mass extinction horizon (Uppermost gigas-Lower triangularis Subzonal interval), or at the
subzonal horizon where an Ir anomaly was reported in the Canning Basin
strata in Australia.

No sharp negative Carbon-13 shift is seen at the biological crisis
horizon, which could be indicative of the collapse of phytoplankton productivity. On the contrary, a sudden increase of Carbon-13 ratios is
seen at the critical horizon, suggesting that either a "bloom" in phytoplankton activity or upwelling occurred in this interval of time. Such
events might have been local phenomena, however, as a comparable Carbon-
13 signature has not yet been found outside of Europe.

No sharp increase in Oxygen-18 is seen, which might have been expected if the latest Frasnian-Famennian period of time were one of global
oceanic cooling, or if upwelling of colder water had occurred.

REFERENCES

BUGGISCH, W. (1972): Zur Geologie und Geochemie der Kellwasserkalke und
    ihrer begleitenden Sedimente (Unteres Oberdevon).- Abh. hess. L.-A.
    Bodenforsch. 62, 68 p.
GELDSETZER; H.H.J.; GOODFELLOW, W.D.; McLAREN, D.J. & ORCHARD, M.J.
    (1985): The Frasnian-Famennian boundary near Jasper, Alberta, Canada.
    Geol. Soc. Amer., Abstr. Progr. 17, 589.
McGHEE, G.R., Jr.; ORTH, C.J.; QUINTANA, L.R.; GILMORE, J.S. & OLSEN,
    E.J. (1986): The Late Devonian "Kellwasser-Event" mass-extinction
    horizon in Germany: No geochemical evidence for a large-body impact.-
    Geology 14, in press.
MEISCHNER, D. (1971): Clastic sedimentation in the Variscan Geosyncline
    east of the River Rhine.- in: MÜLLER, G. (ed.): Sedimentology of part
    of Central Europe. Internat. Sed. Congr. Guidebook 8, 9-43.
PLAYFORD, P.E.; McLAREN, D.J.; ORTH, C.J.; GILMORE, J.S. & GOODFELLOW,
    W.D. (1984): Iridium anomaly in the Upper Devonian of the Canning
    Basin, Western Australia.- Science 226, 437-439.
ZIEGLER, W. (1984): Conodonts and the Frasnian-Famennian crisis.- Geol.
    Soc. Amer., Abstr. Progr. 16, 73.

UPPER FRASNIAN AND LOWER TOURNAISIAN EVENTS AND EVOLUTION OF CALCAREOUS FORAMINIFERA - CLOSE LINKS TO CLIMATIC CHANGES

A contribution to Project

GLOBAL BIO- EVENTS

KALVODA, Jiří *)

Abstract: The extinctions of the calcareous foraminiferal fauna at the Frasnian-Famennian and Famennian-Tournaisian boundaries appear to be connected with more than one event in each case ("episodic gradualism"), 4 of them seeming to be most important - the Kellwasser and Crickites event in the first case and the Hangenberg and sulcata event in the second case. The cumulative nature of extinctions is in contradiction with impact theory. The Famennian is then marked by 2 radiations - in the marginifera and in the expansa Zones. Both extinctions and radiations are closely linked to global changes in carbonate sedimentation and they seem to be best explained by climatic fluctuations of the glacial/ interglacial type which were accompanied by aridity/humidity oscillations.

Introduction

The late Devonian extinctions have been in the centre of attention from seventies when McLaren (1970) formulated its impact theory. Several possible causes that had been suggested were recently summarized by McLaren (1982, 1985). It appears that the immediate cause may have been an anoxic event connected with eutrophication (Walliser 1984). However, there is no general agreement about the processes that were involved.

The new impetus to the study of late Devonian extinctions gave the rise of event stratigraphy in the eighties. The excellent papers by Sandberg et al. (1983) and Johnson, Klapper & Sandberg (1985) correlated sea-level oscillations with fine conodont zonation, Ross & Ross (1985) pointed to the synchronous nature of late Paleozoic depositional events and House (1985) discussed the impact of mid-Paleozoic eustatic oscillations on the evolution of ammonoids.

Our study builts partly on these works and partly on works by Copper (1977), McGhee (1982) and Stanley (1984) that first stressed the importance of climatic changes in the explanation of late Devonian events. The study of the evolution of calcareous benthic foraminifera and the study of changes in carbonate environments that closely influenced the distribution of foraminifera, provides us further tools for the better understanding of processes connected with upper Frasnian - lower Tournaisian events.

*) Mikropaleont. odd., Moravské naftové doly, 695 30 Hodonín, Č.S.S.R.

Lecture Notes in Earth Sciences, Vol. 8
Global Bio-Events. Edited by O. Walliser.
© Springer-Verlag Berlin Heidelberg 1986

Evolution of calcareous foraminifera

Upper Devonian calcareous foraminifera were predominately benthic
organisms, only few forms are supposed to be planctic (thin walled
Parathurammina, Archaesphaera) (Chuvashov 1968, Pojarkov 1979). The
main domain of benthic forms were facies of biodetritic limestones
on the shelf slope in a well aerated nutrient rich environment with
high oxygen content (see Fig. 1). The close link between the distribution of biodetritic limestones and the evolution of calcareous foraminifera is thus apparent (Kalvoda 1986).

The wide extent of flat shallow water carbonate platforms and the
arid climate contributed to the increased salinity in comparison with
basins in the Upper Devonian. According to Kazmierczak, Ittekot &
Degens (1985) such platforms were areas where calcium could concentrate
whereas open seas stayed calcium depleted. Thus it is assumed that
most of the multilocular calcareous foraminifera, especially the highly
organized ones with complicated structure and differentiation of the
wall and often large tests (Multiseptida, Eogeinitzina, Eonodosaria,
Quasiendothyra) thrived in the environment of relatively increased calcium stress and high oxygen concentration.

The most important feature in the study of the evolution of calcareous foraminifera is the fact that the greatest rates of evolution
are shown by complicated multilocular forms, while the evolutionary
rates of primitive unilocular forms are comparatively slow. Consequently we will concentrate our attention only to forms showing greatest
evolutionary rates.

The upper Frasnian was characterized by the occurrence of genera as
Multiseptida, Eonodosaria, Eogeinitzina and partly Nanicella which,

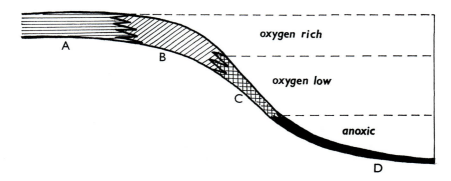

Figure 1. Distribution of four principle Famennian carbonate facies.
A - sphere limestone; B - biodetritic limestones; C - cephalopod limestones; D- black shales (s.l.).

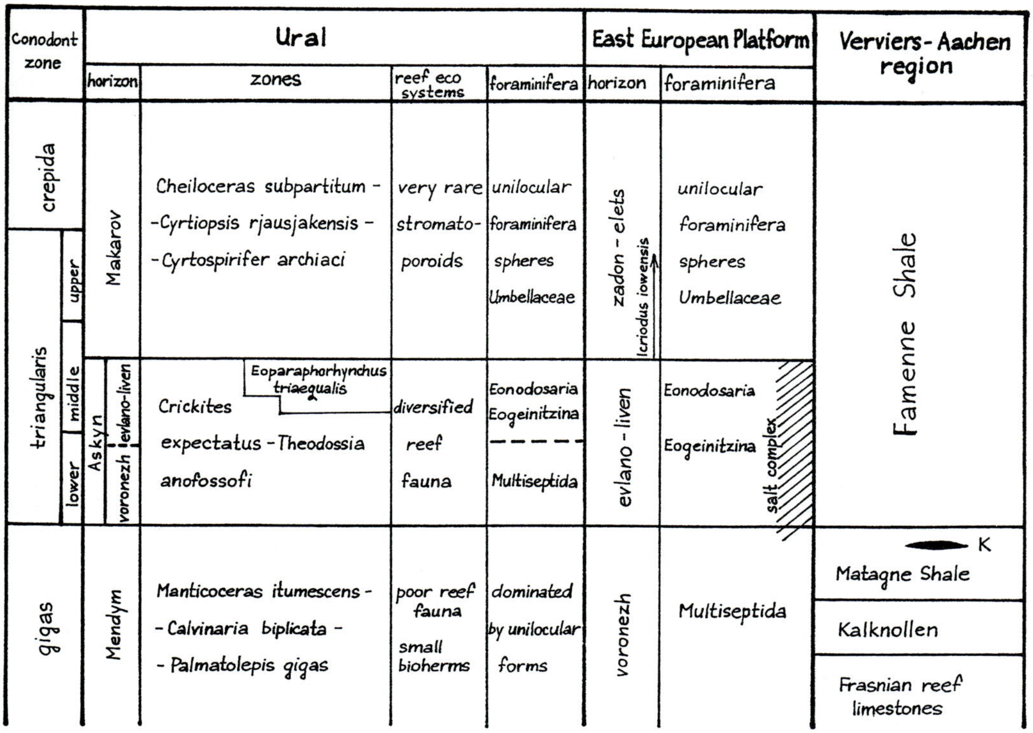

Conodont zone	Ural				East European Platform		Verviers-Aachen region
	horizon	zones	reef eco systems	foraminifera	horizon	foraminifera	
crepida	Makarov	Cheiloceras subpartitum – -Cyrtiopsis rjausjakensis – -Cyrtospirifer archiaci	very rare stromato- poroids	unilocular foraminifera spheres Umbellaceae	zadon - elets Icriodus iowensis	unilocular foraminifera spheres Umbellaceae	Famenne Shale
triangularis (upper / middle / lower)	Askyn (voronezh / evlano-liven)	Crickites expectatus –Theodossia anofossofi	Eoparaphorhynchus triaequalis / diversified reef fauna	Eonodosaria Eogeinitzina Multiseptida	evlano - liven	Eonodosaria Eogeinitzina / salt complex	K Matagne Shale
gigas	Mendym	Manticoceras itumescens – - Calvinaria biplicata – - Palmatolepis gigas	poor reef fauna small bioherms	dominated by unilocular forms	voronezh	Multiseptida	Kalknollen
							Frasnian reef limestones

Figure 2.   Correlation scheme of the Frasnian/Famennian boundary layers in the Eastern and Western Europe. Modified according to Chuvashov (1968), Rzhonsnitskaya (1986), Ovnanatova, Kononova (1984), Manukalova-Grebenyuk (1974), Dreesen et al. (1985).

however, has comparatively greater stratigraphical range. The Multiseptida-Eonodosaria-Eogeinitzina complex was widely distributed. It has been recorded in wide belts including North America, Europe, southern Siberia and China (Kalvoda 1986). The decline of this complex was not abrupt connected with one event but gradual connected with more events, two of them being probably most important. While the decline of this complex in North America seems to be linked with the Kellwasser event, rich assemblages of this complex flourished in Eastern Europe (see Fig. 2) being decimated during the Middle triangularis/Crickites event. In Moravia the decline of this complex was differentiated, at some localities connected with the upper Frasnian event (Zukalová 1982), at some localities the marking genus of this complex (Multiseptida) surviving in the crepida Zone (Kalvoda, in Friáková et al. 1985). The decline of the Multiseptida-Eonodosaria-Eogeinitzina complex was closely linked to the decline of reef ecosystems which is discussed in this volume by Hladil et al.

The lower Famennian (Upper triangularis - crepida Zone) was characterized by the widespread occurrence of unilocular foraminifera, spheres
and charas (Umbellacea). Marking genera of the upper Frasnian foraminiferal complex became extinct (see Fig. 5) while the other important
group, Tournayellidae, the representatives of which first appeared in
the upper Frasnian, was able to survive unfavourable conditions. In the
Eastern Europe the first representatives of the genus Quasiendothyra,
which is characteristic for the Famennian, appear.

In the marginifera Zone we can see the first radiation and wide
dispersal of Tournayellidae as well as Quasiendothyra (Conil, in Dreesen
et al. 1985). The middle Famennian (trachytera, partly postera Zones)
is a period of certain stagnation in the evolution of calcareous foraminifera.

The second major radiation is connected with transgressions of the
expansa Zone and it was marked by the evolution of double layered Quasiendothyra as well as Tournayellidae. In the praesulcata Zone we can
follow the gradual decline of multilocular highly organized forms which
continued in the interval of the sulcata - sandbergi Zone. There exist
only little conodont data about the evolution of foraminifera at the
Devonian/Carboniferous boundary but Quasiendothyra seems to be fairly
abundant at some localities in the praesulcata Zone (Moravia, Dnieper-
Donetz Basin) while at other localities it is very rare or absent (the
Ural) (cf. Reitlinger et al. 1981). The further decline is apparent at
the sulcata event (Moravia, probably also Dnieper-Donetz Basin). During
the uppermost Famennian (praesulcata Zone) and lower Tournaisian we can
follow the gradual evolution and dispersal of Chernyshinellidae (Chernyshinella, Tournayellina). The more abundant occurrence of Tournayellina is apparent already in the praesulcata Zone. The lower Tournaisian
(sulcata-sandbergi Zone) is marked by the domination of primitive Bisphaera and Earlandia as well as by relatively scarce Chernyshinella,
the radiation of which is then characteristic of the middle Tournaisian.
Thus in the lower Tournaisian existed both assemblages with Quasiendothyra and assemblages with Chernyshinella (see Fig. 5) the distribution
of which seems to be facially or perhaps also climatically controlled.

Some trends in the Upper Devonian eustacy and carbonate sedimentation

The most important trend in the carbonate sedimentation at the Frasni-
an/Famennian boundary seems to be the substantial reduction of the belt
of carbonate sedimentation (Heckel & Witzke, 1976). During the upper
Frasnian the carbonate shelves were dominated by reef ecosystems the
distribution of which was continuously restricted, the Kellwasser anoxic

event being the most prominent. It was followed by a widespread regression recognized on the Russian Platform, in the Urals (see Fig. 2),in North America (Sandberg et al. 1983), in Moravia (Hladil et al., this volume) and in Australia (Playford et al. 1984). In this period diversified reef ecosystems are reported in the Urals (see Fig. 2) while on the Russian Platform the deposition of salt seems to be characteristic. In Moravia only impoverished reef fauna has been recorded (Hladil et al., this volume).

The second prominent event that contributed to the decline of mostly already impoverished reef ecosystems as well as foraminiferal fauna (see Fig. 5) seems to be the Crickites event (see Fig. 2).

The lower Famennian seems to be dominated by muddy carbonate facies indicating low oxygen environments. Abundant are spheres, euryfacial unilocular foraminifera and charas (Umbellaceae). Multilocular foraminifera, especially the high organized ones are rare as well as the fauna of reef ecosystems. There an apparent bloom of siliceous organisms - radiolarians (Nazarov & Ormiston 1985) and sponges (McGhee 1982).

The middle part of the Famennian is dominated by regressive tendencies. The widespread occurrence of evaporites is reported from Montana (Sandberg et al. 1983), Pripyat Depression and Dnieper-Donetz Basin (Manukalova-Grebenyuk 1974, Avkhimovich & Demidenko 1985).

During the transgression of the expansa Zone the facies of biodetritic limestones reached its maximal distribution which enabled the widespread radiation of the Quasiendothyra fauna which is reported in the wide belt from Alaska to Australia (Kalvoda 1986), however, it has been found neither in the U.S.A. nor Canada.

The carbonate sedimentation of the Upper expansa Zone is characterized by the occurrence of radiolarian anoxic events. This domanik-type events are reported in the Urals (Kochetkova et al. 1985) as well as in Moravia (see Fig. 3). The culmination of anoxia may be seen in the Lower praesulcata Zone when the similar position of anoxic sediments can be traced in Moravia (see Fig. 3), China and Germany. The middle praesulcata regression is perhaps the most characteristic event in the upper Famennian. It can be traced worldwide - in the Rheinische Schiefergebirge, China, Belgium, Urals, North America, and it seems to be more easily to distinguish than the preceding Lower praesulcata anoxic event (it is preceded by anoxic events in the Upper expansa Zone).

A further important event seems to be the sulcata event. The sediments of the sulcata Zone overlie often unconformably the regressive facies of the praesulcata Zone (middle - upper),or the unconformity is often greater than a part of the sulcata Zone or even the whole zone

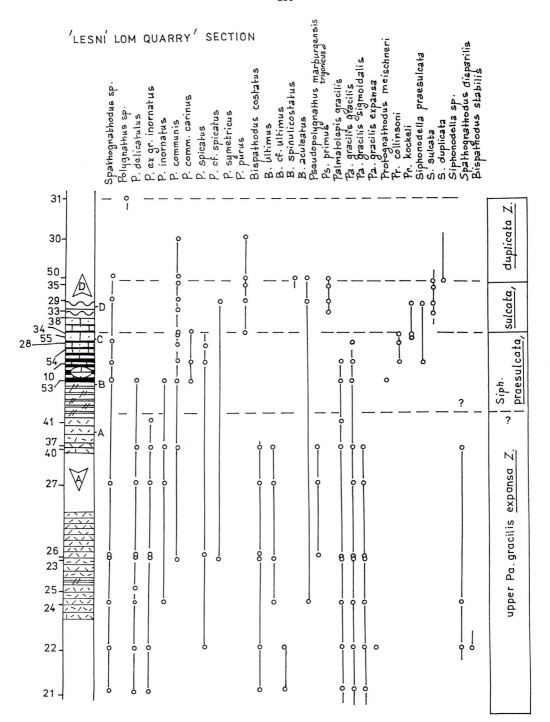

is absent and sediments of the praesulcata or even expansa Zone are
overlain by sediments of the duplicata or even younger zones. Even
though the sea level was at a relatively low stand the sulcata-sandbergi interval is supposed to be transgressive in comparison with preceding upper-middle praesulcata regression.

The sulcata event is marked by a considerable change in the character of carbonate sedimentation. During the sulcata-sandbergi interval the
facies of muddy carbonates became widespread. They are characterized by
abundant radiolarians, thin shelled ostracodes and primitive foraminifera (Earlandia). The environment with more agitated water is characterized by rare multilocular foraminifera and abundant primitive Bisphaera (socalled Bisphaera Beds).

A characteristic feature of the carbonate sedimentation in the
sulcata-sandbergi interval and partly also in the expansa Zone seems
to be signs of fresh water influence (Shcherbakov et al. 1984, Reitlinger et al. 1981).

Upper Devonian climatic changes and their possible impact

Recently Caputo & Crowell (1985) reported glacigenic sediments in South
America in the interval from the lower Famennian to the lower Tournaisian, the stratigraphical position of which is, however, not undisputed
(see Streel, this volume). In our opinion these glaciations in the Famennian contributed to the transition from the generally arid hot Devonian climates to more cool and humid Carboniferous climates and we
assume that they had great impact on eustacy carbonate sedimentation
and evolution of foraminifera and other organisms.

The influence of glacial/interglacial climatic fluctuations accompanied by aridity/humidity fluctuations (see Fig. 4) was complex, it
may have included effect of cold water (Copper 1977), density stratification caused by a low salinity surface layer (cf. Thunnel et al.
1983, Graciansky et al. 1986) or the direct influence of the salinity
decrease.

It was already mentioned that the arid Devonian climate contributed
to higher salinity and calcium concentration on carbonate platforms in
comparison with basins. The greater influx of fresh water during humid

Figure 3.   Schematic profile of the Devonian/Carboniferous boundary
layers in the Lesní lom Quarry section (Brno-Líšeň, Moravia).
A - biodetritic and massive limestones; B - laminite; C - shales with
intercalations of massive to arenaceous limestones, massive limestones,
arenaceous limestones; D - mostly thin bedded micritic limestones, locally nodular.

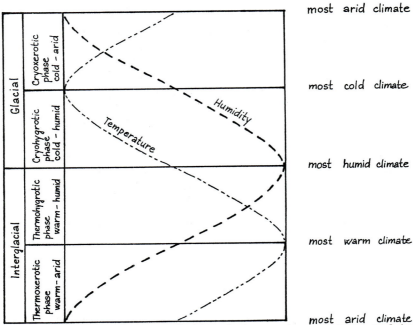

most arid climate

most cold climate

most humid climate

most warm climate

most arid climate

Figure 4. Diagram of the relation between temperature and humidity during glacial/interglacial fluctuations derived from the changes of vegetation. According to Ložek (1973).

periods contributing to the decrease of the salinity and calcium concentration may have been on one side lethal for most of the calcium-stress and relatively higher salinity adapted biota and on the other side it may have favoured the bloom of relatively lower salinity and low calcium concentration adapted siliceous biota migrating from the basinal environment.

We will now try speculative to reconstruct the glacial/interglacial fluctuations in the upper Frasnian - lower Tournaisian interval. It should be stressed the preliminary nature of this reconstruction. It is possible that the refinement of our stratigraphical knowledge may contribute to further subdivisions and modifications.

In our opinion the considerable reduction of the belt of carbonate deposition at the Frasnian/Famennian boundary (Heckel & Witzke 1979) was a consequence of pronounced global climatic cooling. It seems to be apparent especially in North America where it contributed to widespread black shale deposition and rare carbonate occurrence in the Famennian. The influence of cold water reported by Copper (1977) can be distinguished also in calcareous foraminiferal fauna that shows considerable decline in diversity in comparison with the upper Frasnian.

We assume that the cooling trend and accompanying climatic oscilla-

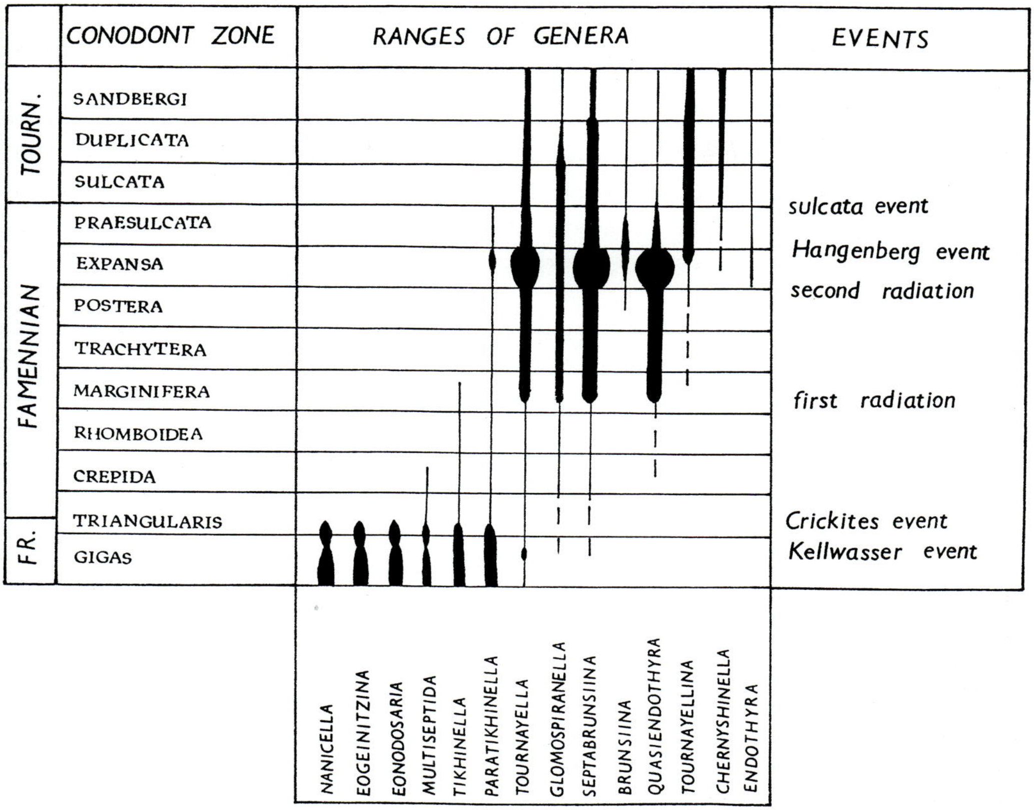

	CONODONT ZONE	RANGES OF GENERA	EVENTS
TOURN.	SANDBERGI		
	DUPLICATA		
	SULCATA		sulcata event
FAMENNIAN	PRAESULCATA		Hangenberg event
	EXPANSA		second radiation
	POSTERA		
	TRACHYTERA		
	MARGINIFERA		first radiation
	RHOMBOIDEA		
	CREPIDA		
FR.	TRIANGULARIS		Crickites event
	GIGAS		Kellwasser event

Genera (columns): NANICELLA, EOGEINITZINA, EONODOSARIA, MULTISEPTIDA, TIKHINELLA, PARATIKHINELLA, TOURNAYELLA, GLOMOSPIRANELLA, SEPTABRUNSIINA, BRUNSIINA, QUASIENDOTHYRA, TOURNAYELLINA, CHERNYSHINELLA, ENDOTHYRA

Figure 5. Suggested distribution of some important genera of calcareous foraminifera in the Late Devonian.

tions may have contributed to anoxic events in the upper Frasnian. The increased humidity would seem to enable the onset of Famennian glaciation in the Lower triangularis Zone. It could be consistent with increased aridity and salt deposition in the Eastern Europe and more arid climate in the Urals which are supposed to be a part of the equatorial region (Heckel & Witzke 1979).

The transgression in the Middle triangularis Zone would seem to be connected with warmer period of an approaching interglacial (see Fig. 4) (Crickites event). The upper triangularis - crepida relatively more humid climate is consistent with increased role of charas, spheres, unilocular foraminifera, radiolarians and sponges and with considerable decline of benthic fauna in the equatorial region.

The middle Famennian with its sea-level lowstand and arid climate would seem to correspond to further glaciation. In the equatorial region of the Eastern Europe ist was marked again by salt deposition. This period is rather long, unfortunately, there is not enough precise data to recognize finer trends.

The lower expansa transgression/strunian transgression -- see Streel this volume -- seems to be connected with approaching interglacial and the Hangenberg anoxic event seems to be connected with humid climate at the glacial/interglacial boundary (see Fig. 4). The pronounced middle-upper praesulcata regression seems then to be compatible with maximum glaciation.

The sulcata event and accompanying changes in the carbonate sedimentation seems, in our opinion, to suggest slight warming contributing to the cool humid nature of the sulcata - sandbergi interval. Streel (this volum), however, supposes that the Tournaisian transgression was connected with tectonic processes.

Conclusions

Were extinctions gradual or episodic? In this respect the problem is similar to other crucial questions of the evolutionary theory -- is the origin of new taxa a gradual or sudden process?

The study of late Devonian calcareous foraminifera suggests that the origin of new taxa seems to be best explained by the theory of "punctuated gradualism" and, similary, the extinctions seem to be best explained by "episodic gradualism".

Global events need not to have the same manifestation in different biotopes and climatic zones, local and global tectonics being other important factors in play. In this respect the process may be very complex with manifold actions, reactions and interactions (Walliser 1986) with different manifestation in different regions. The spatial as well as time scale we take into account is important. The fauna may have become extinct in one or more regions -- it would seem that the extinction was episodic connected with one event, but when we examine the data on larger, e.g. global scale, it will become apparent that the extinction was not sudden connected with one event but it was connected with more events. In global scale then it could be interpreted to be gradual, but the role of true gradualism seems to be comparatively low and what appears to be gradual in global scale is in fact a set of episodes in regional scale ("episodic gradualism").

The nature of the extinction at the Frasnian/Famennian and Famennian/Tournaisian boundary seems to be well compatible with this theory of episodic gradualism. It is apparent that both the extinction at the Frasnian/Famennian and Famennian/Tournaisian boundary can not be attributed to a single event but that more events can be distinguished, some of them as Kellwasser, Crickites, Hangenberg and sulcata event being of greater importance, perhaps because of their worldwide manifestation.

The discussed nature of late Devonian extinctions is in contradiction with the impact theory which also fails to explain why greatest changes were recorded in benthic biota while the planktic biota flourished.

The climatic fluctuations of the glacial/interglacial type seem not only to explain best the discussed complex nature of extinctions but seem also to agree with the trends in the evolution of carbonate sedimentation.

REFERENCES

AVKHIMOVICH, V.I. & DEMIDENKO, E.K. (1985): Biostratigrafiya pogranich-nykh otlozhenii devona i karbona Belorussi (Pripyatskaya vpadina).- Biostratigrafiya pogranichnykh otlozhenii devona i karbona 7, Magadan.
CAPUTO, M.V. & CROWELL, J.C. (1985): Migration of glacial centers across Gondwana during Paleozoic Era.- Geol. Soc. Amer., Bull. 96, 1020-1036.
CHUVASHOV, B.I. (1968): Istoriya razvitiya i bionomicheskaya kharakteri-stika pozdnedevonskogo basseina na zapadnom sklone Srednego i Yuzh-nogo Urala.- Moskva.
COPPER, P. (1977): Paleolatitudes in the Devonian of Brazil and Fras-nian-Famennian mass extinction.- Palaeogeogr., Palaeoclim., Palaeo-ecol. 21, 165-207.
DREESEN, R. et al. (1985): Depositional environment, paleoecology and diagenetic history of the Marbre rouge a crinoides de Baelen (Late Upper Devonian, Verviers synclinorium, Eastern Belgium).- Ann. Soc. Géol. Belg. 108, 311-359.
GRACIANSKY, D.C. et al. (1986): Ocean wide stagnation episodes in the Late Cretaceous.- Geol. Rdsch. 75, 17-41.
HECKEL, P.H. & WITZKE, B.J. (1979): Devonian world palaeogeography de-termined from distribution of carbonates and related lithic palaeo-climatic indicators.- in: HOUSE, M.R.; SCRUTTON, C.T. & BASSETT, M. G. (eds.): The Devonian System. Spec. Pap. Palaeont. 23, 99-123.
HOUSE, M.R. (1985): Correlation of mid-Palaeozoic ammonoid evolutionary events with global sedimentary perturbations.- Nature 313, 17-22.
JOHNSON, J.G.; KLAPPER, G. & SANDBERG, C.A. (1985): Devonian eustatic fluctuations in Euamerica.- Geol. Soc. Amer., Bull. 96, 567-587.
KALVODA, J. (1986a): Lesní lom quarry near Bnro - Líšeň.- in: CHLUPÁČ, I. et al. (ed.): Field Conference Barrandian-Moravian Karst 1986. Excursion Guidebook, 49-54.
-- (1986b): Upper Frasnian - Lower Tournaisian events and evolution of calcareous foraminifera - close links to climatic changes.- in: WALLISER, O.H. (ed.): 5th Alfred Wegener-Conf., 1st Internat. Work-shop IGCP Proj. 216 - Global Bio-events. Summ., 68-76.
KAZMIERCZAK, J.; ITTEKKOT, V. & DEGENS, E.T. (1985): Biocalcification through time: environmental challenge and cellular response.- Paläont. Z. 59, 15-33.
KOCHETKOVA, N.M. et al. (1985): Opornyie razrezy pogranichnykh otlozhenii devona i karbona zapadnogo sklona Yuzhnogo Urala.-Biostratigrafiya pogranichnykh otlozhenii devona i karbona 6. Magadan.
LOŽEK, V. (1973): Příroda ve čtvrtohorách. Praha.

MANUKALOVA-GREBENIUK, M.F. (1974): Foraminifery verkhnedevonshikh otlozhenii Dneprovsko- Doneckoi i Pripyatskoi vpadin.- Moskva.
McGHEE, G.R. (1982): The Frasnian-Famennian extinction event: A preliminary analysis of Appalachian marine ecosystems.- Geol. Soc. Amer.,
Spec. Pap. 190, 491-500.
McLAREN, D.J. (1970): Time, life and boundaries.- J. Paleont. 44, 801-
815.
-- (1982): Frasnian-Famennian extinctions.- Geol. Soc. Amer., Spec. Pap.
190, 477-484.
-- (1985): Ammonoids and extinctions.- Nature 313, 12-13.
NAZAROV, B.B. & ORMISTON, R.A. (1985): Evolution of radiolaria in the
Paleozoic and its correlation with the development of other marine
groups.- Senckenbergiana lethaea 66, 203-216.
OVNANATOVA, N.S. & KONONOVA, L.L. (1984): Koreliatsiya verkhnedevonskikh
i nizhneturneiskikh otlozhenii evropeiskoi chasti SSSR.- Sovetskaya
Geol. 8, 32-42.
PLAYFORD, P.E. et al. (1984): Iridium anomaly in the Upper Devonian of
the Caning Basin, Western Australia.- Science 226, 437-439.
REITLINGER, E.A. et al. (1981): Korreliatsiya pogranichnykh otlozhenii
devona i karbona evropeiskoi chasti SSSR. Biostratigrafia pogranich-
nykh otlozhenii devona i karbona. Magadan.
ROSS, C.A. & ROSS, J.R.P. (1985): Late Paleozoic depositional sequences
are synchronous and worldwide.- Geology 13, 194-197.
RZHONSNITSKAYA, M.A. (1986): Osnovnyie problemy stratigrafii devona
Sovetskogo Soyuza.- Sovetskaya Geol. 3, 53-65.
SANDBERG, C. et al. (1983): Middle Devonian to Late Mississippian geo-
logic history of the overthrust belt region, Western United States.-
Geol. Studies of the Cordilleran Thrust Belt 2, 691-719.
SHCHERBAKOV, O.A. et al. (1985): Biostratigrafiya pogranichnykh otlozhe-
nii devona i karbona v razreze Kosaya Rechka na Srednem Urale.- Bio-
stratigrafiya pogranichnykh otlozhenii devona i karbona 5, 30-67.
WALLISER, O.H. (1984): Pleading for a natural D/C boundary.- Cour.
Forsch.-Inst. Senckenberg 67, 241-246.
-- (1986): Global biological events in earth history. IGCP Project 216.-
5th Alfred Wegener-Conf., 1st Internat. Workshop IGCP Proj. 216.
Inform., 4 p.
ZUKALOVÁ, V. et al. (1982): Závěrečná zpráva z vrtu Jablůnka. 1. MS.
Geofond Praha.

MIOSPORE CORRELATION BETWEEN NORTH AMERICAN, GERMAN AND URALIAN (UDMURTIA) DEEP FACIES THROUGH APPALACHIAN, IRISH AND BELGIAN PLATFORM AND CONTINENTAL FACIES NEAR THE DEVONIAN/CARBONIFEROUS BOUNDARY

A contribution
to Project
GLOBAL
BIO-
EVENTS

STREEL, Maurice *)

At the Devonian/Carboniferous transition, in a rather short timespan that Conodont workers (Sandberg et al. 1986) estimate shorter than three M.y. (from the Lower expansa to the sulcata zones), six interval zones of miospores are available. Each interval zone is based on the first occurrence of one or a few species in assemblages which are on the Devonian side of the boundary, otherwise very similar (the succession of interval zones VCo/LV/LL/LE/LN). At or very near the D/C boundary, the change in miospore assemblages is, on the contrary, rather sharp in all facies except maybe the continental one's (the succession LN/VI). Palynologists work with tenths of thousands of specimens in each sample. This gives some statistical value to their approach on the first occurrence of species, especially when comparing marine sediments. Therefore the lateral correlations are believed to be reliable.

These lateral correlations show that in North America, Western Europe and Uralian (Udmurtia) areas, cycles of low and high levels of sea were synchronous. They are listed below:

(1) "Pre-expansa"- VCo zones of l o w sea-level (Three lick Fm, USA; Lower Evieux Fm in Belgium, Clendening et al. 1980).

(2) "expansa" - LV - LL zones of h i g h sea-level (Cleveland Mbr of Ohio Fm in USA, Winslow 1962; Old Head Sandstone in Ireland, Clayton et al. 1986; Etroeungt Fm in Belgium, Becker et al. 1974; Zavolga beds in Udmurtia, Byvsheva et al. 1984).

(3) "praesulcata" - LE - LN zones of l o w sea-level (Bedford Shales and Berea Sandstone in USA, Eames 1978; Hangenberg Shales in Germany, Higgs & Streel 1984; mudstone part of the Malevka beds in Udmurtia).

(4) "sulcata" - VI zones of h i g h sea-level (Sunbury Shales in USA; Castle Slate Mbr of the Kinsale Fm in Ireland; uppermost Hangenberg Shales and Hangenberg limestone in Germany; limestone part of the Malevka beds in Udmurtia).

The low sea-level (3) is complex: short transgressional movement in the Appalachians, USA (Dennison et al. 1986), very small thickness of rock and structural control in Ireland (Clayton et al. 1986), lack of deposi-

---

*) Paleontology, University of Liège, B-4000 Liège, Belgium.

Figure 1. Paleogeo-
graphical reconstruction
of the Old Red Continent
and other land areas
(shaded) after Scheckler
1986, Fig. 2; slightly
modified. Lined rectan-
gles show studied areas
here compared.

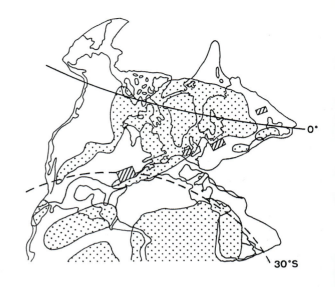

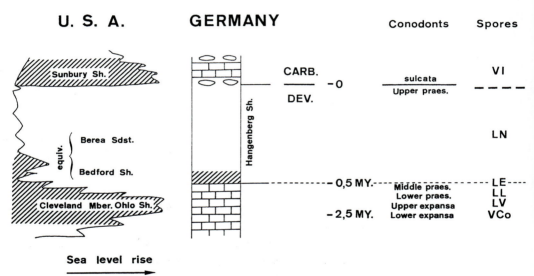

Figure 2. Stratigraphical comparison between America (USA) and
Western Europe (Germany) near the Devonian/Carboniferous limit. Conodonts and time after Sandberg et al. 1986. Lined areas are dominant
black shales.

tion or erosion between the Etroeungt Limestone and the Hastière Limestone in Belgium (Streel 1986), reworking processes in the Hangenberg
Shales of the Oberrödinghausen area in Germany (Higgs & Streel 1984).
Obviously the black shales development at the base of the Hangenberg Fm
in Germany is local or might, at best (see Fig. 2), correspond to the

short transgressional movement within the Bedford Shale equivalents in USA.

The sharp spore-assemblage change (LN/VI) very near the Devonian/ Carboniferous boundary (first Siphonodella sulcata entry) corresponds to the high sea-level (4) of the Sunbury Shale in Eastern USA and of the Castle Slate Mbr of the Kinsale Fm in Ireland. It corresponds to the end of the supply of terrigenous material in deeper facies in Germany and near the Ural. This floral change however has not the same characteristics when one compares the tropical realm of Eastern USA and Western Europe with the equatorial realm at the East of the Russian Platform (see Byvsheva et al. 1984).

Moreover, the floral change is less obvious in the "inland"-continental beds (see the Middle Sandstone and Shale Mbr of the Pocono Fm in Streel & Traverse 1978).

A suggestion is here made that the Devonian/Carboniferous sea-level rise (4) might well have affected the coastal vegetation more drastically than the inland vegetation.

REFERENCES

BECKER, G.; BLESS, M.J.M.; STREEL, M. & THOREZ, J. (1974): Palynology and ostracode distribution in the Upper Devonian and basal Dinantian of Belgium and their dependence on sedimentary facies.- Meded. Rijks Geol. Dienst, N. ser. 25, 9-99.
BYVSHEVA, T.V.; HIGGS, K. & STREEL, M. (1984): Spore correlations between the Rhenish Slate Mountains and the Russian platform near the Devonian-Carboniferous boundary.- Cour. Forsch.-Inst. Senckenberg 67, 37-45.
CLAYTON, G.; GRAHAM, J.R.; HIGGS, K.; SEVASTOPULO, G.D. & WELSH, A. (1986): Late Devonian and Early Carboniferous Paleogeography of Southern Ireland and Southwest Britain.- Ann. Soc. géol. Belgiques 109, 103-111.
CLENDENING, J.A.; EAMES, L.E & WOOD, G.D. (1980): Retusotriletes phillipsii n. sp., a potential Upper Devonian guide palynomorph.- Palynology 4, 15-22.
DENNISON, J.M.; BEUTHIN, J.D. & HASSON, K.O. (1986): Latest Devonian-Earliest Carboniferous Marine transgressions Central and Southern Appalachians, USA.- Ann. Soc. géol. Belgique 109, 123-129.
EAMES, L.E. (1978): A palynologic interpretation of the Devonian-Mississippian boundary from northeastern Ohio, USA (abst.).- Palynology 2, 218-219.
HIGGS, K. & STREEL, M. (1984): Spore stratigraphy at the Devonian-Carboniferous boundary in the northern "Rheinisches Schiefergebirge", Germany.- Cour. Forsch.-Inst. Senckenberg 67, 157-179.
SANDBERG, C.A.; GUTSCHICK, R.C.; JOHNSON, J.G.; POOLE, F.G. & SANDO, W. J. (1986): Middle Devonian to Late Mississippian Event Stratigraphy of Overthrust Belt Region, Western United States.- Ann. Soc. géol. Belgique 109, 205-207.
SCHECKLER, S.E. (1986): Old Red Continent facies in the Late Devonian and Early Carboniferous of Appalachian North America.- Ann. Soc. géol. Belgique 109, 223-236.

STREEL, M. (1986): Miospore contribution to the Upper Famennian-Strunian Event Stratigraphy.- Ann. Soc. géol. Belgique 109, 75-92.
WINSLOW, M.R. (1962): Plant spores and other microfossils from the Upper Devonian and Lower Mississippian rocks of Ohio.- U.S. Geol. Surv. Prof. Pap. 364, 1-93.

PERMIAN FUSULINACEAN FAUNAS OF THAILAND – EVENT CONTROLLED EVOLUTION

INGAVAT-HELMCKE, Rucha *) & HELMCKE, Dietrich **)

Abstract: The evolution of the fusulinacean faunas of Thailand during Permian times is controlled by bioevents. The first bioevent happened in the upper Lower Permian and is characterized by the disappearance of the Arctic-Tethyan elements. This bioevent can be correlated with the closure of the Urals and a worldwide regression. The second bioevent happened at the boundary Middle/Upper Permian to lower Upper Permian (Midian). It is characterized by the extinction of approximately 90 % of the fusulinaceans. This bioevent is controlled by the orogenic evolution of the Kunlun and the Petchabun fold and thrust belt, which again led to a worldwide regression. The third bioevent happened at the boundary Permian/Triassic. It is characterized by the final extinction of the fusulinaceans and might be controlled by the isostatic uplift of the newly formed mountain belt.

Introduction

Traditionally the Permian of Thailand and adjacent regions was regarded as a time of stable conditions during which tectonic quiescence prevailed. The most characteristic sediments deposited during this time are shallow marine platform-carbonates ("Ratburi Limestone") which are extensively distributed and form scenic landscapes. Beside the carbonates also clastic sequences are quite widely spread. Especially the Permian carbonates contain rich fauns with bryozoans, brachiopods, corals, pelecypods, ammonites and -- most important from a stratigraphical point of view -- fusulinids.

It was said that the stable conditions of the Permian came to a close with the onset of deformation during the Lower Triassic and the main deformation during the Upper Triassic "Indosinian" orogeny. Since platetectonic scenarios were applied the view was widely accepted that the central parts of mainland Southeast Asia were created by continent/continent collision during the Upper Triassic after closure of the "Paleotethys". It was postulated by many authors that at least the western plate involved ("Shan-Thai Craton", Bunopas & Vella 1978) should be regarded as a displaced terrane of Gondwana origin.

During the last years the evolution of the Permian foraminiferal faunas of Thailand was studied in some detail (Toriyama et al. 1978, Ingavat et al. 1980, Ingavat 1984) with special emphasis on the corre-

---

*) Geological Survey Division, Department of Mineral Resources, Bangkok 10400, Thailand.

**) Institut für Geologie und Dynamik der Lithosphäre, Universität Göttingen, D-3400 Göttingen, F.R.G.

Lecture Notes in Earth Sciences, Vol. 8
Global Bio-Events. Edited by O. Walliser
© Springer-Verlag Berlin Heidelberg 1986

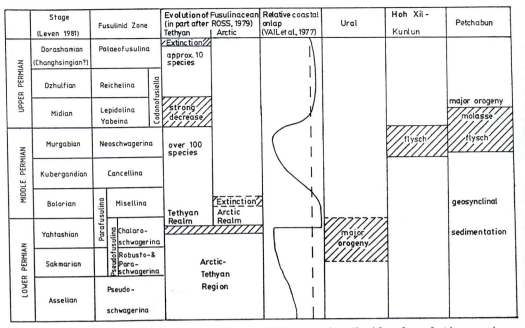

	Stage (Leven 1981)	Fusulinid Zone	Evolution of Fusulinacean (in part after ROSS, 1979) Tethyan	Arctic	Relative coastal onlap (VAIL et al., 1977)	Ural	Hoh Xil - Kunlun	Petchabun
UPPER PERMIAN	Dorashamian (Changhsingian?)	Palaeofusulina	Extinction approx. 10 species					
	Dzhulfian	Reichelina						major orogeny
	Midian	Lepidolina Yabeina	strong decrease					molasse
MIDDLE PERMIAN	Murgabian	Neoschwagerina	over 100 species				flysch	flysch
	Kubergandian	Cancellina						
	Bolorian	Misellina	Tethyan Realm	Extinction Arctic Realm				geosynclinal
LOWER PERMIAN	Yahtashian	Chalaro-schwagerina				major orogeny		sedimentation
	Sakmarian	Robusto-& Para-schwagerina	Arctic-Tethyan Region					
	Asselian	Pseudo-schwagerina						

Figure 1. Permian Fusulinacean Bioevents in Thailand and the mechanisms which possibly control the events.

lation between the western, central and eastern provinces of Thailand. These studies revealed

-- that the foraminiferal assemblages of the western, central and easte: zones of Thailand are related and that certain differences are due t paleoecology and biofacies: the eastern province had a more favourab environment for certain genera (Ingavat 1984, p. 99)

-- and that a remarkable paleobiogeographic change took place after the early Permian: while the early Permian foraminifera faunas of Thaila and Malaysia suggest that the domains of continental Southeast Asia still had a connection with the Arctic sea, indicate the late Early Permian faunas that a separation between the two realms took place (Toriyama et al. 1978, p. 110).

Explanation for Tables 1 - 3 (comp. Toriyama et al. 1978, Ingavat et al 1980).
Element I   = Cosmopolitan
        II = Arctic-Tethyan
            II_0 = Arctic only
            II_1 = Arctic and western Tethyan
            II_2 = Arctic and Tethyan
        III = Tethyan
        IV = narrower Tethyan
        V   = western Tethyan (V' = predominantly western Tethyan)
        VI = eastern Tethyan (VI' = predominantly eastern Tethyan)
        VII = endemic (only Thailand and Malaysia)

Element	I	II			III	IV	V'	V	VI'	VI	VII	No. of species
		II_0	II_1	II_2								
Ozawainellidae											4	4
Schubertellidae	1					1			1	5	5	13
Schwagerinidae	4	3	3	9	7	1	1	8	7	18	31	92
Verbeekinidae					3			3		3	1	10
Neoschwagerinidae											1	1
Staffellidae										1	3	4
Thailandinidae											2	2
No. of species	5		15		10	2	1	11	8	27	47	126
Percentage	4.1		11.9		8.1	1.6	0.8	8.9	6.5	21.9	38.2	

Table 1. Paleogeographic distribution of Early Permian Fusulines in Thailand-Malayan realm.

Element	I	II			III	IV	V'	V	VI'	VI	VII	No. of species
		II_0	II_1	II_2								
Ozawainellidae										1	3	4
Schubertellidae					1	3	1	1	1	1	3	11
Schwagerinidae	1(+4)				1			3	4	10	5	24
Verbeekinidae					4	1		1	2	2	8	18
Neoschwagerinidae					3	2	1	5	2	15	7	35
Staffellidae									1	3	2	6
Thailandinidae										5		5
No. of species	1(+4)				9	6	2	10	10	32	33	103
Percentage	1.0				8.7	5.8	1.9	9.7	9.7	31.0	32.0	

Table 2. Paleogeographic distribution of Middle Permian Fusulines in Thailand-Malayan realm.

Element	I	II			III	IV	V'	V	VI'	VI	VII	No. of species
		II_0	II_1	II_2								
Ozawainellidae										2	1	3
Boultonidae										4	1	5
Schwagerinidae										1		1
Neoschwagerinidae										1	3	4
No. of species										7	6	13

Table 3. Paleogeographic distribution of Late Permian Fusulines in Thailand-Malayan realm.

Recently it became apparent that the traditional view which regarded the Permian as a time of tectonic quiescence in Thailand and adjacent regions is not compatible with the facts nature created. Chonglakmani & Sattayarak (1978) described for the first time a Permian section from Petchabun (Central Thailand) which they called "geosynclinal". In the following years Helmcke and coworkers (Helmcke & Kraikhong 1982, Helmcke & Lindenberg 1983, Helmcke 1983, Altermann et al. 1983, Winkel et al. 1983) could show that the Permian strata found in the Petchabun fold and thrust belt (Wielchowsky & Young 1985) comprise the whole sequence of pre-flysch, flysch and molasse sediments as characteristic of the pre-orogenic, syn-orogenic and late-orogenic stage of an ongoing orogeny. This orogeny has to be dated therefore as Permian in age and should be labelled a Late Variscan orogenic event.

The recent discovery that the geodynamic evolution of Thailand is governed by an orogenic event of Permian age raises the question as whether the evolution of the fusulinacean faunas might be correlatable with this abiotic evolution or not, i.e. whether a control or at least a strong influence of the orogenic evolution on the biosphere might be rightly postulated for the Permian extinctions which led to the greatest biologic revolution in the Phanerozoic history of the earth (Sheng et al. 1984).

Outline of the fusulinacean evolution in Thailand during the Permian

During the Lower Permian (Table 1) fusulinacean faunas are most conductive in Thailand and adjacent areas. About 47 % of the whole fusulinaceans found belong to the Lower Permian. Among them the Schwagerinidae are extensively. The Arctic-Tethyan elements which have been found are 15 species of Schwagerinidae, including these genera:

Pseudoschwagerina	Rugosofusulina
Sphaeroschwagerina	Chusenella
Triticites	Monodiexodina
Pseudofusulina	Robustoschwagerina

Of the 15 species mentioned, 3 species are found only in the Arctic realm during the Asselian. The rest are Arctic-Tethyan and cosmopolitan elements (Ingavat et al. 1980), which range in age up to the Yahtashian.

The faunal composition shows a rather remarkable change during the upper Lower Permian (Yahtashian) by the decrease of the Arctic-Tethyan and Western Tethyan elements and the increase of the Eastern Tethyan elements.

In the Middle Permian (Table 2) Arctic-Tethyan elements were no longer settling. The Verbeekinidae and Neoschwagerinidae occur dominantly, especially the Eastern Tethyan species, which are about 56 % of the whole Middle Permian species. The cosmopolitan and endemic species have also been found associated.

The percentage increment of the Eastern Tethyan elements suggests that during the Middle Permian Thailand had a closer paleobiogeographic relationship with the Eastern Tethys than in the Lower Permian.

During the Upper Permian the fusulinacean fauna became limited both in diversity and distribution. The decrement in number of families and species is distinctly, only about 10 genera of Boultonidae and Ozawainellidae evolved to the Dorashamian (Table 3). Advanced species of Neoschwagerinidae - Lepidolina are known from the Midian. Paleofusulina sinensis, the youngest and most characteristic species of the Upper Permian in the Eastern Tethys region has only been found in two areas in the east of nothern Thailand. The strata containing these are overlain unconformably by Lower Triassic shales with bivalves and ammonites.

This faunal association is known from the Tethyan region from southern Europe eastward into central Asia, southwestern China and Southeast Asia where the biogeographic boundaries are closely related to those of the Mesozoic Tethyan realm. (In this paper the terms "Tethyan region" or "Tethyan realm" are used to describe a paleobiogeographic province -- it does not indicate that this region is underlain by oceanic crust).

This short outline of the Permian fusulinacean evolution in Thailand clearly shows remarkable changes (bioevents). The first bioevent happened during the upper Lower Permian (Yahtashian). This event is characterized by the disappearance of the Arctic-Tethyan elements. The second event can be dated approximately boundary Middle Permian/Upper Permian to lowermost Upper Permian (Midian). This event is characterized by the strong decrease in fusulinacean diversity. The third event finally is characterized by the extinction of the fusulinaceans at the Permian/Triassic boundary.

The mechanisms which may control the Permian fusulinacean bioevents in Thailand

We like to suggest that the described bioevents are controlled or at least strongly influenced by geodynamic events which happened during the Permian. The bioevent during the upper Lower Permian (Yahtashian) which led to increased provincialism (Ross 1979) indicates that an innerasiatic seaway (comp. Ross 1967, Toriyama et al. 1978, Helmcke 1984), which linked the Arctic with Thailand closed. Regional geology tells us that the Urals were folded approximately during the upper Lower Permian (Nalivkin 1973). The relative coastal-onlap chart (Vail et al. 1977) -- in Fig. 1 reproduced mainly after Wielchowsky & Young (1985) -- shows a

sharp regression in the upper Lower Permian. The following equation may
describe the interactions between the geosphere and this bioevent:

major orogeny = worldwide regression = increase of provincialism.

The bioevent approximately boundary Middle Permian/Upper Permian to
lowermost Upper Permian (Midian) which led to a strong decrease in fusulinacean diversity might be caused by the orogenic evolution in Thailand
and China. During this time the Petchabun geosyncline is filled with
flysch (Winkel et al. 1983, Ingavat-Helmcke et al., in prep.) and molass
strata (Altermann et al. 1983). According to Chinese information (Jin
Yu-gan 1981, Wang Yu-jing & Mu Xi-nan 1981) there stretches a flysch bel
all the way from the Kunlun mountains (Karamiran pass) to the eastern
banks of the Jinsha river in Yunnan. This flysch contains few members of
Neoschwagerina fauna. Therefore this flysch is in Chinese terminology
upper Lower Permian (Maokouan), which equals approximately upper Middle
Permian (Murgabian) in the terminology we use. Again the relative
coastal-onlap chart (Vail et al. 1977) shows a pronounced regression
approximately during this time. The following equation may describe the
interactions between the geosphere and this bioevent:

major orogeny = worldwide regression = decrease in diversity.

The bioevent at the Permian/Triassic boundary which led to the
extinction of the fusulinaceans might also be controlled by a geodynamic
process. In the Alps there ellapses an interval of approximately 3 - 5 m
between the meso-alpine deformation and the uplift (Trümpy 1985, p. 38).
If it is a general rule that there ellapse always some millions of years
between the folding of an orogen and the uplift caused by isostasy of th
newly formed mountain belt, then this might be the reason for this event
due to the isostatic uplift the sea withdrew from the last remaining
basins. The following equation may describe the interactions between the
geosphere and this bioevent:

isostatic uplift = closure of remaining basins = extinction.

If the described scenario is correct then we can conclude that the
greatest biologic revolution in the Phanerozoic history of the earth
(Sheng et al. 1984) is closely related to the Late Variscan orogenies
which caused the final assembly of the Permo-Triassic Pangaea (Trümpy
1982).

REFERENCES

ALTERMANN, W.; GRAMMEL, S.; INGAVAT, R.; NAKORNSRI, N. & HELMCKE, D. (1983): On the Evolution of the Paleozoic Terrains bordering the Northwestern Khorat Plateau.- Conf. Geol. Mineral Resources of Thailand, November 1983, Bangkok, 5 p., preprint.

BUNOPAS, S. & VELLA, P. (1978): Late Paleozoic and Mesozoic Structural Evolution of Northern Thailand a Plate Tectonic Model.- in: NUTALAYA, P. (ed.): Proc. Third Regional Conf. Geol. Mineral Resources SE Asia, Bangkok, 133-140.

CHONGLAKMANI, C. & SATTAYARAK, N. (1978): Stratigraphy of the Huai Hin Lat Formation (Upper Triassic) in northeastern Thailand.- in: NUTALAYA, P. (ed.): Proc. Third Regional Conf. Geol. Mineral Resources SE Asia, Bangkok 739-774.

HELMCKE, D. (1983): On the Variscan Evolution of Central Mainland Southeast Asia.- Earth Evol. Sci. 4/1982, 309-319.

-- (1984): The orogenic Evolution (Permian-Triassic) of central Thailand. Implications on paleogeographic Models for Mainland SE-Asia.- Mém. Soc. géol. France, N.S., 147, 83-91.

-- & KRAIKHONG, C. with a contribution by LINDENBERG, H.-G. (1982): On the Geosynclinal and Orogenic Evolution of central and northeastern Thailand.- J. geol. Soc. Thailand 5, 52-74.

-- & LINDENBERG, H.-G. (1983): New Data on the "Indosinian" Orogeny from central Thailand.- Geol. Rdsch. 72, 317-328.

INGAVAT, R. (1984): On the Correlation of the Permian Foraminiferal Faunas of the western, central and eastern Provinces of Thailand.- Mém. Soc. géol. France, N.S. 147, 93-100.

--; TORIYAMA, R. & PITAKPAIVAN, K. (1980): Fusuline Zonation and Faunal Characteristics of the Ratburi Limestone in Thailand and its Equivalents in Malaysia.- Geol. Palaeont. SE Asia 21, 43-56.

INGAVAT-HELMCKE, R.; STROBEL, C. & HELMCKE, D. (in prep.): Fossil Evidence of upper Middle Permian Age of the Flysch Sequence in the Petchabun Fold and Thrust Belt.

JIN YU-GAN (1981): On the Paleoecological Relation between Gondwana and Tethys Faunas in the Permian of Xizang.- in: Geological and Ecological Studies of Qinghai-Xizang Plateau 1, Sci. Press, Beijing, 171-178.

NALIVKIN, D.V. (1973): Geology of the U.S.S.R.- Oliver & Boyd, Edinburgh, 855 p.

ROSS, C.A. (1967): Development of Fusulinid (Foraminifera) Faunal Realm.- J. Paleont. 41, 1341-1354.

-- (1979): Evolution of Fusulinacea (Protozoa) in Late Paleozoic Space and Time.- in: GRAY, J. & BOUCOT, A.J. (eds.): Historical Biogeography, Plate Tectonics, and the Changing Environment. Oregon State Univ. Press, 215-226.

SHENG JIN-ZHANG et al. (1984): Permian-Triassic Boundary in Middle and Eastern Tethys.- J. Fac. Sci. Hokkaido Univ., Serv. IV 21, 133-181.

TORIYAMA, R.; PITAKPAIVAN, K. & INGAVAT, R. (1978): The Paleogeographic Characteristics of Fusuline Faunas of the Rat Buri Group in Thailand and its Equivalent in Malaysia.- in: NUTALAYA, P. (ed.): Proc. Third Regional Conf. Geol. Mineral Resources SE Asia, Bangkok, 107-111.

TRÜMPY, R. (1982): Das Phänomen Trias.- Geol. Rdsch. 71, 711-723.

-- (1985): Die Plattentektonik und die Entstehung der Alpen.- Naturforsch. Ges. Zürich 187. Neujahrsblatt, 47 p.

VAIL, P.R.; MITCHUM, R.M., Jr. & THOMPSON, S. III. (1977): Seismic Stratigraphy and Global Changes of Sea Level, Part 4: Global Cycles of relative Changes of Sea Level.- in: PAYTON, C.E. (ed.): Seismic Stratigraphy - Applications to Hydrocarbon Exploration. Amer. Assoc. Petrol. Geologists, Mem. 26, 83-97.

WANG YU-JING & MU XI-NAN (1981): Nature of the Permian Biotas in Xizang and the northern Boundary of the Indian Plate.- in: Geological and Ecological Studies of Qinghai-Xizang Plateau, 1, Sci. Press, Beijing, 179-185.

WIELCHOWSKY, C.C. & YOUNG, J.D. (1985): Regional Facies Variations in
    Permian Rocks of the Petchabun Fold and Thrust Belt, Thailand.-
    in: Proc. Conf. Geol. Mineral Resources Development NE, Thailand,
    Kohn Kaen Univ., 41-55.
WINKEL, R.; INGAVAT, R. & HELMCKE, D. (1983): Facies and Stratigraphy of
    the Lower - lower Middle Permian Strata of the Petchabun Fold-Belt in
    central Thailand.- in: Proc. Workshop Stratigraphic Correlation
    Thailand & Malaysia, 1, Haad Yai, 293-306.

TRIASSIC BRYOZOA AND THE EVOLUTIONARY CRISIS OF PALEOZOIC STENOLAEMATA

SCHÄFER, Priska *) & FOIS-ERICKSON, Elisabetta **)

A contribution
to Project
GLOBAL
BIO-
EVENTS

Triassic Bryozoa occur in the Arctic and Western N-America, in Mexico, in the Western (European) Tethyan Region, in the NW-Caucasus and Pamir, in the Himalaya, in Tibet, S-China and Japan, in the Maritime Province and SE-Siberian Jakutzkaya Region, USSR, and in New Zealand and New Caledonia (Fig. 1).

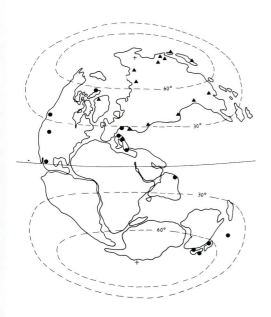

Figure 1. Map with Triassic bryozoan localities; position of continents according to plate tectonic model;

● : studied faunas;
▲ : faunas compiled from literature.

Most Triassic Bryozoa are Paleozoic holdovers belonging to the orders Trepostomata, Cryptostomata, Cystoporata and Fenestrata. Many species derive from Middle to Upper Paleozoic genera extending into the uppermost Triassic; however, others seem to originate and be restricted to the Triassic exclusively. Among the 4 orders the Trepostomata are the most diverse and abundant.

The Upper Permian is still characterized by a highly diverse Tethyan fauna during the Kazanian being clearly distinctive from the low diverse

*) Institute of Geology, University of Marburg, D-3550 Marburg, F.R.G.
**) Institute of Geology, University of Milano, I-20133 Milano, Italy.

Lecture Notes in Earth Sciences, Vol. 8
Global Bio-Events. Edited by O. Walliser.
© Springer-Verlag Berlin Heidelberg 1986

non-Tethyan fauna (Ross 1978). Bryozoans are poorly presented in the Dzhulfian, since most Permian lineages become extinct at the end of the Kazanian or during Early Dzhulfian. However, a few genera persist throughout the Dzhulfian and provide initial populations for a new radiation of Paleozoic stenolaemates in the Early to Middle Triassic.

The extinction of most Paleozoic lineages in the Late Permian can be clearly related to plate tectonics. Among the manifold interacting factors are the most important: 1.) final accretion of continents to form Pangaea during Late Permian; 2.) climatic extremes; 3.) decrease of environmental stability for benthic organisms; 4.) worldwide regression; 5.) reduction of shelf areas; 6.) reduction of areas for colonization; 7.) decrease of separation possibilities.

The reorganization of Paleozoic bryozoans in the Early Triassic coin cides with initial rifting indicating the impending break up of Pangaea. Transgression occurs and shelf seas increase again, although the coheren shelf seas especially of the Inner Tethys still counteract a rapid radia tion due to still extreme continental climatic conditions. Thus a few Scythian taxa including one new genus have been found in the Arctic Sea, while only one is known from the Tethys. Several genera (most likely all

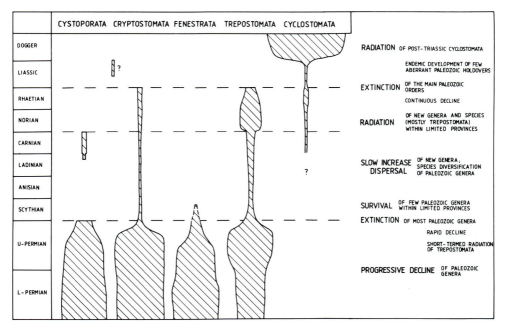

Figure 2.   Diagram showing Paleozoic stenolaemate orders during Permia to M-Jurassic time and evolutionary events affecting diversity patterns of the class across the Paleozoic/Mesozoic boundary; data based on genus level.

Fenestrata) are finally extinct at the end of the Scythian.

Since the Anisian the development of extended shallow carbonate platforms and the proliferation of benthic communities on the Tethyan shelf and surrounding Panthalassa result in a stepwise re-occurrence and gradual generic increase of bryozoan taxa (Fig. 2, 3). Still most of the Middle Triassic Bryozoa belong to Late Paleozoic trepostomate genera; they are widely known from the Central and Eastern Tethys. However, also the number of newly originating genera increases with more provincial species in the Central and Western Tethys and in the West-Cordillera of N-America.

In the Carnian bryozoans are known from most paleogeographic provinces. They occur with several cosmopolitan trepostomate genera throughout the Tethys but also in the Circumpacific (Japan, Maritime Province, SE-Siberia, West-Cordillera and Mexico) and with cryptostomates in the Central to Eastern Tethys. Endemic occurrences of Cystoporata in the Western Tethys and Mexico are restricted to the Carnian and have no known ancestors in the Middle and Lower Triassic.

In the Norian and Rhaetian bryozoans again are less cosmopolitan and seem to be restricted to certain provinces. Most shallow water reef biotopes by this time are dominated by scleractinian corals associated

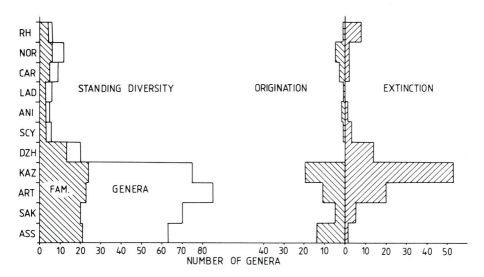

Figure 3.   Standing diversity of genera and families of Paleo-zoic Stenolaemata during the Permo-Triassic. Ratio of generic origination to extinction of Paleozoic Stenolaemata displaying two evolutionary cycles in the Permo-Triassic: at the beginning of each cycle extinction lags behind generic origination, catches up with origination in the middle and dominates origination at the end. Data for Permian from Ross 1978.

with sphinctozoans, sclerosponges and calcareous algae. Bryozoans seem to be lacking even as scarce cryptic reef dwellers. Lagoonal environments, however, have been unfavorable for bryozoans throughout the Triassic.

Thus provincial, but in part highly diverse bryozoan faunas of Noria to Rhaetian age occur in the Murihiku Supergroup of New Zealand and in New Caledonia interpreted as the Australian margin of Gondwana, and have been reported also from the Jakutzkaya Region of SE Siberia. Both provinces are supposed to have had subpolar to boreal climatic conditions. Thus most likely the bryozoans did not have to compete against the scleractinians that otherwise florished in the tropical Tethys. The Gondwana fauna of New Zealand and New Caledonia comprises several genera that presumably originated in the Early Norian (Fig. 2, 3). Some of them show morphologic characters that obviously have newly evolved but otherwise had signified Late or even Middle Paleozoic taxa. Others display characters that are unique and seem to have no counterparts in earlier stenolaemates. Upper Triassic Bryozoa are less frequent but occur with new taxa also in the Torlesse Supergroup of New Zealand commonly interpreted as a displaced terrane with alpine facies.

In general our recent knowledge of Triassic stenolaemate faunas indicates a rapid radiation of new taxa during the Latest Triassic that affected mainly the Trepostomata and less distinctly the Cryptostomata (Fig. 2). This remarkable but short-termed radiation of stenolaemates within isolated provinces at the dawn of the Triassic has by now no conclusive explanation. It may be comparable and be attributed, however, to the same reasons that caused the short-termed radiation and extinction in other phyla.

By the end of the Triassic all Paleozoic stenolaemates become extinct except of a few, tubuliporid Cyclostomata that occur both in the Upper Triassic in the Western Tethys as well as in the Lower Jurassic epicontinental seas of Central Europe. Only two forms have been reported from the Liassic of British Columbia and from Chile, which obviously display characters of several of the Paleozoic orders simultaneously and are explained here as a blind ending evolutionary branch of Paleozoic holdovers. Several other species assigned to post-Triassic cerioporid Cyclostomata have been reported from the Upper Triassic of several localities (Austria; Dolomites, Italy; Hungary; SE-Siberia; Timor); however, their assignment to bryozoans in general and to the Cyclostomata particularly seems questionable and needs restudy.

The final extinction of most Paleozoic stenolaemate stocks again corresponds in time with plate tectonic events, although cause/effect relations are less understood. The rifting in the Tethys due to the opening of the first Atlantic segment caused a worldwide transgression and the formation of the Penninic Ocean. Deep water environments with pelagic and basinal sedimentation prevailed in the Tethys; black shale formation occurred on the European epicontinental shelf. Thus potential biotopes for bryozoans and other sessile-benthic organisms were diminished. Where shallow water carbonate sedimentation is known to have persisted into the Liassic (i.e. Pantokrator Lm., Greece; Zanskar Region, Himalaya), it was dominated by unfavorable intertidal and by innershelf to interior platform conditions respectively.

Early post-Triassic radiation of modern stenolaemates might be expected theoretically in areas not or less affected by the Tethyan oceanic stage (i.e. SE-Siberian platform, Circumpacific or New Zealand); however, no bryozoans except of a few obscure forms have been reported from some of these areas that might document the radiation of the Cyclostomata already during the Liassic. The explosive radiation of post-Triassic Cyclostomata seems to have taken place not before the Early Dogger within the epicontinental shelf seas of Central Europe.

R E F E R E N C E

ROSS, J.R.P. (1978): Biogeography of Permian Ectoproct Bryozoa.- Palaeontology 21, 341-356.

BIOLOGICAL EVENTS IN THE EVOLUTION OF MESOZOIC OSTRACODA

WHATLEY, Robin *)

A contribution
to Project
GLOBAL
BIO-
EVENTS

Abstract: Plotting the stratigraphical distribution of all ostracod species through the Mesozoic reveals major fluctuations in their evolutionary activity and diversity. These fluctuations, in many instances, correlate well with major Mesozoic boundaries and global geological events. At other times enhanced evolutionary activity seems to correlate with advances in, e.g., carapace articulation and sensorial ability. The major features of the evolution of Mesozoic Ostracoda seem therefore to be the product of a complex interaction of both extrinsic and intrinsic factors.

This contribution is based on a paper presented to the 9th International Symposium on Ostracoda held at Shizuoka, Japan, 1985. Data has been abstracted from 1,267 references on Mesozoic Ostracoda. This must represent more than 95 % of such references and is up to date to April, 1985. Great care has been taken to eliminate synonyms and homonyms.

A total of 6,797 species belonging to 739 genera (excluding subgenera) are recorded from the Mesozoic. Their distribution by stage and other divisions is plotted throughout the Mesozoic and the same is done for various of the major groups of Ostracoda.

Although the data for each stage/division has not been normalised with respect to the duration of the divisions, a number of features are immediately apparent in the following figures. For example, the diversity of all Ostracoda is shown to be much higher in the Cretaceous, particularly the Upper Cretaceous, than elsewhere in the Mesozoic and that Triassic diversity is higher than Jurassic. Also, high extinction levels can be seen to occur immediately prior to major Mesozoic division boundaries and are almost immediately succeeded by high origination levels in either absolute or percentage terms. High rates of turnover of ostracod taxa correlate with these major stratigraphical boundaries, but are these boundaries the product of geological or biological events -- or both? Not all of the diversity changes nor rates of turnover of taxa can be correlated with geological events. For example, the Middle Jurassic diversity increase in the Cytheracea can possibly be correlated with the acquisition of a more complex (entomodont) hinge and increased tactile sensory ability anteriorly and the Upper Cretaceous diversity increase is surely not unrelated to the advent of the amphidont hinge.

However, many of the boundaries preceded by high extinctions and followed by high originations can be correlated with more or less major global geological events. Among these are:

---

*) Micropalaeontology Division, Department of Geology, University College of Wales, Aberystwyth, Dyfed SY23 3DB, U.K.

Lecture Notes in Earth Sciences, Vol. 8
Global Bio-Events. Edited by O. Walliser
© Springer-Verlag Berlin Heidelberg 1986

1) The ending of the generally restricted marine conditions of the
   Triassic by the transgression of the Rhaetian/Liassic.
2) The widespread creation of non-marine aquatic environments in the
   Neocomian in a major global regressive phase initiated in the late
   Tithonian.
3) The virtual eradication of "Wealden" environments in most parts of
   the globe (except China and Mongolia) by the Aptian transgression.
4) The Cenomanian onset of the Upper Cretaceous transgression.

These four events seem to have a more than coincidental correlation
with enhanced evolutionary activity in Mesozoic Ostracoda.

A number of stratigraphical comprises, such as the use of tripartite
division of the Triassic, the non use of Tithonian and the use of the
Senonian, have been forced on the author by virtue of the prepordenance
of records in the literature. In the paper the relationship between
fluctuating diversity, rates of evolution and both extrinsic and intrinsic events are discussed in detail. An attempt is also made to recognize
anoxic and other events and to relate the history of the development
of Mesozoic Ostracoda to them. The anomalous nature of the Turonian data
seems certainly related to a late Cenomanian event. The paper also contains a section in which the evolution of Mesozoic Ostracoda is discusse
in terms of both the Red Queen and stationary models of macro-evolutionary theory.

Readers are asked to send any written comments to the author who
will also be glad to send reprints of the paper when it is eventually
published (late 1986-1987).

Figures 1 & 2.   All species in the L. Triassic are assumed to be new.
Origination rates almost always notably increase in the first stage afte
a major geological boundary (i.e. L. Trias, Hettangian, Aalenian , Berri
asian, Cenomanian), this is followed by a dramatic decline to much lower
origination levels, (M. Trias, Sinemurian, Bajocian, Valanginian, and
Turonian) this low level being maintained until the major boundary.
Figs. 1 & 2 illustrate a complex relationship between the number of
species per stage and the % of those which are new. For example, the
L. Trias has the highest number/% of new species but the lowest number
of spp. for the Trias. Similarly the low number of total spp. in the
Hettangian and Aalenian does not match the high % originations for these
stages. Despite the high number of new species in the Cenomanian, the
total number of spp. in the stage is, for the Upper Cretaceous, relative
ly low. Only in the Berriasian does high numbers of new spp. go with
the highest number of spp. for the Neocomian.

Fig. 1 also clearly shows the considerable degree by which the simple
species diversity of Ostracoda fluctuated in the Mesozoic. In simple
terms it was higher in the Upper Triassic than at any time in the Jurassic and considerably higher during all the Cretaceous stages than hitherto in the Mesozoic. The mean simple species diversity of the Neocomian
is, for example, 70 % higher than that of the Triassic while that of the
Maastrichtian is X10.5 greater than that of the Turonian, which is the
lowest for the Mesozoic.

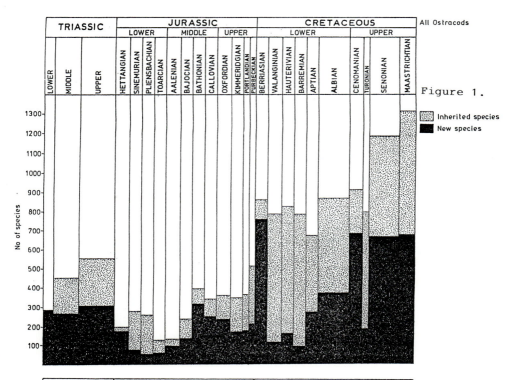

Figure 1.

Inherited species

New species

All Ostracods

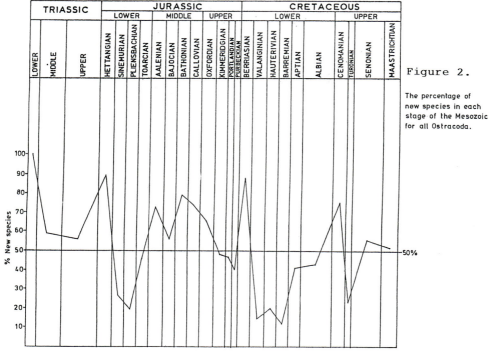

Figure 2.

The percentage of new species in each stage of the Mesozoic for all Ostracoda.

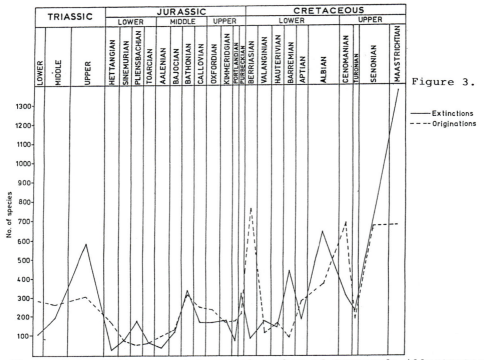

Figure 3.

Figure 3.   With respect to total number of Ostracoda illustrates the relationship between originations and extinctions of spp. by stage/division.

In the following 12 stages, extinctions exceed originations:

Upper Triassic	+274	Purbeckian	+110
Sinemurian	+ 1	Valanginian	+ 61
Pliensbachian	+128	Barrêmian	+343
Toarcian	+ 5	Albian	+270
Bathonian	+ 20	Turonian	+ 43
Kimmeridgian	+ 10	Maastrichtian	+694

The (certainly invalid) assumption is made that all Mesozoic species became extinct in the Maastrichtian.

In the following 13 stages, originations exceed extinctions:

Lower Triassic	+182	Portlandian	+ 98
Middle Triassic	+ 74	Berriasian	+672
Hettangian	+153	Haaterivian	+ 17
Aalenian	+ 62	Aptian	+ 91
Bajocian	+ 15	Cenomanian	+375
Callovian	+ 79	Senonian	+ 50
Oxfordian	+ 69		

As Figs. 1 and 2 showed that high origination levels took place immediately after major geological boundaries, so Fig. 3 shows that in absolut or percentage terms high extinction rates immediately preceed these boundaries.

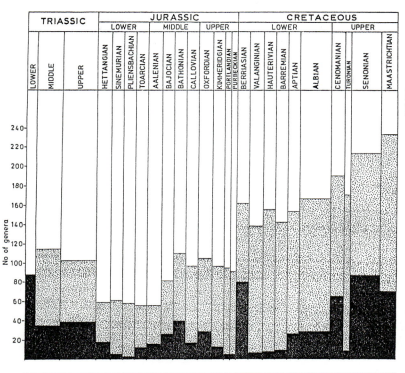

Figure 4.

Distribution of all
ostracod genera.

Inherited genera

New genera

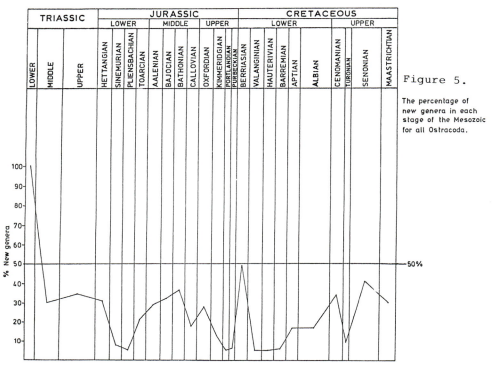

Figure 5.

The percentage of
new genera in each
stage of the Mesozoic
for all Ostracoda.

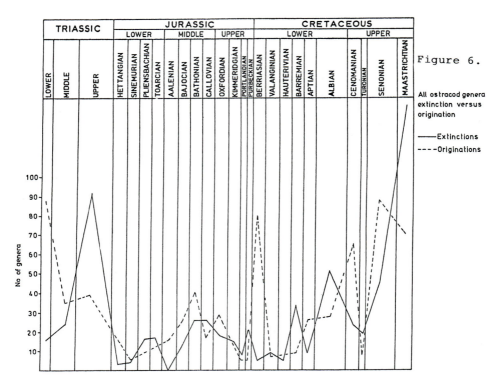

Figure 6.

All ostracod genera
extinction versus
origination

———Extinctions
----Originations

Figures 4, 5, 6. These Figs. illustrate for genera what Figs. 1-3 did for species. These Figs. largely mirror those for species but are influenced by the greater longevity of genera.

Figures 7a, b - 9a, b. Illustrate graphically the distribution of the major groups of Mesozoic Ostracoda in terms of simple species diversity. The most important factor to note is in Fig. 7b. The principally marine superfamily Cytheracea dominate the principally (from Bathonian onwards) non-marine Cypridacea, except in the Neocomian-Aptian interval. This is associated with the large scale availability of "Wealden" environments at this time. The persistence of such environments in China and Mongolia into the Upper Cretaceous is responsible for the sustained high level of the Cypridacea at this time.

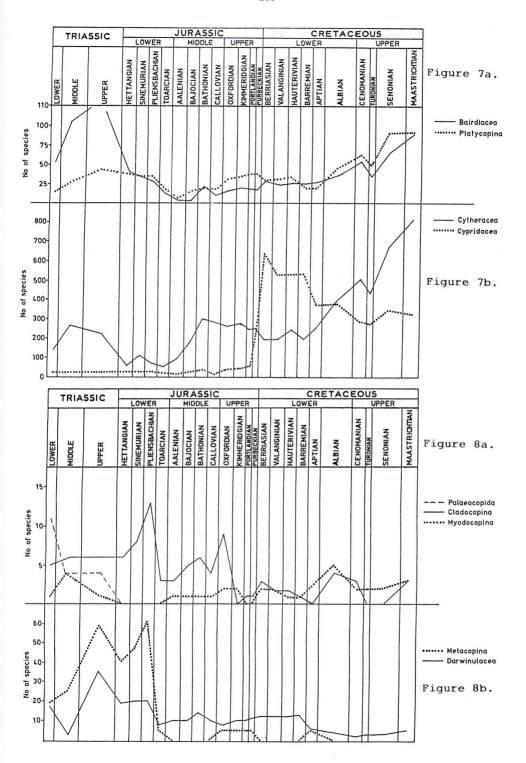

Figure 7a.

Figure 7b.

Figure 8a.

Figure 8b.

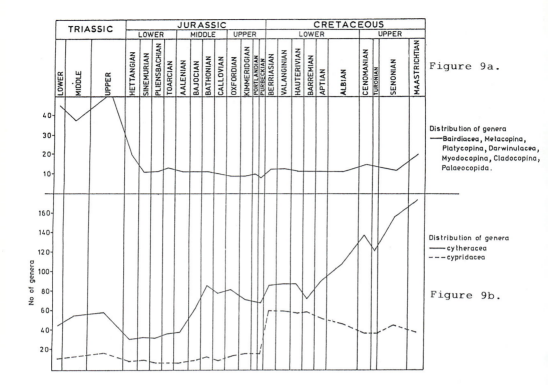

Figure 9a.

Distribution of genera
—Bairdiacea, Metacopina,
Platycopina, Darwinulacea,
Myodocopina, Cladocopina,
Palaeocopida.

Distribution of genera
——cytheracea
- - -cypridacea

Figure 9b.

			SPECIES			GENERA		
CRETACEOUS	UPPER	Maastrichtian	1311 (678)			233 (70)		
		Senonian	1181 (664)	1048		213 (87)	202	
		Turonian	794 (188)	(2211)		171 (8)	(230)	
		Cenomanian	907 (681)	(32.5%)		190 (65)	(31.1%)	173
	LOWER	Albian	864 (371)		899	167 (28)		(388)
		Aptian	669 (275)	154	(3985)	154 (26)	154	(52.5%)
		Barremian	786 (91)	799	(58.6%)	143 (9)	(158)	
		Hauterivian	825 (163)	(1774)		156 (8)	(21.4%)	
		Valanginian	789 (117)			139 (7)		
		Berriasian	860 (757)	(26.1%)		162 (80)		
JURASSIC	UPPER	Purbeckian	512 (211)			91 (5)		
		Portlandian	367 (176)	396		95 (5)	97	
		Kimmeridgian	346 (170)	(793)		97 (13)	(52)	
		Oxfordian	360 (236)	(11.7%)		105 (29)	(7.0%)	
	MIDDLE	Callovian	340 (250)			97 (17)		
		Bathonian	396 (316)	276	296	110 (40)	86	81
		Bajocian	235 (133)	(794)	(1950)	81 (26)	(99)	(189)
		Aalenian	131 (95)	(11.7%)	(28.7%)	56 (16)	(13.4%)	(25.6%)
	LOWER	Toarcian	126 (59)			56 (12)		
		Pliensbachian	257 (51)	214		58 (3)	59	
		Sinemurian	277 (76)	(363)		61 (5)	(38)	
		Hettangian	199 (177)	(5.3%)		59 (18)	(5.1%)	
TRIASSIC		Upper	552 (308)	431	431	103 (39)	102	102
		Middle	456 (268)	(862)	(862)	115 (35)	(162)	(162)
		Lower	286 (296)	(12.7%)	(12.7%)	88 (88)	(21.9%)	(21.9%)
			No. of species and No. of new species per stage	Mean No. of species. No. of new species and % of new species per division	Mean No. of species. No. of new species and % of new species per system	No. of genera and No. of new genera per stage	Mean No. of genera, No. of new genera and % of new genera per division	Mean No. of genera, No. of new genera and % of new genera per system

Table 1.   Number of species and genera occurring and appearing for the first time in each stage. (First appearance of taxa in brackets).

EFFECTS AND CAUSES IN A BLACK SHALE EVENT -- THE TOARCIAN POSIDONIA SHALE OF NW GERMANY

RIEGEL, Walter, LOH, Hartmut, MAUL, Bernd & PRAUSS, Michael *)

A contribution to Project GLOBAL BIO-EVENTS

Abstract: Black shale formation in the Toarcian of NW Germany is associated with a major turnover in phytoplankton assemblages interpreted as the response to lowered salinities in surface waters of the epicontinental sea. The resulting halocline leeds to anoxia of varying degrees at the sediment surface and in the water column controlling the extent of benthic life. The bulk of organic matter concentrated in the Posidonia Shale occurs as thin bituminous laminae considered to be the remnants of microbial mats stabilizing the redox boundary at the sediment surface. Possible links with oceanic circulation may provide a basis for correlation with contemporaneous events in other regions.

Introduction

The formation of the Lower Toarcian Posidonia Shale represents a black shale event of regional scope extending across several basins in north-

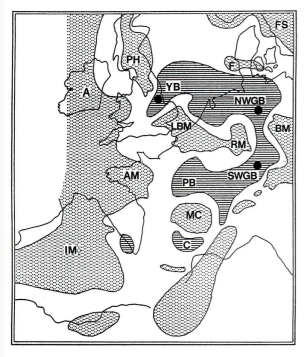

Figure 1. Distribution of land areas and bituminous facies in the Upper Liassic (Toarcian) of northwestern Europe (Adapted from Hoffmann 1968 and Ziegler 1982).

A   = "Atlantis"
PH   = Pennine High
F   = Funen High
FS   = Fennoskandia
LBM   = London-Brabant-Massif
RM   = Rhenish Massif
BM   = Bohemian Massif
AM   = Armorican Massif
MC   = Massif Central
IM   = Iberian Massif
YB   = Yorkshire Basin
NWGB = Northwest German Basin
SWGB = Southwest German Basin
PB   = Paris Basin
C   = Chalhac

▨ Land areas   ▤ Black shales

---

*) Institut und Museum für Geologie und Paläontologie, Universität Göttingen, D-3400 Göttingen, F.R.G.

Lecture Notes in Earth Sciences, Vol. 8
Global Bio-Events. Edited by O. Walliser
© Springer-Verlag Berlin Heidelberg 1986

western Europe (Fig. 1) and leading to the accumulation of one of the foremost hydrocarbon source rocks in this region. Sections from widely separated locations often exhibit a surprisingly close similarity of lithologic succession within the Posidonia Shale indicating a common underlying control of its formation in different areas.

Few attempts have previously been made to reconstruct the sources of primary productivity, the pathways of maceral input and the early dia-genetic processes and to integrate them into models of depositional environments. This is the aim of our study of section through the Posidonia Shale near Marienburg about 4 km SE of Hildesheim, NW-Germany and material from adjacent drill cores.

The sections (Fig. 2)

The underlying claystones of the Domerian contain few fossils but are partially bioturbated, carbonates are restricted to sideritic concre-

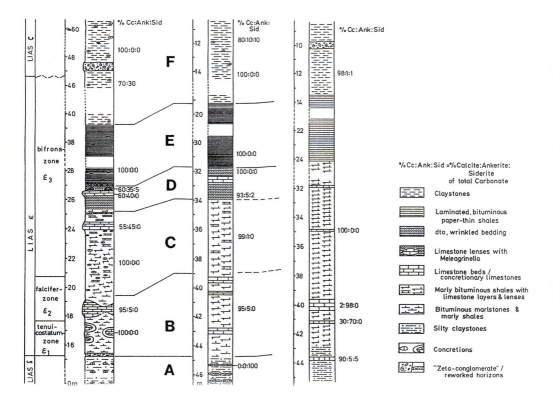

Figure 2.   Sections through Posidonia Shale in the Marienburg trench (left), borehole Hildesheim 4 (middle) and borehole SPD 18 showing ammonite zones and ratios of carbonate species (from LOH et al. 1986).

tions. The Toarcian begins with a horizon of reworked concretions and belemnite rostra embedded in a glauconitic matrix. The marly clays above are unbedded before showing the characteristic lamination of the Posidonia Shale. Higher up carbonate is concentrated in large concretions typically zoned in the tenuicostatum Zone and showing a grey calcitic core surrounded by an ankeritic/sideritic mantle (ockre-yellow in weathered sections, "Ockergeoden"). Concretions of the falciferum Zone are 1 to 2 m in diameter ("Laibsteine", "whalestones") and vary in their composition from 95 % calcite to 100 % ankerite. Their growth must have taken place late in diagenesis since laminations cross concretion boundaries without deformation and shells are consistently compacted.

The bivalve assemblage of the upper part of the tenuicostatum Zone and the falciferum Zone is composed of Meleagrinella, Inoceramus and Steinmannia. Normally, juvenile stages predominate, but in the upper part of section C adult specimens may densely cover bedding planes. This bivalve assemblage is suddenly replaced by the Bositra community in section D. Among cephalopod belemnites are commonly present throughout. Harpoceratoid ammonites are replaced by Dactylioceras in the lower part of the bifrons Zone.

The "Monotis-Bank" interval (upper part of section C and section D) shows evidence of considerable current action in the carbonate beds and the belemnite layer ("Belemnitenschlachtfeld") above. Wrinkled bedding within this interval is due to the dense covering of bedding planes by Bositra shells. Pyritic concretions are typical for the entire bifrons Zone, but they are particularly frequent in the lower part of section C.

Section E is characterized by papery grey marly shales and differs from section F by the total lack of bioturbation and higher organic content. Occasional small-sized Bositra shells are the only fossils. The bituminous facies of the Posidonia Shale is terminated somewhat prior to the occurrence of a marked erosional event ("Zeta-conglomerate") made up of belemnite rostra, sideritic and phosphatic nodules and representing three ammonite subzones.

Primary production in the Posidonia Shale

Fig. 3 shows the distribution of the major organic-walled plant protist groups, of spores and gymnosperm pollen across the Posidonia Shale interval. Of particular interest for the interpretation of Posidonia Shale formation and its underlying causes is the great turnover in phytoplankton associations at the Posidonia Shale boundaries.

Phytoplankton assemblages in the Pliensbachian are dominated by

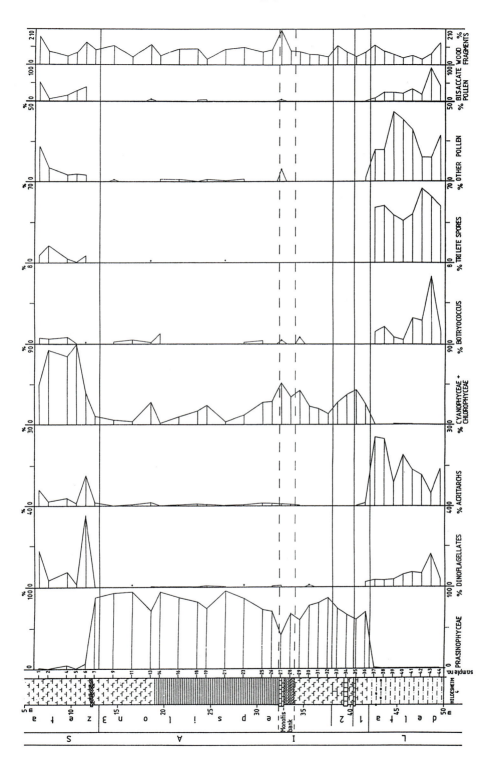

Figure 3. Distribution of palynomorph groups in the Posidonia Shale, core Hildesheim 4 (from Loh et al. 1986).

acritarchs and three species of early dinoflagellates. Within the Posi-donia Shale, acritarchs are greatly reduced in number, while Nannocera-topsis gracilis is the lone occasional survivor of dinoflagellates. Instead, phytoplankton assemblages are vastly dominated by a diversity of prasinophyte species, which are exceedingly rare to absent outside the Posidonia Shale. In addition small spherical bodies (probably chlo-rophyta) and clusters of flaky organic material (Cyanophyta ?) are well represented in the Posidonia Shale.

Immediately above the terminal Zeta-conglomerate dinoflagellates return not only with great frequency, but also with increased diversity in the Upper Toarcian (22 species), while prasinophytes nearly dis-appear. It is obvious from this that during the dinoflagellate black-out in the Posidonia Shale facies dinoflagellates have flourished else-where. Similar trends of phytoplankton distribution have been observed throughout NW Europe, but with differing degrees of intensity. In SW Germany f.i. several dinoflagellate species occur well within the Posi-donia Shale (Wille 1982). Thus, the cause for the dinoflagellate black-out in NW Germany has been effective but less pronounced in SW Germany.

If it is accepted that acritarchs and dinoflagellates represent the phytoplankton associated with normal marine conditions and prasinophytes are less restricted with regard to salinity requirements, it may be reasonable to conclude, that lowered salinities in the surface water account for both, the replacement of phytoplankton groups in the photic zone and the establishment of a pyncocline in the water column leading to anoxia at the basin floor.

The role of coccolithophorids in the Posidonia Shale is difficult to assess due to uncertainties regarding their ancient habitat and their actual proportion in the phytoplankton. Considering their susceptibility to dissolution coccolithophorids along with schizopheres are a potential-ly important component of the phytoplankton. In analogy to their recent representatives (Tappan 1980) Jurassic coccolithophorids may have been able to thrive in greater depth under reduced light intensities.

Organic petrology

Maceral classification provides the most refined means of particulate organic matter identification. It also allows for a broad assignment of organic input to sources: Macerals of the vitrinite and inertinite group (except micrinite) as well as some liptinites (sporinite, cutinite, resinite) are derived from terrestrial sources, while other liptinites (bituminites, various alginites) and in part micrinite have to be assigned to aquatic sources.

The distribution of important maceral types or groups in the Marien-burg section shows that terrestrial woody material (vitrinites and fusi-nites) represents a minor constituent. Since only finely dispersed wood fragments have been recorded and large driftwood has not been included, the total vitrinite content is potentially higher. Micrinite in this section is apparently generated by thermal alteration of algal material (bituminite, alginite), since its peak begins to appear just above a burnt zone.

By far the most abundant macerals are the aquatically derived algini tes and bituminites. The bulk of alginite A in the Posidonia Shale in NW Germany consists of prasinophyte phycomata. Alginite B occurs as thin stacked laminae best visible by their light yellow fluorescence. Bitumi-nite varieties have a similar mode of occurrence but differ in their reduced fluorescence. They are considered to be thermally or bacterially matured alginite laminae. There is a distinct negative correlation in the distribution of bituminites, the vastly dominant organic constituent in the lower part of the Posidonia Shale up to the belemnite layer, and alginites showing a sharp increase in section E. The interpretation of laminar alginite B, and consequently of bituminites with similar modes of occurrence, as remnants of microbial mats (Hutton et al. 1980) grants support to Kauffman's claim (1979, 1981) for the existence of such a mat at the sediment surface and its control of the redox boundary in the Posidonia Shale.

Discussion and conclusions

Low concentrations of pyrite, the lack of calcitic cements and the abundance of sideritic concretions indicate a quick passage of Domerian sediments through the sulfate reduction zone into the fermentation zone due to the combination of high sedimentation rates and lack of easily digestible organic material. Most significant for reconstructing the mechanisms of anoxia initiation in the "Posidonia Shale sea" is the recognition of widespread reworking at the base of the Toarcian (Jenkyns 1985). Above this, the initial presence of oxygenated bottom waters as indicated by bioturbation is gradually replaced by anoxia. The consisten succession within short intervals of a conglomeratic horizon, bioturbate marlstones grading into laminated bituminous shales may be explained by currents introducing oxygen depleted high density oceanic water into the warm or less saline epicontinental seas initiating a pronounced pycnocline there. This leads to the rise of the redox boundary from within the sediment surface and into the water column within an estimate 2000 to 4000 years (Loh et al. 1986).

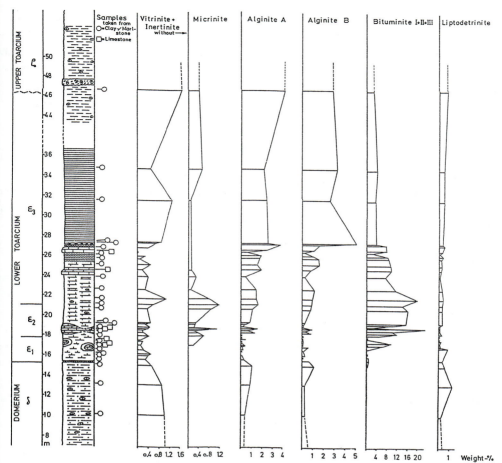

Figure 4. Maceral distribution in the _Posidonia_ Shale (from Loh et al. 1986).

Early bituminous stages represented by section B show a characteristic succession of ironrich carbonate concretions. The zoned "Ockergeoden" may be considered to be a transitional type between the sideritic concretions below and the ankeritic/calcitic limestone beds (whalestones) formed in bituminous facies above. The lack of ankerite and the abundance of pyritic concretions in the greater part of section C points towards a long residence time of these sediments in the sulfate reduction zone. This section also shows the prolific growth of the bituminite precursor, a microbial mat that may have provided a firm substrate for epibenthic bivalves like _Meleagrinella_ and _Inoceramus_ and the food source for gastropods (_Coelodiscus_).

While the Posidonia Shale sequence up to 24 m in the Marienburg section indicates progressive development of anoxic conditions and an organic mat combined with continued lowering of sedimentation rates, the "Monotis-Bank" interval (upper part of section C and section D) is characterized by turbulence. Frequent holomixis and subsequent build-up of the redox boundary may have caused repeated mass mortality, the most pronounced of which has formed the belemnite layer.

The change from the turbulent events of section D to the monotonous succession of finely laminated shales in section E may be explained by an increase in density difference between surface and bottom waters stabilizing the pycnocline and preventing turbulences reaching the sediment surface. As a result, the redox boundary rose into the water column precluding the existence of oxygen requiring benthic life and aiding in the formation of a bottom suspension zone. In this situation microbial populations may no longer form competent mats but disintegrate into discontinuous films "floating" around various chemoclines. The sediment surface has then become highly permeable for the diffusion of sea water and the escape of bacterially produced CO_2 and H_2S resulting in the precipitation of total iron as FeS_2.

Section F represents the gradual return from the fully anoxic to nearly normal marine conditions similar to those assumed for section A, indicated by the progressive appearance of bioturbation. The Zeta-conglomerate topping the Lower Toarcian does not generate any recognizable

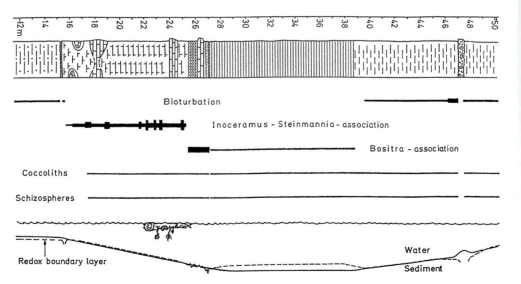

Figure 5. Section showing distribution of major fossil associations and model of suggested migration of redox boundary around the sediment/water interface during Posidonia Shale formation (from Loh et al. 1986)

change in sea bottom ventilation despite its intensity of reworking.

Fig. 5 shows our suggested reconstruction for the migration of the redox boundary around the sediment surface during Posidonia Shale deposition. In contrast to the assumption of Jenkyns (1985) we suggest a model (Fig. 6) by which conditions of Posidonia Shale formation result from a spill-over of the oceanic oxygen minimum zone into the epicontinental sea and the subsequent water column stratification. In this way Posidonia Shale formation, despite its regional aspect, may be linked with broader oceanic conditions and eventually correlated with facies situations in other regions attributable to a common global cause. It may also serve as an example to demonstrate the effects of anoxia sustaining systems on primary production, organic sediment input and early diagenetic alteration that may also be applied to global anoxic events.

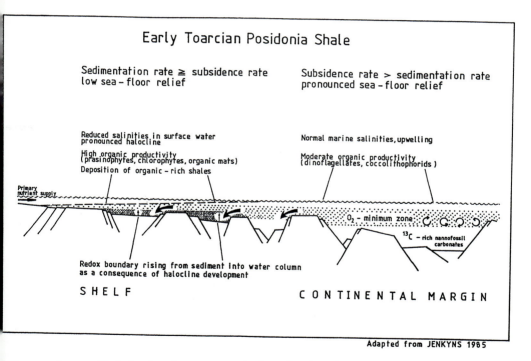

Figure 6. Model showing suggested relation between Posidonia Shale formation and ocean margin conditions.

REFERENCES

HUTTON, A.C.; KANTSLER, J.J.; COOK, A.C. & McKIRDY, D.M. (1980): Organic matter in oil shales.- A.P.E.A. J. <u>20</u>, 44-66.

JENKYNS, H.C. (1985): The early Toarcian and Cenomanian-Turonian anoxic events in Europe: comparisons and contrasts.- Geol. Rdsch. <u>74</u>, 505-518.

KAUFFMAN, E.G. (1981): Ecological reappraisal of the German Posidonien-schiefer (Toarcian) and the stagnant basin model.- in: GRAY, J.; BOUCOT, A.J. & BERRY, W.B.N. (eds.): Communities of the past.- 311-381.

LOH, H.; MAUL, B.; PRAUSS, M. & RIEGEL, W. (1986): Primary production, maceral formation and carbonate species in the Posidonia shale of NW Germany.- in: DEGENS, E.T.; MEYERS, P.A. & BRASSEL, S.C. (eds.): Biogeochemistry of black shales. Mitt. Geol.-Paläont. Inst. Univ. Hamburg <u>60</u>, in press.

TAPPAN, W. (1980): The paleobiology of plant protists.- 1028 p., Freeman & Co.

WILLE, W. (1982): Evolution and ecology of Upper Liassic dinoflagellates from SW Germany.- N. Jb. Geol. Paläont. Abh. <u>164</u>, 74-82.

CRETACEOUS

HIGH-RESOLUTION EVENT STRATIGRAPHY: REGIONAL AND GLOBAL CRETACEOUS BIO-EVENTS

KAUFFMAN, Erle G. *)

A contribution to Project GLOBAL BIO-EVENTS

Introduction

High-resolution event stratigraphy (HIRES) involves the careful documentation of sedimentary deposits and depositional surfaces representing
geologically instantaneous to short term (1000 years or less) physical,
chemical, thermal, and biological events (Fig. 1); the primary purpose
of HIRES is the integration of these diverse data into a highly refined,
applied system of chronostratigraphy with regional to global correlation
potential. Collection of data requires field observation to the cm or
less scale on fresh outcrop surfaces or cores. Common types of events
recorded in the stratigraphic record are volcanic ash falls, mass flow
deposits, oxygen overturns, desalination events resulting from giant
storms in shallow seas, Milankovitch climate cycle boundaries, and mass
mortalities or mass extinction steps. In the Western Interior Cretaceous
Seaway of North America, from which examples will be drawn in this paper,
integration of event-deposit data has allowed chronostratigraphic resolution to 50,000 year or less event-bounded intervals (Fig. 2), and
correlation over hundreds to thousands of square kilometers within and
between major marine facies. Interbasinal correlation to the Gulf Coast
and Caribbean can be achieved at the 100,000 year level in many cases.
This level of resolution, in turn, allows a new generation of integrated
basin analyses and three-dimensional modeling of basin dynamics within
narrow intervals of time.

Biological events are the least understood and most complicated
processes in Earth history, and especially at such a fine scale. The
concept of geologically instantaneous bio-events, such as mass mortalities or catastrophic extinctions, has met with considerable opposition
from the biological and paleontological community. This mainly reflects
awareness of the ecological resiliency that modern species populations,
communities, and broader ecosystems demonstrate today under fluctuating
environmental conditions.

But the present is atypical of the geological past, being essentially
a glacial interval and thus representing less than ten percent of geological time; with the modern Tropics abnormally constricted, a higher
percentage of living organisms are more broadly adapted to environmental

---
*) Department of Geological Sciences, University of Colorado, Boulder,
   Colorado, 80309, U.S.A.

Lecture Notes in Earth Sciences, Vol. 8
Global Bio-Events. Edited by O. Walliser
© Springer-Verlag Berlin Heidelberg 1986

variations than in the past, when most of the global marine biota evolve
into a predominantly equable, warm marine world without permanent ice
at the poles or cold climatic zones. Ancient marine organisms, and especially those of warm intervals associated with high sealevel stand (i.e.
the Cretaceous), must have had narrower adaptive ranges with regard to
temperature, marine chemistry, and biological competition, and thus woul
have responded more rapidly and at a larger scale to diverse environmental perturbations than most of their modern counterparts. It is this
factor, more than any other, that makes the study and stratigraphic
application of bio-events feasible. In addition, modern bio-events such
as mass mortalities, rapid widespread immigration events, and population
bursts that might be recorded in the fossil record are not a strong focu
of contemporary biological research, and have not been summarized for
the paleobiologist; however, meticulous documentation of some modern bio
events does exist in diverse journals. We have just begun to realize the
profound impact of bio-events on the evolution of species and communities; the extremely rapid, punctuated evolution of new adaptive types
within a few hundred thousand years of Cretaceous and Tertiary mass extinction events comprises an obvious example. Such events may, in fact,
be the most important driving mechanism of evolution.

The well studied stratigraphic sequence of the Cretaceous Western
Interior Seaway and adjacent Gulf Coastal Plain of North America provide
a unique opportunity to document and test the temporal aspects of divers

Figure 1. Schematic model of the components of a high-resolution even
stratigraphy (HIRES), plotted against an upward-fining, eustatic rise-
transgressive sequence. Key: left column, C_{org} = organic carbon analysis
(AE = Anoxic event; CS = chemical spike ^{org} useful in event-correlation); this analysis and stable isotope data (not shown) of ^{18}O and ^{13}C
are common chemical event data used in HIRES. Stratigraphic column;
Arrows indicate abundance of points at which HIRES data typically occur
in a sequence; based on actual data in Pratt, Kauffman & Zelt (1985).
PE column = Physical Event Data, typically including CC (climate cycle
beds), CH (concretion horizons), D (some widespread disconformity or
bypass surfaces), NH (phosphatic, siliceous, or ferruginous nodule horizons), MF (mass flow deposits), SB (storm beds), VA (volcanic ash beds,
or bentonites), VF (volcanic flows), etc. CE column = Chemical Event
Data, typically including AE (Anoxic Events), CS (short-term chemical
spikes, or excursions above background in data), CC, CH, and NH (above)
(where chemically precipitated). BE column = Biological Event Data,
typically including CE (rapid colonization events), EE (emigration
events), IE (immigration events), ME (mass extinction steps or events),
MM (mass mortality events) and PE (productivity events). IES column =
integrated event stratigraphy matrix from Physical, Chemical, and Biological data, in which each line essentially represents an isochronous
surface, and event-bounded intervals (numbered) average about 50 Ka in
Cretaceous examples from the Western Interior of North America.

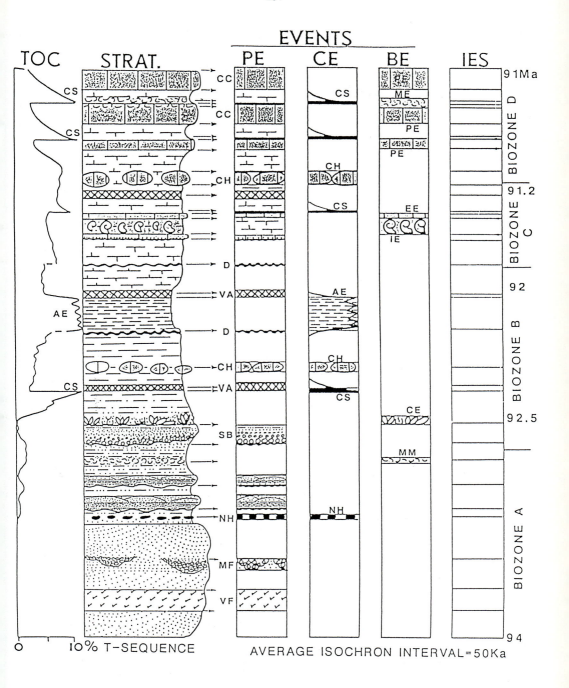

Figure 1

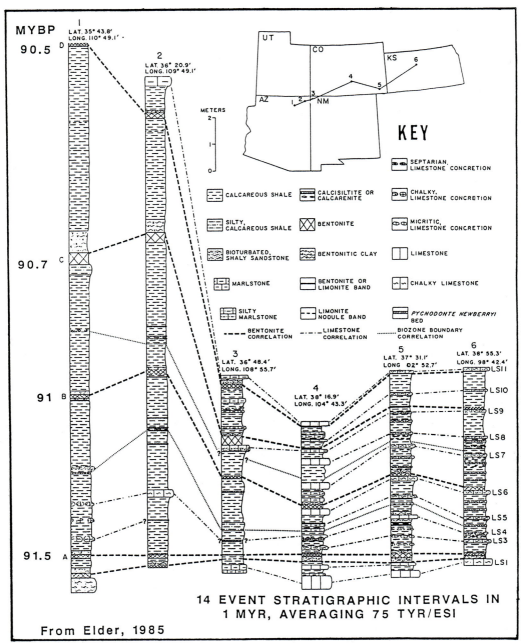

Figure 2.   A field example of high resolution event stratigraphic correlation, utilizing volcanic ash (bentonite) beds and climate cycle bedding surfaces, in Cenomanian-Turonian boundary strata across the Western Interior basin of North America (From Elder 1985). Event correlation lines are compared against molluscan biostratigraphic boundaries (heavy lines), showing close but not isochronous correlation of biozone boundaries. Note radiometric scale; event bounded intervals average 50-100 Ka in duration.

bio-events of regional and global scale, and especially those with very short time spans. This reflects our ability to regionally compare individual bio-events against other essentially isochronous surfaces, in particular volcanic ash falls (bentonites) and the boundaries between Milankovitch climate cycle deposits. The results of these analyses, essentially completed now for the Late Albian through basal Campanian part of the Western Interior sequence, demonstrate that short-term biological events are pervasive throughout the stratigraphic record, and may be used as an important tool in regional chronostratigraphy. Their study provides important insights into the dynamic processes which have changed the global biota through time, and leads to a better understanding of biological evolution during typical intervals of Earth history. In this paper, characteristic examples of North American Cretaceous bioevents have been selected from a very large data bank to demonstrate diverse kinds of rapid biological responses to dynamic changes in a marine system.

The Cretaceous Western Interior Seaway; Environmental setting

Shallow epicontinental seaways such as that which invaded the Western Interior region of North America during middle and Late Cretaceous time were extremely sensitive to various allocyclic forcing mechanisms related to global and regional tectonics, eustacy, climate, and ocean history, far more so than deeper, more resilient oceanic settings. Largescale perturbations (impacts, explosive volcanism, giant storm events, etc.) invoked similar response. Consequently, physical, chemical, ther-al mal, and biological factors of the environment were highly dynamic in epicontinental settings, and short-term changes affecting hundreds to thousands of square kilometers of shallow warm Cretaceous seas were common, and predictable. Further, most Cretaceous shallow water biotas were narrowly adapted to warm epicontinental environments, and underwent extensive rapid changes in response to environmental perturbations of all levels, enhancing the record of bio-events.

Kauffman (1977, 1984a, 1985b, and references therein) has summarized the climatic geological, oceanographic, and biological evolution of the Western Interior Cretaceous Seaway, based on the work of many researchers (cited). Thus, only a brief summary of basin history is presented below as background for understanding the common and diverse bio-events that characterized this marine system.

Tectonically, the Western Interior Basin was an elongate, complex, eastward-migrating foreland basin developed first in Jurassic time in response to uplift and eastward thrusting along the great north-south

trending Cordilleran Geanticline (Fig. 3). Development of this orogenic
belt was in direct response to rapid convergence and subduction along
the Pacific Coast of North America, which was greatly accelerated during
the middle and Late Cretaceous (Kauffman 1984a, 1985b). Much evidence
suggests rapid tectonic response within the Cordilleran orogenic belt
and adjacent foreland basin complex, to intervals of rapid seafloor
spreading (as measured in detail from eustatic rise events) and Pacific
margin subduction; little lag time apparently existed between the
forcing mechanisms (Pacific margin convergence, subduction) and tectonic
response in the Cordillera and foreland basin (intrusion, volcanism,
thrusting, basin subsidence, and isostatic rebound) during basin evolution. Consequently, a two phase model of basin evolution is suggested
(Fig. 4): (1) During third- and fourth-order eustatic rise events,
reflecting volumetric displacement of sealevel upward with active seafloor spreading and ridge building, Pacific convergence and probably
subduction rates were relatively high; most tectonic and orogenic
activity within the Cordillera and adjacent foreland basin occurred
during these intervals of rapid subduction and convergence. The initiation of most intrusions and all but one dated major thrusting event
was associated with eustatic rise and regional epicontinental transgression; 85 percent of all volcanism, as measured from a data base of
more than 1,000 bentonite and volcanic ash layers, was also associated
with eustatic rise (active spreading, subduction) events, as was most
active basin subsidence and adjacent block uplift along forebulge and
tectonic hinge zones. Deep water environments above rapidly subsiding
mid-basin axes, and extensive water stratification leading to widespread
oxygen restriction in bottom waters, characterized eustatic rise and
active tectonic intervals.

(2) Episodes of global eustatic fall, reflecting relative plate
tectonic quiescence, were associated in the Western Interior Cretaceous
Basin with little tectonic activity along the Cordilleran orogenic belt,
slowing or cessation of tectonically driven subsidence and rebound, and
by basin filling and extensive progradation of thick clastic wedges from

Figure 3. Generalized tectonic cross-section (A) and map (B) of
major tectonic zones and peak transgressive marine facies of the Cretaceous Western Interior Seaway of North America (From Kauffman 1984a)
Key: HZ - Hinge zone of block faulting separating the stable craton to
the east from the east slope of AB - the axial basin and deepest part
of the seaway. FRB - Forebulge Zone of line at basement uplifts along
the zone of rebound marginal to the FB - Foreland Basin produced by
thrust and sediment loading from the Western Cordilleran thrust belt.
The various tectonic zones were predominantly active during eustatic
rise in the Cretaceous.

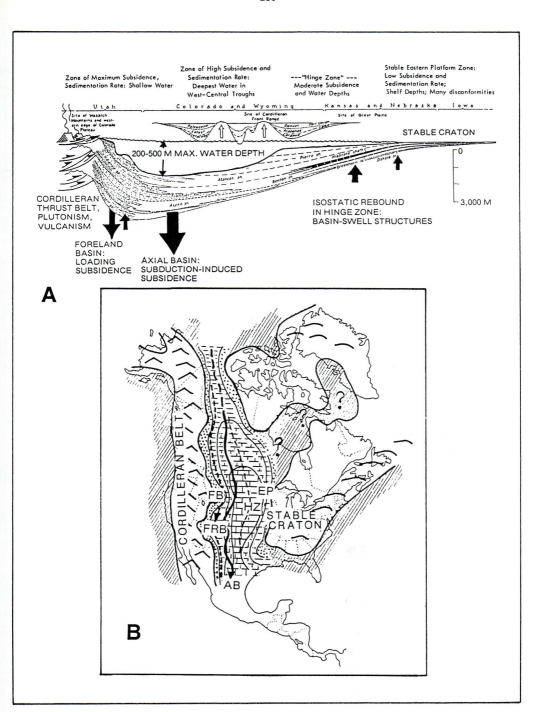

Figure 3

286

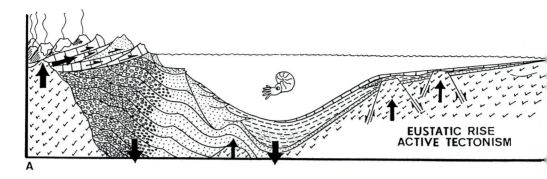

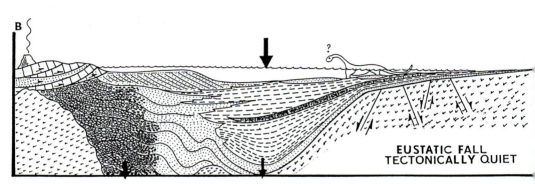

Figure 4. Two phase model of evolution of the Western Interior Cretaceous Basin, North America. <u>A.</u> The great majority of plutonic emplacement, volcanism, thrusting, synorogenic and thrust loading, foreland and axial basin subsidence, and both forebulge and hinge zone deformatio are associated with rapid plate movements, eustatic rise and paleobathymetrically deep phases of the seaway. <u>B.</u> Times of tectonic quiescence in the basin are associated with slow plate spreading and eustatic fall, sediment filling and shoaling of the Western Interior Seaway, and basinward progradation of major clastic wedges from the western margin.

the Western tectonic margin far into the basin. Shallow active water environments were associated with basin filling during eustatic fall, breaking down water stratification and oxygenating the benthic zone. Figure 5 shows correlation of major tectonic, volcanic, and eustatic events within the Western Interior Basin, as determined by high-resolution event- and biostratigraphy. High-resolution stratigraphic correlation suggests that many major tectonic events took place within less than a million years, and some within a few tens of thousands of years; turbidity, paleobathymetry, and sedimentation parameters changed rapidly These changes were capable of generating rapid biological response within the marine realm. In addition more than 1,000 explosive volcanic events, many resulting in very thick (10 cm - 5 m or more), laterally extensive ash and bentonite deposits provide a unique set of isochronous

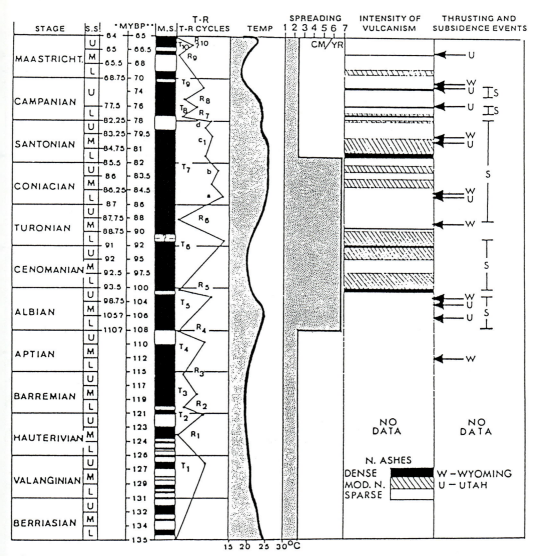

Figure 5. Albian-Maastrichtian tectonic history of the Western Interior Basin, United States, compared to seafloor spreading and eustatic curves (modified from Kauffman (1984a). Columns from left to right are: Cretaceous stages; substages (SS); Western Interior (*) and Van Hinte (**) radiometric scales (add 2.5 % for new decay constants) (MYBP); Paleomagnetic standard (MS); Kauffman's global epicontinental sealevel curve (1977) (T-R CYCLES); an average Cretaceous Temperature curve (TEMP.); a generalized half-spreading rate curve; a measure of intensity of volcanism in the Western Interior Basin (based on number and volume of bentonites), with solid dark lines representing most intense, cross-striped areas moderately intense, and white areas least intense volcanic intervals; and in the far right column, arrows indicating initiation and main movement on Wyoming-Montana-Idao thrusts (W) and Utah-Arizona thrusts (U), matched against times of rapid foreland and axial basin subsidence (vertical S bars). Note close correlation of most tectonic and volcanic activity to sealevel rise intervals.

surfaces for correlation and analysis of the timing of other events in
the basin, even during eustatic fall when they comprised only 15 percent
of the total number and volume of volcanic deposits. Further, many ash
falls had a direct effect on the biotas, causing mass mortality and, subsequently, the development of unique benthic colonizing surfaces; volcanic events were very closely linked to widespread biological events
in the Western Interior Cretaceous Seaway.

A second major external (allocyclic) forcing mechanism for biological response in the Western Interior Cretaceous Seaway was ocenaography,
and especially eustatic fluctuations directly linked to relative plate
tectonic activity and basin deformation, watermass movement (immigration
and emigration of watermasses, changes in current systems and their velocity, etc.), and density stratification due to thermal and chemical
(especially salinity) differences between watermasses; the latter
strongly affected oxygen levels in the lower water column and at the sea
floor, and thus the development of regional anoxic or dysaerobic events
(Kauffman 1984a, 1985b; papers in Pratt, Kauffman & Zelt, Eds. 1985).

Five orders of global tectonoeustacy have been documented in the
Western Interior Basin (see Kauffman 1985b, for discussion and references), following and logically extending the scheme of Vail et al.
(1977, 1978) (Fig. 6). First and second order cycles, defined by overall
Late Jurassic through Paleocene flooding of the North American Craton,
are too large in scale to be considered in event stratigraphy. The basic
eustatic-stratigraphic unit of the Western Interior Cretaceous is the
third-order cyclothem, averaging 9-10 Ma in duration (Fig. 5). Five such
cyclothems, of Late Albian-Middle Maastrichtian age, characterize the
sedimentary record of the entire seaway; additional, older Cretaceous
cyclothems defined by Caldwell (1984) are mainly known from northern and
central Canadian settings. Mid-basin third-order cyclothems have sub-
symmetrical transgressive-regressive facies suites, whereas marginal
basin cyclothems are asymmetrical preserving only regressive facies abov
disconformities. Global event correlation of third-order cycles is restricted to the points of maximum eustatic rise and maximum lowstand
(sequence boundaries), as determined from regionally correlative peaks
of deepening or transgression and shallowing or regression of epicontinental and shelf seas; large scale diachronism marks all transgressive-
regressive history between these points. "Peak" third-order eustatic
rise or fall events (sequence boundaries) are commonly marked by slow
rates of reversal or stillstand intervals spanning tens of thousands of
years, and the precise time of eustatic reversal may be difficult to
define in the stratigraphic record. Fourth-order eustatic cycles,
averaging 0.5-3 Ma in duration, are more useful in regional event stra-

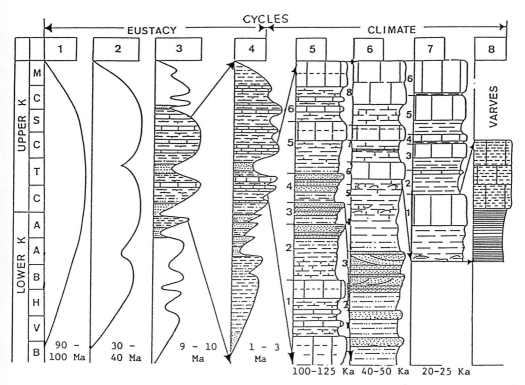

Figure 6. Generalized classification of eustatic and climatic cycles for the Cretaceous, showing average durations for the North American record. Scheme follows classification of Vail et al. (1977, 1978), as modified and extended by Kauffman (1985b). Fifth order cyclicity may be either eustatic or climatic, or both. Climate cycles approximate Milankovitch cyclicity. Typical Western Interior Cretaceous facies patterns utilized to show cycles in this diagram.

tigraphy because of their narrower time constraints, but may be more difficult to precisely correlate globally because they are more subtly defined in stratigraphic sequences. Criteria for recognition of third-order eustatic cycles and their peak reversals (event units) are closely similar for those utilized for the definition of fourth-order cycles in the Western Interior Basin; only the scale is fundamentally different. Fourth-order cyclothems may be caused either by small scale, short term eustatic fluctuations or by stillstand events within eustatic rise or fall sequences.

A fifth-order of eustatically regulated cyclothems averaging 100-125 Ta in duration has been proposed (Kauffman 1985b) and would incorporate glacioeustatic fluctuations which are so useful in global Pleistocene correlation. During the Cretaceous, however, without permanent ice at the poles, depositional cycles at this scale have been recognized primarily as small scale reversals of the strand line producing regionally correlative progradational clastic sequences 1-25 m thick, within very short time intervals (100 Ka or less). These are possibly related to short term eustatic stillstand events, but are also similar to climatically-controlled sedimentary cyclothems reflecting wet-dry Milankovitch cycles. In this hypothesis, increased weathering erosion and run off during wet cycles would significantly increase coarser clastic sediment input at the shoreline, causing nearly synchronous regional progradation of strand sequences. Wet periods would also significantly increase silt and clay input to more offshore facies, producing large-scale (0.5-3 m) rhythmites (limestone-calcareous shale; clay shale-silty shale; shale-siltstone or fine sandstone bedding alternations). Whereas such small cyclothems and bio-events associated with them are an important component of regional event stratigraphy within individual basins like the Western Interior Cretaceous Seaway, they are difficult to correlate between basins of the world without other event markers as reference because of their similarity to one another through time.

Sequence boundaries representing eustatic highstand/lowstand peaks provide, along with magnetostratigraphy (Fig. 5) and impact deposits, the most reliable and commonly represented means of truly global event-stratigraphic correlation. Eustatic fluctuations further comprise one major factor affecting the development of regional to global bioevents. Recognition of eustatic fluctuations of all lower levels (third- through fifth-order cyles) is thus critical to event stratigraphy, and may be done in epicontinental basins using regionally correlative fluctuations of marine strandline deposits and/or paleobathymetric fluctuations determined from sedimentary cyclothems.

Eustatic fluctuations strongly influenced both oceanographic and climatic trends during the Cretaceous, producing many of the short term physical, chemical, and biological events that comprise the major components of high-resolution event stratigraphy in North America. Three major types of oceanographic phenomena produced synchronous to short-term, stratigraphically recorded events in the Western Interior Seaway: (1) Anoxic or dysaerobic events reflecting rapidly induced intervals of water stratification, or immigration of oxygen minimum zones (Fig. 7); (2) rapid ventilation (oxygenation) and erosion/bypass events reflecting giant storm intervals or rapid changes in current regime with changes in basin topography or water depth; and (3) rapid changes in water temperature associated with major climate shifts or relatively abrupt immigration and emigration of surface watermasses from the Caribbean across

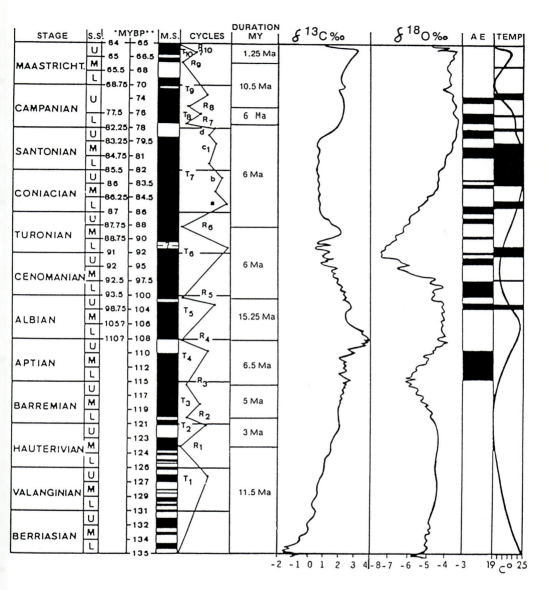

Figure 7. Albian-Maastrichtian oceanographic history of the Western Interior Seaway, North America. Plotting base same as in Fig. 5. Generalized oceanic and epicontinental changes in $\delta^{13}C$ and $\delta^{18}O$ ratios averaged in central two columns; AE column shows, in black, the most prominent levels of organic carbon enrichment, defining dysaerobic and anoxic events in the seaway; TEMP column shows generalized ocean temperature curve and, in black, intervals of abrupt regional warming of the seaway, biologically and chemically defined. Note strong correlation of dysaerobic/anoxic and abrupt thermal warming events with eustatic rise and peak highstand, represented by regional transgression in the Western Interior Seaway and tectonic deepening of the basin.

the partially silled southern aperture of the seaway (Fig. 7). The last strongly influenced plankton and pelagic carbonate production.

Regional anoxic or dysaerobic events in the Western Interior Seaway were very numerous, and a few appear to be world-wide in extent (i.e. the Bonarelli Event; Cenomanian-Turonian boundary interval) (Fig. 7). They are normally represented by finely laminated, non- to sparsely bio-turbated, pyritic and organic carbon-rich shales with depleted benthic faunas. Kauffman (1984a) noted that all major Cretaceous anoxic events in the Western Interior seaway occurred during intervals of eustatic rise, peak rise, and earliest eustatic fall, when the basin was being extensively flooded and, at the same time, was rapidly subsiding in re-sponse to tectonic loading and increased subduction under the western edge of North America. Thus, there exists a close relationship between the frequency and intensity of regional anoxic and resultant biological events, and paleobathymetry of the seaway favoring development of a stratified water column. Rapid basin filling during tectonically more passive intervals associated with eustatic fall soon brought bottom waters into the mixing zone, disrupting stratification and benthic to midwater oxygen depletion. For many eustatic cycles, there seem to be four major intervals during which anoxic events occurred (Fig. 8): (a) early in eustatic rise due to salinity stratification in deeply embayed   shallow seas with high internal freshwater runoff; (b) middle to late transgression and early regression at times of therma stratification of the seaway between warm southern and cool northern currents; and (c) during incursion of oceanic oxygen-minimum zones near peak eustatic rise and global warming (Fig. 7). Each "anoxic event" was a composite of rapidly fluctuating anoxic and dysaerobic intervals that reflected rapid changes in climatic or oceanographic forcing mechanisms and that abruptly placed high stress on the resident biotas, causing diverse bioevents.

Regionally, the onset of these anoxic events appears to have been very rapid to nearly sychronous when compared to biostratigraphic and other event-stratigraphic units (e.g. ash falls). This suggests that their forcing mechanisms were very pervasive, as in the case of immi-grating oxygen minima zones. It may also suggest that Interior Cretaceou seas in North America and elsewhere were already chemically stressed -- broadly dysaerobic, thermally and/or chemically stratified, and poorly circulated -- to the extent that relatively small changes in the degree of density stratification, water chemistry, temperature, or other factor may have driven the marine system rapidly to the anoxic state over broad areas. The predominance of low diversity, low equitability benthic com-munities dominated by eurytopic taxa like the Inoceramidae (Bivalvia)

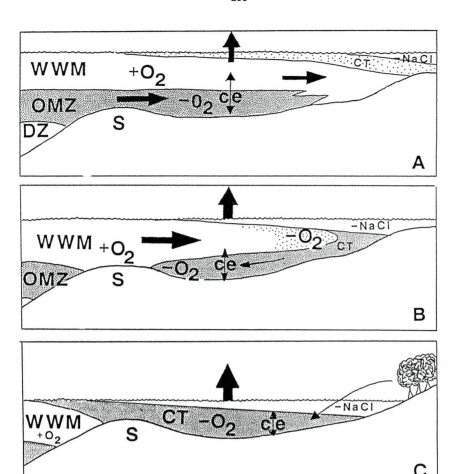

Figure 8. Models of the development of dysaerobic to anoxic events, and related bioevents, in the Western Interior Seaway associated with eustatic rise (<u>A</u>, <u>B</u>), peak rise and regional transgression (<u>C</u>) and earliest eustatic fall (B). Key: <u>CE</u> - fluctuating anoxic boundary during climate cycles; <u>CT</u> - Cool Temperate northern water mass; DZ - Dysaerobic bottom waters below oxygen minimum zone; ——<u>NACL</u> - subnormal (brackish) marine salinity layer forming cap on seaway during times of high internal runoff; $\pm O_2$ - normal (+) or depleted (-) oxygen in water column; <u>OMZ</u> - oxygen minimum zone in Caribbean and lower part of seaway during eustatic highstand; <u>WT</u> - Warm Temperate to Subtropical southern water mass; First dysaerobic to anoxic event (A) forms during early transgression in response to sluggish circulation in deeply embayed seaway and brackish water cap on seaway creating density stratification; terrestrial and marine organic material is plentiful and preserved in dysaerobic to anoxic waters below salinity stratified water column; (B) second type of anoxic event forms in response to broad mixing of warm southern and cool waters and establishment of thermal stratification and deoxygenation of lower water column late in eustatic rise and regional transgression, and again early in eustatic fall and regional regression; (C) The third type of anoxic event develops when the Caribbean oxygen minimum zone expands upward and floods northward into the Western Interior Basin where it comes in contact with the benthic zone.

in Cretaceous epicontinental seas and ocean basins suggests that this
might have been the case. Such chemically perched marine systems would
have greatly enhanced the development of all kinds of events, but especially bio-events relating to oxygen depletion, in the Cretaceous.

High-resolution stratigraphic analysis of Cretaceous oxygen-depletion
events shows that they were internally dynamic; finely laminated, organic-rich layers representing anoxic benthic conditions alternate with
dysaerobic intervals characterized by microbioturbated and, rarely,
macrobioturbated layers containing event colonization communities of
Chondrites and/or Planolites and/or thin (1-5 cm) intervals containing
abundant epifaunal bivalves (Inoceramidae, Pteriidae, Pectinidae,
Ostreidae) within a meter of section. Similarly, analyses of ^{18}O, ^{13}C,
and organic carbon through these oxygen-depleted intervals show rapid,
large-scale changes in thermal and chemical parameters (e.g. Pratt 1985),
further suggesting a dynamic system. Factors affecting small scale environmental oscillations of an already unstable lower water column include Milankovitch climate cycles, large storm perturbations, changes
in deep benthic circulation patterns, temperature fluctuations, variation in productivity of organic carbon at the surface and its subsequent
rate of accumulation in the benthic zone, etc. Each small-scale fluctuation of benthic oxygen produced a rapid regional imprint on the sedimentation, geochemistry and benthic biota of the seafloor, and thus
small-scale event surfaces or thin intervals with great potential in
regional correlation.

Anoxic or dysaerobic events and their biologic signature were terminated almost as rapidly as they began in the Western Interior Seaway by
recirculation events bringing increased oxygen to the seafloor. Slightly
coarser-grained, more bioturbated strata with abundant and diverse benthic molluscs (colonization bio-events) characterize recirculation
events. Commonly, detailed stratigraphic observations indicate a narrow
gradational interval of recovery to normal benthic conditions a few cm
thick -- still within the range of high-resolution event stratigraphic
data. Recirculation events may have had many causes: oxygen overturn,
giant storms, changes in current patterns and intensity, or rapid break-
down of water stratification in the seaway.

A final major oceanographic phenomenon controlling the development
of regional to global bio-events were rapid, widespread shifts in the
water temperature of shallow epicontinental seas. Data for such shifts
are drawn from isotopic analyses of ^{18}O (Pratt 1985) and from rapid biogeographic and stratigraphic shifts in the distribution of typical Tropical-Subtropical Cretacoues biotas (e.g. rudistid bivalves, Tethyan

ammonites, nerineid and actaeonellid gastropods, thick shelled echinoids,
diverse planktonic foraminifera and nannoplankton, etc.). Kauffman
(1984a) documented several such events in the Western Interior Seaway
during Late Albian - Middle Maastrichtian time, mainly associated with
peak eustatic highstand, epicontinental transgression, and rapid northward immigration of Caribbean watermasses and biotas as rising sealevel
breached tectonic-sedimentologic bariers at the southern end of the
Western Interior Seaway (Fig. 7). These warm water immigration intervals
variably lasted from a few hundred thousand to nearly two million years;
however, their initial rate of immigration was unusually rapid, based
on event-stratigraphic correlations, 100 Ka or less over as much as
1 000 -2 000 Km distance. Emigration intervals of Tropical biotas were
longer, up to 0.25 Ma. Rapid immigration and somewhat slower emigration
events, marked by abrupt changes between predominantly Tropical-Subtropical, and predominantly Temperate biotas, comprise important regional
and possibly global bio-event markers.

The last major allocyclic forcing mechanism for event-stratigraphic
phenomena in shallow epicontinental seas is climate. Climate forcing
produces two kinds of short-term biological and sedimentological responses: (1) Perturbations reflecting, for example, short-term intervals or extraordinary storm phenomena; and (2) cyclic sedimentation and
biotic response reflecting climate cycles, and especially Milankovitch
orbital cycles of (today), 21 Ka, 42 Ka and 100 Ka durations.

Barron & Washington (1982) recently modeled Cretaceous climates and
noted that, especially during eustatic highstands, it was probable that
Subtropical storm tracks shifted north and south so that they lay over
the southern portions of North America and Eurasia, and over North
Africa and northern South America, where they would have directly affected broad shallow epicontinental seas. The sedimentologic evidence for
large storms in the Western Interior seaway of North America is extensive. Proximal offshore and lower shoreface facies are characterized
by thick to thin storm beds with scoured bases, planar high-flow regime
basal beds overlain progressively by hummocky cross-stratified and
ripple laminated bedding with sharp clay drapes. High-resolution stratigraphic analyses of such sands, comparing them against volcanic ash
beds, suggest that some may be regionally correlated over a few hundred
kilometers parallel to the shoreline, and upt to 100 km offshore. Storm
events created bio-event horizons such as mass mortality-escape burrow
horizons at the base of the beds, transported mass mortality horizons
within them, and recolonization surfaces on top of them. Ongoing studies
show that storm beds commonly occur in bundles of several closely-

spaced units separated by intervals with sparse storm beds, suggesting possible control on their distribution by climate cycles, or at least climatically disturbed intervals.

Milankovitch and other climate cyclicity is variably expressed as bedding rhythms within fine-grained facies in epicontinental seas like the Western Interior, and has recently been analyzed by Baron, Arthur & Kauffman (1985). Calculation of sedimentation rates, and the number of chalk-marl, limestone-shale, etc. bedding rhythms between well-dated bentonite/ash layers strongly suggest three levels of cyclicity with approximately 20-25 Ka, 40-50 Ka, and 100-125 Ka durations during the Cretaceous. The last is the most prominent bedding interval, and the 40-50 Ka cycle the most consistently represented and commonly preserved interval in stratigraphic sequences. These intervals generally reflect the predicted Milankovitch cyclicity, and thus suggest alterinating dry and wet intervals, possibly associated with warm and cool climate fluctuations, respectively. Several types of Cretaceous sedimentary alternations thought to reflect Milankovitch cycles in the Western Interior Basin and elsewhere are shown in Figure 9, i.e. bedding alternations of (in dry phase - wet phase order): chalk-argillaceous chalk: chalk-marl; limestone - calcareous shale; calcarenite or calcisilt - calcareous shale; calcareous shale - clay shale; clay shale - silty shale: silty shale - siltstone: and possibly sandy shale - sandstone or storm bed bundling in nearshore facies. In general, these are asymmetrical bedding cycles with very sharp to narrowly transitional tops and upward grading facies from the base. Figure 10 models a typical limestone-shale Milankovitch cycle in which varied sedimentologic, biologic, and geochemical data suggest a climatic forcing mechanism; the example is taken from the Upper Cenomanian-Lower Turonian Bridge Creek Limestone Member, Greenhorn Formation, in Colorado. In these cycles, organic carbon increases in basal clay units, representing the most rapid sedimentation within the cycle, and is commonly associated with an unusual negative excursion of ^{18}O, low biotic diversity and low abundance. These data collectively suggest stratification of the water column by short-term emplacement of a brackish water lens on top of the seaway during wet climatic phases and increased internal runoff. Such a lens would inhibit normal populations of calcareous plankton, diluting carbonate input and the stratification would restrict downward movement of oxygenated waters to the seafloor, depleting benthic faunas and allowing increased accumulation of organic carbon. Benthic mass mortality bio-events brought on by the rapid onset of such conditions commonly mark the base of these cycles. Upward-grading of micro-facies representing improving

FACIES MODELS FOR CRETACEOUS MILANKOVITCH CYCLES

OFFSHORE ← ————————————————————————————— → ONSHORE

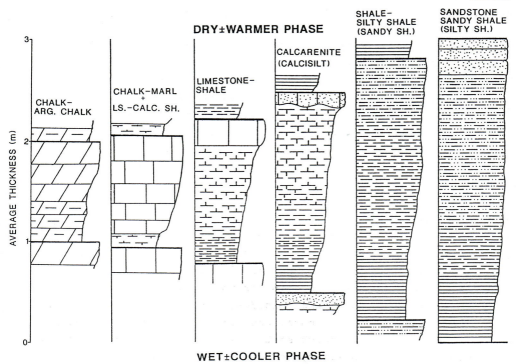

Figure 9.   Various lithologic expressions of 20-25 Ka, 40-50 Ka and 100-125 Ka bedding cycles thought to reflect Milankovitch climate cycles, in typical Western Interior Cretaceous facies.

marine conditions characterize the middle of each cycle. The top of each cycle has normal marine ^{18}O values, low organic carbon levels, high planktonic productivity and carbonate production, low sedimentation rates with little terriginous clay reaching the basin center, and normal marine benthic and planktonic biotas. These data depict oxygenation of the seafloor resulting from breakup of the brackish water lens and destruction of the density stratification in the water column. Event communities occupy the tops of such cycles within a few thousand years, over large areas.

In summary, the sedimentologic and biologic records of the Western Interior Cretaceous Seaway of North America are demonstrably controlled by regional and global allocyclic forcing mechanisms such as plate and regional tectonics, large scale volcanism, oceanography, and climate.

Physical, chemical, and biological events of many types result from allocyclic forcing, and are thus predictable on a regional and global scale. Whereas autocyclic forcing mechanisms - local tectonic, sedimentologic, oceanographic and climatic phenomena - also produce event-stratigraphic units, they are rarely extensive enough to be useful in regional chronostratigraphic correlations or in understanding the dynamics of regional to global bio-events. Event stratigraphy based on predictable, allocyclic influences on sedimentation is further enhanced by regional to global punctuational events - massive environmental perturbations related to impact by extraterrestrial objects, giant storms, giant volcanic explosions, and similar phenomena.

A wide variety of short-term to geologically instantaneous events of 1000 year or less duration result from allocyclic and punctuational phenomena, forming the basis for high-resolution event stratigraphy and, especially, the study of diverse bio-events on a regional (basin or continent-wide) scale. These include physical events (e.g. impacts, volcanic eruptions, rapid thrusting and synorogenic sedimentation intervals, storm and mass flow sedimentation); oceanographic events related to rapid immigration and emigration of watermasses, dysaerobic or anoxic intervals, overturn and mixing episodes, current distribution and velocity changes, desalination events, and short-term eustatic changes; chemical events sensed through stable isotope and elemental analyses reflecting changes in salinity, oxygen, trace elements, temperature, and the carbon cycle; climate events of both punctuated (storm events, temperature changes) and predictable (Milankovitch) aspect; and short-term biological events related to abrupt environmental changes produced by physical, chemical, thermal, and climatic events. These bio-events are discussed in detail, with Cretaceous examples from the Western Interior of North America, in the following pages.

Biological events

Short-term biological events are very common in the fossil record, but are overlooked for the most part because of the coarse resolution of stratigraphic observation that characterizes much field work even today. The observational format of high-resolution event stratigraphy, down to a cm or less for typical analyses in the Cretaceous, is well suited for detecting common biological event data and has led to great increases in the number and diversity of bio-events recognized in the last decade. Biological events are those in which some dynamic aspect of past life - rapidly occurring death, immigration, colonization, emigration, population size change, biofacies changes and punctuated evolutionary events

- is primarily responsible for the resulting stratigraphic deposit. Obviously, the differentiation of bio-events from certain physical and chemical events is not always easy, and they may grade into one another in, for example, mass mortality caused by storm events or oxygen overturn events in marine systems.

In high-resolution event stratigraphy, biological events are classified into three groups based on their regional extent and lateral persistence, and thus their use in precise correlation: (1) **Local Bio-events** which effect only portions of a sedimentary basin and are thus of limited value in correlations exceeding 100 km; (2) **Regional Bio-events** which effect most or all of a large sedimentary basin, or which extend between basins within and peripheral to a major craton; and (3) **Global to Intercontinental Bio-events** with near-synchronous to synchronous expression in strata of two or more major regions (continents and/or ocean basins) of the world. Regional and intercontinental to global bio-events are important components of high-resolution event stratigraphic matrices. Local bio-events might result from diverse causes such as floods and their effect on marginal marine environments, isolation and stagnation of a lagoon by barrier bars, an outbreak of "Red Tide", or burial of benthic infauna by a turbidite deposit, etc. Regional and global bio-events have more pervasive environmental causes such as rapid basin-wide shifts in water chemistry and temperature resulting from oceanographic, tectonic, or regional climatic phenomena, volcanic explosions producing widespread atmospheric dust, giant storm events of hurricane or monsoon force, and the effects of extraterrestrial impacts on Earth. Regional and global bio-events are mainly produced by allocyclic forcing mechanisms, the origins of which are external to any specific sedimentary basin in which they influence stratigraphic response.

Another classification of bio-events in the ancient record relates to the type of deposit and its causes; this is a more dynamic and potentially detailed approach to the discussion of biological events. The following types of bio-events are common in the Phanerozoic record, and are the major biological components of high-resolution Cretaceous stratigraphic systems developed in the Western Interior of North America, from which typical examples are taken.

(1) **Punctuated evolutionary events**, both at the character and the species/subspecies level, that occur among taxa with rapid dispersal mechanisms (e.g. mobile adult stages or planktotrophic larvae) result in the near-isochronous distribution and first appearance of key biostratigraphic taxa. Outstanding examples occur among Cretaceous ammo-

nites, inoceramid bivalves, and calcareous microplankton on both a regional and global scale. For example, in the latest Cenomanian and Early Turonian of Europe, America, North Africa, and elsewhere, the first appearance, species level evolution and near-global distribution of the inoceramid bivalve genus Mytiloides, and the successive species M. submytiloides (Seitz), M. opalensis (Böse), M. mytiloides (Mantell) and subsequently, M. labiatus (Schlotheim) seem to occur each within a few thousands of years. Similarly, the evolution and dispersal of the cosmopolitan ammonite Mammites nodosoides, the planktonic foraminifer Praeglobotruncana helvetica, and the calcareous nannofossil Quadrum gartneri, seem to represent widespread punctuated evolutionary bio-events during the Early Turonian Stage.

(2) Population bursts, especially among planktonic and nektonic taxa, occur over broad areas in response to rapid development of highly favorable watermass conditions, e.g.: (a) a temperature change that crosses some critical threshold level for food production and/or reproductive activity; (b) rapid oxygenation following overturn of a formerly anoxic basin; (c) changes in salinity in response to regional climate shifts favoring greater evaporation or rainfall which quickly attain optimal levels for certain species; or (d) biological responses to short term climate cycles. These may be well-defined by abrupt increases followed by abrupt decreases in population size and abundance of certain taxa, in some cases reflected in lithologic changes (e.g. from calcareous shale to pelagic limestone during population bursts of calcareous plankton species). In the Western Interior of North America, Upper Cenomanian strata of the Lincoln and Hartland members, Greenhorn Formation, contain abundant varves around 0.5-2 mm thick capped by very thin light gray bands composed of one or a few species of coccoliths that probably represent short term population bursts in response to very short climate cycles (10 years?); these strata also contain regionally persistent layers up to a few mm thick of globigeriniid foraminiferal sand which probably represent population bursts among single species of planktonic foraminifera. At best, such event deposits are of regional (basinal) scale; similar bursts may occur at different times between basins with the variation in response time being mainly controlled by the oceanographic dynamics of the basin itself. Acmezones, or epiboles based on these population bursts similarly are best applied only to regional or basinal biostratigraphic correlation.

(3) Productivity events are an extension of (2) above, but involve synchronous increase in population size of, mainly, diverse planktonic and

nektonic taxa. Major changes in lithology may result from such events. A common example of productivity events is found in the cyclic alternation of shale and calcareous shale, shale and limestone, and marl and chalk facies in Cretaceous Milankovitch climate cycle deposits (Barron et al. 1985, Fisher et al. 1985). Figure 10 provides a model of these cycles in the Western Interior Cretaceous of North America, based mainly on data from the Upper Cenomanian-Lower Turonian Bridge Creek Limestone, and the Lower Coniacian Fort Hays Limestone. Dark, laminated, organic-rich shales and calcareous shales near the base of each small bedding cycle contain little or no body and trace fauna, but high organic carbon of pelagic origin. This represents the wet and possibly cool part of the climate cycle, high rates of weathering, erosion, and transport of clay to the basin center, relatively high sedimentation rates, and brackish water as a surface layer over much of the seaway. This greatly restricts calcareous plankton in the water column (Watkins 1985) but favors high productivity among opportunistic, organic-walled algae, dinoflagellates, etc. Salinity stratification in the water column restricts flow of oxygen to the benthic zone and enhances preservation of organic carbon, whereas the lack of benthic oxygen restricts macrofaunas to, at best, sparse low-oxygen tolerant bivalves (Inoceramidae) and detritus-feeding burrowers (Chondrites, Planolites). Low benthic diversity and population size results. As climatic conditions change through time toward the dry and possibly warmer part of the Milankovitch cycle, carbonate content increases, rates of (clay) sedimentation decrease, and salinity stratification breaks down allowing benthic oxygenation and proliferation of populations among diverse normal marine benthic taxa. A second productivity event within each cycle takes place near the top (pure chalk and limestone facies), under warm normal marine conditions which favor great proliferation of pelagic calcareous plankton, mainly coccoliths and planktonic foraminifera. Whereas the upward gradation of pelagic carbonate production prevents a clear event boundary from being drawn in some cases (but not all), and diagenetic enhancement of the cycles occurs (Ricken 1986), the upper contact between pelagic carbonate and black organic-rich shale facies in adjacent cycles, reflecting two different kinds of productivity events, is abrupt. These surfaces form persistent, productivity bio-event marker beds (e.g. Hattin 1971, 1985; Elder & Kirkland 1985; Fisher et al. 1985) with high-resolution potential for correlation over areas a few hundred to over a thousand kilometers in north-south and east-west extent.

Productivity bio-event deposits in the Cretaceous are also associated with upwelling regions, although most of these do not have great lateral

extent and thus have limited utility in high-resolution event stratigraphy. The initiation and termination of upwelling episodes, and
dramatic changes from low to high to low productivity deposits (especially among siliceous and organic-walled micro-plankton) can be very
rapid and thus stratigraphic contacts between such deposits have high
potential in bio-event correlation. Bedded radiolarian chert - laminated shale deposits of deep marine seetings may, in part, reflect such

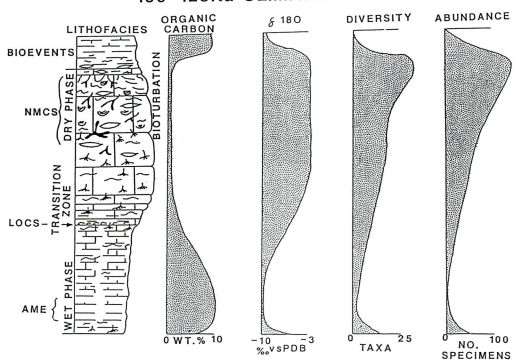

100-125Ka CLIMATIC CYCLICITY

Figure 10. Model of numerous chemical, biological and lithological
studies of Milankovitch climate cycle deposits (e.g. limestone-shale
rhythms), showing association of clay enrichment, low levels or no bioturbation, low benthic faunal diversity and abundance, and negative $\delta^{18}O$
values indicating a brackish water influence, as a surface layer, during
lower or wet (and cooler?) part of the cycle, changing upward to carbonate-enriched, slowly deposited, highly bioturbated, faunally rich
deposits formed in normal marine conditions (as per $\delta^{18}O$ values) during
dry (and warmer?) part of cycle. Various types of bioevents associated
with these cycles are coded as follows: AME - Anoxic mass mortality
event, LOCS - low oxygen event colonization surface during first break-
down of anoxic bottom conditions; NMCS - normal marine climate-cycle
regulated, colonization event surface on top of bedding rhythm. Modeled
from Lower Turonian Bridge Creek Limestone, and Lower Coniacian Fort
Hays Limestone, in the Western Interior Cretaceous Seaway.

upwelling-related productivity cycles throughout the Phanerozoic.

More extensive productivity events in shallow Cretaceous epicontinental seas are related to rapid watermass movements, and in particular
to rapid short-term immigration of warm waters from the Tethyan Realm
into epicontinental seas of America, Europe, and North Africa during the
Cretaceous (Kauffman 1984a, Eicher & Diner 1985). In the North American
Interior Seaway, warm Subtropical waters along the north edge of the
Caribbean Sea were held at bay during eustatic fall and early eustatic
rise by a structurally and sedimentologically defined sill or barrier.
This was rapidly breached during Late Albian, latest Cenomanian-Early
Turonian, Coniacian, Santonian, Early Campanian and Late Campanian
eustatic highstand events, abruptly introducing warm normal marine waters
northward for over 1000 km. Each such warm water incursion, in some cases
a series of smaller short term events, was characterized by very rapid
increase in pelagic carbonate production by plankton (Kauffman 1984a,
Eicher & Diner 1985). These highstand events lasted for thousands to a
few million years, producing thick intervals of foraminifer-coccolith
enriched strata; the initiation of these productivity events, when
plotted against widespread bentonite marker beds, seems to have taken
place across much of the basin within only a few thousands of years at
best, forming a valuable bio-event surface for regional correlation.
Within these carbonate-enriched intervals representing eustatic highstand, smaller scale productivity bio-events representing Milankovitch
climate cycles are especially well defined by shale-limestone and marl-
chalk bedding couplets, 1 m or less thick, of 20-100 Kyr duration
(Figs. 9, 10) (Fisher et al. 1985).

(4) Immigration and emigration events are bio-events characterized by the
rapid regional immigration and proliferation of new taxa throughout much
of a sedimentary basin in association with regional changes in watermass properties (temperature, chemistry, or circulation) and equally
rapid emigration of such taxa as conditions within the seaways returned
to their former state. In dated examples from the Western Interior Seaway of North America, such events take place within a few thousand to
100,000 years over regions exceeding 1,000-2,000 Km among taxa with
mobile adult and/or planktotrophic larval stages (Kauffman 1984a, Eicher
& Diner 1985, give examples); first and last appearances of these immi-
grant biotas may form valuable, nearly synchronous event stratigraphic
surfaces. Typical examples from the Western Interior Cretaceous, representing different causal mechanisms, are as follows: (a) During the Late
Albian extensive epicontinental seaways independently invaded the
Western Interior basin from the north (predominantly) and south, but

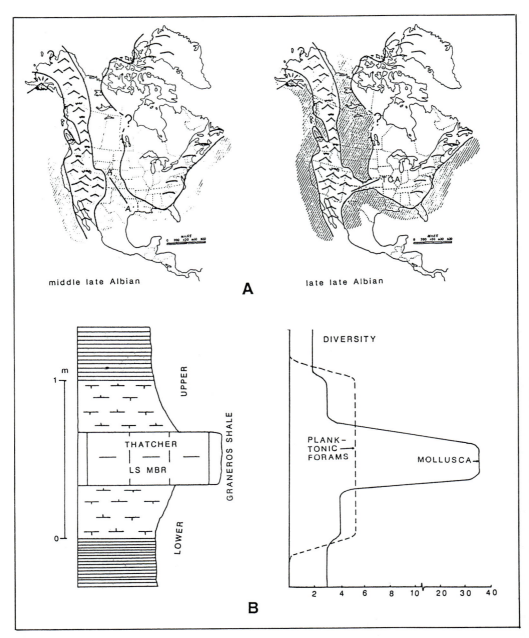

Figure 11.   A.- Paleogeographic maps showing pre- and post-breaching
of the Transcontinental Arch during Late Albian sealevel rise. Bar A-A'
denotes extent of rapid northward immigration and colonization by
southern Warm-Temperate taxa, and rapid southern immigration of Cool
North Temperate taxa, during brief connection (< 0.5 Ma). B. Stratigraphic distribution of planktonic foraminifer (PF) and warm temperate
molluscan diversity patterns during the Middle Cenomanian Thatcher
immigration-emigration bioevent. Data combined from Colorado and norther
New Mexico.

were separated by a tectonic barrier, the Trans-Continental Arch; this arch was breached during Late Albian eustatic highstand (Fig. 11) for approximately a half million years, allowing northward and southward migration of biotas from Subtropical-Warm Temperate and Cool Temperate paleobiogeographic provinces for hundreds of Km; migration was near-instantaneous in both directions, probably taking a few thousand years at most; Subtropical Gulf Coast ammonites (e.g. Engonoceras) and bivalves (Texigryphaea; Trigonia) reached northern Colorado within the same short interval that Boreal Inoceramidae reached southern Kansas, north Texas, and southern New Mexico. In both directions, these immigrations are manifest in fossils within a meter or less of strata; early regression associated with eustatic fall quickly terminated the marine connection.
(b) Another short-term bio-event, termed the Thatcher Bio-event by Kauffman (1985a), occurs within the lower Middle Cenomanian Calycoceras (Conlinoceras) gilberti Biozone in the Middle Graneros Shale and equivalents of New Mexico and Colorado. The interval probably spans no more than 10,000-25,000 years, and includes 0.25-1 m of calcareous shale containing a thin persistent limestone or concretion horizon. In contrast to low diversity Temperate molluscan faunas and rare to absent planktonic foraminifera in surrounding Graneros Shale, the Thatcher interval shows abrupt regional immigration and emigration of moderately diverse planktonic foraminifera and nearly 50 species of molluscs, including small rudistid bivalves, of Warm Temperate to Subtropical origin (Fig. 11). The immigration and emigration boundaries of this rapid warm water Subtropical watermass are essentially isochronous when compared to surrounding bentonite deposits (volcanic ash falls) over a distance of 500 to 750 Km along outcrop, and probably further south in the subsurface.
(c) Larger scale Cretaceous immigration-emigration bio-events in the Western Interior Cretaceous Basin have been documented by Kauffman (1984a) and Eicher & Diner (1985) in association with eustatic sealevel fluctuations. These authors have mapped rapid northward incursions of Warm Temperate and Subtropical molluscs and foraminifera as much as 2500 Km into the Western Interior Seaway from the Caribbean and Gulf Coast. This immigration was associated with peak eustatic highstand, breaching of the structural-sedimentologic sill at the southern end of the seaway, and rapid northward movement of warm Caribbean water masses into the basin. Figure 12 compares biogeographic distributions of warm water biotas of the Caribbean Province and Southern Interior Subprovince during eustatic lowstand (A) with peak Lower Turonian (B) and Coniacian-Santonian (C) eustatic highstands. Although the immigrations of these watermasses and biotas are somewhat diachronous over long distances, they

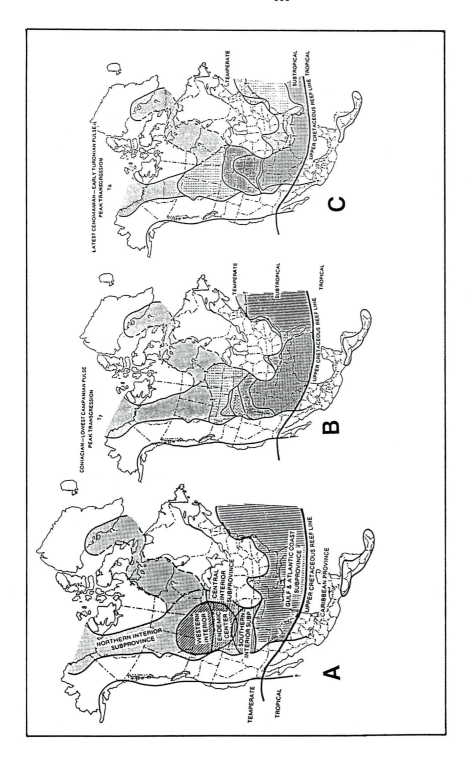

Figure 12. Paleobiogeographic maps showing the comparison of (A) average distribution of cool, mild, and warm temperate subprovinces during mid to Late regression, and early to mid-transgression of the Western Interior Seaway, and (B) (C) the rapid northward immigration of Subtropical taxa into the Temperate zone during Early Coniacian (B) and Early Turonian (C) eustatic highstand. From Kauffman (1984a).

appear to represent essentially isochronous event surfaces over hundreds of Km and are further characterized by abrupt change from non- or slightly calcareous to highly calcareous strata.

(5) Ecostratigraphic events are characterized by abrupt widespread changes in the community structure of one pervasive facies caused by regional and/or allocyclic forcing mechanisms (oceanography, climate, regional volcanism, etc.), as well as by evolutionary and biogeographic changes in community structure (e.g. introduction of a new, more efficient competitor).

Outstanding examples of ecostratigraphic events, in addition to those previously cited for immigration and emigration events, are the dynamic community changes associated with the onset and demise of regional to global "oceanic anoxic events (OAE)". These oxygen depletion events result from basin-wide water stratification (Kauffman 1984a, 1985b; Sageman 1985) and/or the incursion of oxygen minima zones into cratonic epicontinental seas (e.g. Frush & Eicher 1975, Eicher & Diner 1985, Kauffman 1984a). Anoxic or dysaerobic events mainly occur during eustatic rise and a peak eustatic highstand (e.g. the Bonarelli event and its global expression at the Cenomanian-Turonian boundary) (Kauffman 1984a, Sageman 1985) in epicontinental settings of North America and Europe, as do larger-scale global anoxic events (Jenkyns 1980). Elder (1985) and Kauffman (1984b) provided detailed taxonomic and biostratigraphic data (Fig. 13) showing abrupt regional change in the molluscan faunas at initiation of the Bonarelli Global Anoxic Event in the Western Interior Basin of North America. Highly diverse molluscan faunas of the global Sciponoceras gracile Biozone, totaling nearly 175 species in the Western Interior Basin (Koch 1980, Kauffman 1984b, and subsequent data) and representing normal marine benthic conditions in epicontinental seas, abruptly change to low diversity (N = 5-30 species), inequitable communities dominated by low-oxygen tolerant Inoceramidae (Bivalvia) and rare other molluscs with initiation of the Bonarelli OAE (Elder 1985, Kauffman 1984b). Event-stratigraphic ash markers and biostratigraphy indicate that this ecostratigraphic event took place almost simultaneously across the Western Interior Basin and in Western Europe. Termination of the Bonarelli OAE abruptly gave rise to, first, population bursts of the inoceramid bivalve Mytiloides, and secondly to moderately diverse benthic and pelagic biotas of the Mammites nodosoides biozone (Fig. 13, Kauffman 1984a). This pair of regional ecostratigraphic events was associated with withdrawal of the oxygen minimum zone from epicontinental seas recirculation and oxygenation of the benthic zone followed by abrupt warming of shallow seas with peak eustatic highstand in the middle Early Turonian. Among Foraminifera, Eicher & Diner (1985, Fig. 1) have noted a regionally correlative, near-synchronous, benthic-foraminiferal population burst and diversity bio-event in the latest Cenomanian (Sciponoceras gracile Biozone), which abruptly terminated just below the Cenomanian-Turonian extinction boundary and onset of the Bonarelli OAE. This, and the previously cited Thatcher planktonic foraminiferal event (see Immigration-Emigration Bio-events) comprise typical ecostratigraphic bio-events among the microplankton in the Western Interior Basin. For other examples the interested reader is referred to papers in Pratt, Kauffman & Zelt (Eds. 1985) for the Western Interior Basin of North America. Ecostratigraphic events clearly merge with, and are characteristic of, other types of regional to global bio-events.

(6) Regional colonization bioevents refer to near-simulataneous coloni-zation of the benthic zone by one or related communities as a result of a rapid regional change in benthic environments usually initiated by allocyclic forcing mechanisms. Three major types of colonization events characterize Cretaceous epicontinental seas like that of the Western Interior of North America. (a) Oxygen-related colonization events: regional stagnation followed by recirculation of oxygen in the benthic zone caused widespread initial colonization by low-oxygen-tolerant, opportunistic edge species among bivalves like the Inoceramidae (or Jurassic Posidonia and Bositra: Kauffman 1981) during brief dysaerobic intervals bounding benthic anoxic intervals. These may be asociated with both global (Elder 1985, Kauffman 1984a) and regional anoxic event

Figure 13.   Composite plot of molluscan species ranges in the Western Interior United States having last occurrences in the Cenomanian-Turoni boundary interval. Data are plotted against the Western Interior standa reference section (left) at Rock Canyon Anticline near Pueblo, Colorado and δ ^{13}C organic and δ ^{18}O carbonate versus PDB values from the same section (after Pratt 1985). Species ranges are a graphic composite of data from 10 widely scattered stratigraphic sections analyzed by Elder (in progress), and from published range data of Koch (1975, 1980), Cobban and Scott (1972) and Cobban (1985). Biostratigraphic zones and subzones, in ascending order, are: Vascoceras diartianum (V. d.) and Euomphaloceras septemseriatum (E. s.) subzones of the Sciponoceras gra-cile Biozone (SCIP); the Vascoceras gamai (V. g.), Neocardioceras juddi (N. j.), and Nigericeras scotti (N. s.) subzones of the Neocardioceras juddi Biozone (NEOCARD.); the Pseudaspidoceras flexuosum (P. f.) and Vascoceras birchbyi (V. b.) subzones of the Watinoceras Biozone; the Mammites nodosoides Biozone; and the Collignoniceras woollgari (C. w.) Biozone. Numbered horizontal lines or intervals mark the stepwise extinctions noted in the text. Approximately 15 additional, mostly in-faunal taxa (not plotted) are known to range through the entire inter-val; and numerous other species are either too rare or their taxonomy is too poorly known to accurately determine their ranges.

309

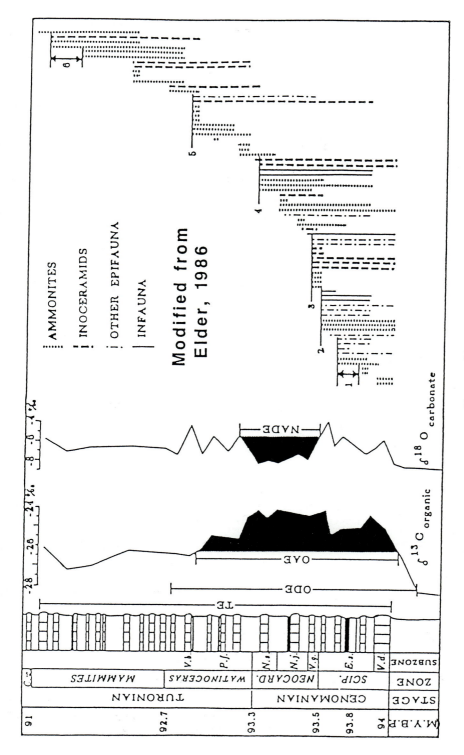

Figure 13

for example the Hartland Anoxic Event of the Western Interior Cretaceous Basin (Sageman 1985). Similarly, the environmental expression of global Milankovitch climate cycles of 20-25,000, 40-45,000, and 100-125,000 year duration in epicontinental basins involves anaerobic to dysaerobic benthic conditions during wet phases due to resultant density strati- fication and emplacement of temporary fresh to brackish water lenses on top of the seaway. These low-oxygen intervals are abruptly followed by reoxygenation of the benthic zone and normal marine conditions during dry phases of these cycles. Three basic colonization events are common- ly associated with each regionally expressed Milankovitch cycle (Fig. 14): an inoceramid bivalve-dominated epibenthic assemblage without trace fossils during lowest oxygen conditions (dysaerobic) at or near the base of each cycle; a mixed inoceramid bivalve, epibenthic mollusc, detritus- feeding trace fossil assemblage (e.g. <u>Chrondites</u>, <u>Planolites</u>) during initial re-oxygenation phases in mid-cycle; and a diverse mollusc - trace fossil - benthic foraminifer assemblage during dry phases at the

COLONIZATION AND MORTALITY BIO-EVENTS

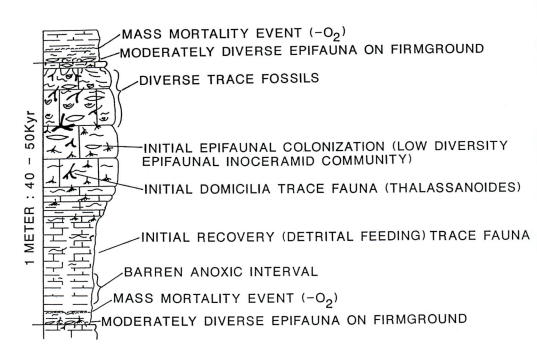

Figure 14. Model of a typical shale-limestone Milankovitch climate cycle showing distribution of various regional bioevent surfaces.

top of each sedimentary cycle (e.g. limestone-shale bedding rhythms; see Barron et al. 1985). These colonization events can be traced at a small scale over hundreds of kilometers in the Western Interior Cretaceous Basin of North America, especially the diverse assemblages marking the tops of each Milankovitch bedding cycle.

(b) Event-sedimentation related colonization surfaces: in many cases benthic biotas are regionally excluded from habitation of substrates by chemistry at the sediment-water interface (e.g. excessive H_2S seepage, low Oxygen, etc.). Locally, storm deposits and turbidites provide short-term, physically and chemically favorable substrates for essentially synchronous colonization of the benthic zone, providing a basis for local bio-event correlation, but regionally, bio-event correlations can be established utilizing such colonization surfaces related to volcanic ash layers (represented by bentonites). Preliminary studies of Cenomanian strata from the Western Interior show that large ash falls, extending over thousands of square Km in the seaway, commonly have short-lived, diverse benthic foraminifer and mollusc communities developed on top of them over large regions because the ashes temporarily seal off the seepage of toxic pore waters from underlaying organic-rich sediments to the sediment-water interface. Figure 15 shows a typical example from the middle Graneros Shale (Middle Cenomanian) of the Colorado Front Range outcrop belt. Some of the most outstanding examples of this phenomenon are the _Ostrea_ _beloiti_ biostromal beds associated with the regional X-Bentonite Marker Bed (Hattin 1965, Sageman & Johnson 1985). _Ostrea beloiti_ appears abruptly in the Western Interior Seaway in association with the boundary between the Middle Cenomanian Graneros Shale and Lincoln Limestone Member of the Greenhorn Formation, initiation of calcareous (planktonic) sedimentation, and deposition of ash related to large-scale explosive volcanism producing the regional X-Bentonite marker bed; the oysters form extensive lenticular biostromes primarily in strata directly or closely overlying the X-Bentonite and stabilization of the substrate related to it. The biostromes extend as a near-isochronous zone from Wyoming and South Dakota south to the Texas-Mexican border, possibly farther, providing one of the best examples of a colonization bio-event surface in the North American Cretaceous.

(c) A third common type of colonization surface is related to change in regional background sedimentation over large areas of an epiconti-nental basin. The physical characteristics of volcanic ash falls (see above) can obviously produce such a regional effect and, coupled with the role of volcanic ash in sealing off detrimental benthic chemistry at the sediment-water interface,can provide a significant shift in

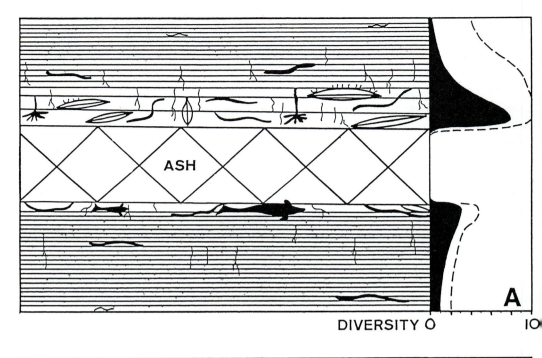

DIVERSITY O   IO

A

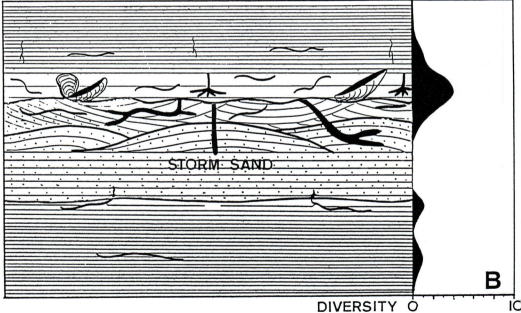

DIVERSITY O   IO

B

Figure 15.   Model of (<u>A</u>) volcanic ash fall and (<u>B</u>) regional storm event
showing regional mass mortality and recolonization bio-events associated
with these short-term depositional events. Data from numerous Cenomanian-Turonian examples in Colorado, Western Interior Cretaceous Seaway.

physical substrates. Another common example is related to fluctuations in clay input related to pelagic carbonate (foraminifera, coccolith) production, for example in marine varves and climate cycle deposits which result in alternating limestone and shale, or chalk and marl deposits over a major portion of a sedimentary basin. Calcisilts and foraminiferal sands (calcarenites) produced by these events result in more stable colonization surfaces, and allow widespread shifts in short-term event communities from soft organic-rich substrate-adapted taxa, to those dependent on firm substrates (byssate and cemented bivalves, trace fossil taxa building permanent domicilia, etc.). Kauffman (in Barron et al. 1985) provides an example from limestone-shale Milanko-vitch climate cycle deposits in the Western Interior Cretaceous Seaway in which large inoceramid bivalves, cemented oysters, and domicile burrows are primarily associated with winnowed pelagic carbonates on slow sedimentation or bypass surfaces at the top of the cycle (purest limestone, dry climate phase, slowest sedimentation). Figure 14 shows a more generalized model. Where such surfaces actually show interrup-tions in sedimentation (regional bypass) and early cementation, forming firmgrounds, borings by worms and lithophagid bivalves may also be associated with these event-colonization-communities which extend, like the climate cycles themselves, over thousands of square kilometers.

A related kind of colonization event surface is associated with abrupt regional shifts from clay to bioclastic benthic substrates during small-scale eustatic fluctuations and/or regional changes in benthic current patterns. Increased benthic current energy during these inter-vals winnow clays over broad areas, leaving behind foraminiferal-mollusc shell fragment sands and silts. Sageman (1985) and Sageman & Johnson (1985) give examples from the Middle and Upper Cenomanian, Lincoln and Hartland members of the Greenhorn Formation in the Western Interior. Such sediments become short-term event colonization surfaces for byssate inoceramid, pteriid, pectinid, and cemented ostreid bivalves, for a few tens to hundreds of years, over very large areas of the Cretaceous seafloor. These form regional bio-event correlation surfaces.

(7) **Mass mortality bio-events** refer to regional isochronous surfaces characterized by mass mortality (but not necessarily extinction) deposits of benthic or pelagic taxa in the fossil record, usually reflecting some regional perturbation or allocyclic causal mechanism. Possible causes in shallow marine epicontinental seas are: (a) Large volcanic ash falls with both their chemical and physical effects; (b) oxygen overturn events as stagnant water moves through the water column; (c) rapidly induced oxygen restriction due to short-term density strati-

fication events, or incursion of oxygen minimum zones from oceanic sources; (d) rapid shifts in light and temperture related to atmospheric dust clouds of volcanic or impact origin; (e) giant storm events that tear up shallow water benthic communities and form extensive shoreface and downslope shell deposits; (f) desalination and hypersalinity events that result from abnormally wet and dry climatic intervals, respectively; and (g) disease or "red tide" intervals.

The Western Interior Cretaceous sequence of North America contains some outstanding examples of regional mass mortality bio-event deposits with high correlation potential. Several thick Lower and Middle Cenomanian ash (bentonite) deposits have been sampled within the basal cm of the ash and immediately underlying shales; macro- and microfaunal content, when compared to background biofacies, suggest a modest increase in planktonic foraminifers and fish scales - a pelagic mass mortality signal - and termination of benthic molluscan communities over very large areas, followed by unfossiliferous ash-enriched sediment (Fig. 15 provides a summary model). Near the Albian-Cenomanian boundary, from northern Alberta, Canada to central Colorado, a remarkable mass mortality deposit known as the "Fish Scale Marker Bed" occurs in the same stratigraphic position in virtually all sections, suggesting a regional oxygen overturn bio-event within the anoxic to dysaerobic Mowry Sea. The abrupt extinction of 75-80 percent of Cenomanian molluscan species at the top of the Sciponoceras gracile biozone in the Western Interior (Koch 1975; Kauffman 1984a, b; Elder 1985) is largely a result of rapid incursion of the oceanic oxygen minimum zone from the Caribbean into the Western Interior Seaway and, in part, resultant mass mortality as this watermass first reached broad areas of the benthic zone (Fig. 13). These are but a few of many examples of regional mass mortality beds, forming valuable bio-event correlation units, that have been noted in our high-resolution analyses of Western Interior Cretaceous strata.

(8) **Extinction bio-events** refer to synchronous or short-lived, regional to global extinctions of specific organisms (background extinction events), or in the case of mass extinction, encompassing 50 or more percent of the existing biota worldwide. Examples are the cyclic mass extinctions recognized by Raup & Sepkoski (1984, 1986); most abrupt extinction bio-events, however, are of lesser magnitude, and have been widely used in biostratigraphy. Many examples could be given where both event and biostratigraphic data suggest that certain individual taxa or taxa-set extinction events are very short lived or essentially synchronous worldwide, e.g. the global extinction of the planktonic foraminifer Rotalipora and, later, Sciponoceras gracile (ammonite) and

Inoceramus _pictus_ (bivalve) in the Late Cenomanian (Elder 1985, Eicher & Diner 1985, Leckie 1985) associated with abrupt changes in temperature, water chemistry, and marine oxygen leading up to the Cenomanian-Turonian boundary (Arthur, Schlanger & Jenkyns 1985, Pratt 1985, Elder 1985). Ecological sensitivity and limited adaptive range of diverse Cretaceous taxa during warm, equable intervals associated with eustatic highstand, caused them to be highly susceptible to extinction during even small regional or global perturbations in light, temperature, chemistry, and niche parameters. These conditions commonly lead to development of nearly synchronous extinction bio-event horizons coincident with regional biostratigraphic zone boundaries in the Western Interior Cretaceous Seaway, as demonstrated by regional comparison of extinction levels with isochronous event stratigraphic deposits (e.g. volcanic ash and climate cycle deposits (Hattin 1971, 1985; Elder & Kirkland 1985, for the Cenomanian-Turonian boundary interval utilizing graphic correlation). Clearly, other taxa extinctions can be shown to be diachronous using the same test, and it is logical to assume that most taxa have somewhat dia-chronous extinction records due to varying intensity of causes of ex-tinction, varying resistance among populations to those causes, varying longevity of favored environments from place to place, differential preservation, and the existence of refugia for long periods of time following extinction of a taxon (taxa) over most of its preferred range.

Mass extinctions, as defined by Kauffman (1984a, 1985c, in manu-script) are of greatest interest to us as global bio-events and major tools of correlation, and have received considerable attention in recent years. Mass extinctions involve the loss of 50 or more percent of ecologically and genetically diverse taxa, worldwide, within a narrow interval of geological time (4 Myr or less, Kauffman 1984a, 1985c). Some contain, within their history, isochronous, catastrophic bio-events which form unparalleled global correlation surfaces. As reported in the literature, the Frasnian-Famennien (Devonian) invertebrate ex-tinction (McLaren 1985), and the extinction of calcareous microplankton and a number of other typical Cretaceous invertebrates associated with the bold impact at the Cretaceous-Tertiary boundary (Alvarez et al. 1984) represent such synchronous global extinction events.

There is a tendency among workers today to regard all such mass extinctions as synchronous global bio-events with singly driving mechanisms; i.e. to regard them as catastrophes. Whereas this paper is not the proper format for a review of either the history of this debate between catastrophists and gradualists on extinction, nor the myriad proposed causes and theories of mass extinction on a regional

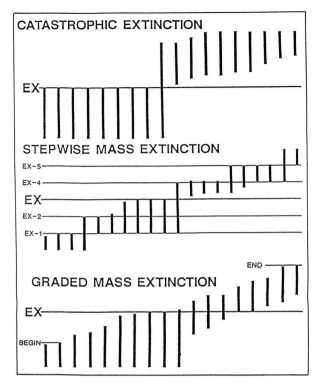

Figure 16.   Models of extinction patterns of taxa (vertical lines) within the framework of the three major theories for mass extinction. EX represents the level of either maximum taxonomic loss or of the extinction level of certain characteristic taxa (e.g. biostratigraphic. indices). EX-1, EX-2, etc. define levels of stepwise extinction as components of a single mass extinction event (Kauffman, in manuscript).

to global scale (see papers in Science and Nature over the past decade: in Christensen & Birkelund 1979, in Geol. Soc. America Special Paper 190, 1982, etc.) it is appropriate to review the main theories, and evidence, for the dynamics of mass extinction and associated regional to global bio-events from stratigraphically well studied mass extinction intervals.

Three concepts of mass extinctions now exist in the literature (Fig. 16): (a) **Graded mass extinction** theory suggests that during mass extinction, the average rate of extinction is significantly higher than normal background rates, but the disappearances of individual taxa are evenly to randomly spread through this interval. Graded mass extinction may proceed along an ecological gradient from initially more highly stenotopic groups to more eurytopic groups. This extinction pattern is ususally attributed to accelerated intervals of envrionmental change resulting from Earthbound causes (such as major eustatic fall of sea level, climate changes including greenhouse events, intense volcanic episodes, etc.) which produce global ecological crises. A corollary of the graded mass extinction concept is that mass extinctions involve the same processes as background extinctions, but at an accelerated rate. However, Jablonski (1986) has shown that the ecological characteristics of mass extinctions are very different from those of background extinctions.

(b) **Catastrophic mass extinction** theory views world-wide mass extinction among genetically and ecologically diverse organisms as occurring simultaneously, within days, months, or a few years, as a direct or indirect result of some enormous perturbation of global environments. Such events have been attributed to extraterrestrial causes, notably to asteroid or comet impacts (Alvarez et al. 1980).

(c) **Stepwise mass extinction** theory (Kauffman 1984b, c, and in Hutt et al. 1986, in review) proposes that mass extinction episodes occur as a series of discrete steps spread over 1-4 Myr; each step consists of a very short interval of highly accelerated or catastrophic extinction affecting only a portion of the global biota. In an idealized model of stepwise mass extinction, the steps are also ecologically graded, first affecting the most environmentally sensitive (stenotopic) and/or Tropical organisms, and proceeding through a series of intermediate steps involving progressively more tolerant groups, to the final extinction event in which many Temperate Zone, eurytopic, ecological generalists disappear, including surviving eurytopic species of once diverse, dominantly stenotopic groups.

In all of these hypotheses, ecological generalists and taxa living in protected habitats survive to seed the ecosystem recovery following a mass extinction event. The recovery follows a brief (0.05-0.5 Myr) interval of little evolution, characterized by low diversity among small-size, eurytopic taxa, and is itself characterized by rapid radiation of new biotas into the vacant ecospace formerly occupied by the victims of the mass extinction. A growing body of carefully collected paleontological evidence strongly supports the concept of stepwise mass extinction as the predominant pattern during biotic crisis on Earth.

Only the Cenomanian-Turonian (C-T), Cretaceous-Tertiary (K-T), and Eocene-Oligocene (E-O) boundaries provide detailed evidence for the evaluation of these theories at present, with the global data synthesized. In the following summaries, the reader is referred to listed references for detailed data and discussions, and to summary paper (Hut et al. 1986) currently being revised for publication in S c i e n c e .

The Cenomanian-Turonian mass extinction event. Raup & Sepkoski (1984, 1986) recognized the Cenomanian-Turonian boundary interval as a second-order mass extinction event based on their analyses of ordinal, familial and generic data. The long recognition by biostratigraphers that this stage boundary was one of the most clearly defined in the Cretaceous seems to support these observations. Collectively, estimates of 70-85 percent species loss across the boundary have been made from, primarily, North American data (Koch 1975, Kauffman 1984, Elder 1985 and in press). But there are two unusual aspects to this extinction event/boundary interval which make its study especially important. First, unlike most extinction events, it occurs during near-eustatic highstand with asso-ciated, broadly ameliorated warm global climates representing seemingly optimal conditions for life and evolutionary diversification. Secondly, the stage boundary is defined above the major extinction of typical Cenomanian taxa, suggesting a more complex history than implied by catastrophic extinction theory.

In recent years, Koch (1975), Kauffman (1984b) and especially Elder (1985, and in press) have attempted to document the detail of the Ceno-manian-Turonian mass extxinction event in North America. These environ-mental and paleontological data have recently been summarized in Hut et al. (1986, in review) and the following is a condensation of our collective observations and interpretations, with relevant references.

To summarize, the Cenomanian-Turonian mass extinction event occurred over a period of 2.5 Ma just prior to and during eustatic highstand of the Greenhorn Marine Cycle -- the greatest global sealevel rise of the Cretaceous (Kauffman 1984a). It is associated with several major en-vironmental perturbations clustered around the boundary and otherwise warm, equable, moist global climates, as follows: (a) A global anoxic event characterized by incursion of oxygen minimum zone watermasses into epicontinental marine basins (Jenkyns 1980; Arthur, Schlanger & Jenkyns 1985) the Bonarelli Anoxic Event. In the Western Interior Seaway this event was represented by fluctuating anaerobic and dysaerobic conditions in the lower water column between 93.2 and 93.6 Mybp (Fig. 13), producing laminated to microbioturbated organic-rich shales basin-wide. (b) A regional desalination event, possibly representing increased Tropical storm activity for long periods of time over central North America and high levels of internal runoff, is seemingly represented by a major negative excursion of ^{18}O isotopic data (Pratt 1985, Fig. 13 herein), reaching values of -8 to -14 ‰ vs. PDB in the middle of an interval otherwise characterized by near normal marine ^{18}O values (-1 to -3 ‰ vs PDB) at peak eustatic highstand. This unusual event lasted from 93.5 Ma, below the C-T boundary, to 93.2 Ma in the Early Turonian (Fig. 13). (c) A global positive ^{13}C isotopic excursion (Pratt 1985) with basin-wide expression in the Western Interior Seaway, suggesting an interval of enhanced depletion of 12-carbon due to increased photosynthetic activity and/or enhanced preservation of organic matter in the benthic zone. This event started below the C-T boundary at approximately 94.1 Ma, and extended into the Early Turonian (92.9 Ma). (d) Major changes in global marine temperatures also characterized this interval, including abrupt rise of 2-5° C through a series of rapid pulses around the boundary in Temperate epicontinental seas, based on biogeographic evidence (Kauffman 1984a) (Fig. 13); this suggests rapid incursion of Subtropical biotas from the Tethyan Realm to the north and south Temperate climatic belts around the C-T boundary. At the same time, a summary of oceanic thermal data (Fig. 7) suggests a broad drop of 4-5° C through the Cenomanian and across the boundary (Kauffman 1984a). Collectively, (a-d) above define a 1.5 million year interval of massive oceanic and atmospheric desta-bilization (ODE; Fig. 13) spanning the C-T boundary, with maximum rates and magnitude of chemical fluctuation centered around the boundary inter-val in association with the Bonarelli OAE. During this interval, major isotopic and chemical fluctuations in the oceans and epicontinental seas occurred in approximately 50-100,000 year duration intervals -- suffi-ciently fast to be a major cause of mass extinction among highly steno-topic, normal marine, warm water -- adapted organisms. Large scale tectonism and volcanism was also associated with the Cenomanian-Turonian boundary interval in North America (Kauffman 1984a, 1985b). Although no Iridium or other rare element spike has been confirmed to date around the C-T boundary (Asaro et al. 198?), three major impact craters (Steen River, Boltysh, and Logish; Grieve 1982, Alvarez & Muller 1984) have age ranges which overlap the C-T boundary mass extinction interval, suggesting a possible role of extraterrestrial impact in climate/ocean destabilization. This interval of environmental destabilization includes the major portion of the genera and over 70 percent of the species of marine invertebrates (North American data; Elder, in Hutt et al. 1986, in review).

Kauffman (1984b) and Elder (1985, in press) (collectively in Hut et al. 1986, in review) have detailed the patterns of loss of macro-invertebrates across the C-T boundary interval in the Western Interior and Gulf Coast of North America, and Eicher (1969), Eicher & Worstell (1970), Eicher & Diner (1985) and Leckie (1985) have documented extinc-tion patterns among Foraminifera. Figure 13 is a summary of Elder & Kauffman's data on molluscan extinction patterns in the central seaway (see also Kauffman 1984b, Figs. 8-10, 8-11). In every case, careful stratigraphic analyses of the C-T mass extinction boundary shows it to be stepwise in nature, with each major and some minor steps closely correlated to major shifts in the stable isotope records of ^{18}O and ^{13}C; this suggests a cause -- effect relationship between chemical and thermal changes in the marine realm and extinction among stenotopic taxa adapted to stable, equable marine environments. Of importance to this paper is the fact that each of six major and some minor steps (bio-events) within the C-T extinction interval is abrupt and regionally isochronous or of very short duration under the most careful stratigraphic scrutiny, i.e. when compared to isochronous ashfall event desposits in the Western Interior Seaway. Similar patterns of extinction are beginning to emerge from Western European C-T data although precise correlations to North American extinction steps remains to be tested. The main part of the C-T mass extinction (steps 1-5; Fig. 13) takes place around the C-T boundary within the 1.5 Myr interval defined by major stable isotope and organic carbon flucutations (see also Pratt 1985); whereas it appears that these environmental fluctuations are the direct cause of C-T mass extinction, a relationship of climate/ocean perturbations to one or multiple impacts around the C-T boundary interval (based on clustered crater ages) cannot be ruled out.

The Cretaceous-Tertiary mass extinction event. This is the most studied mass extinction because of its highly probable association with a large extraterrestrial impact event at the K-T boundary (Alvarez et al. 1984), based on diverse physical, chemical, and biological evidence. However, biological data for this boundary interval are mainly derived from marine plankton studies in open oceanic facies. Detailed marine macrofaunal analyses are mainly from outer shelf carbonate sites in Denmark and Spain, and from middle to inner shelf siliciclastic facies in Texas

Figure 17.   Summary of stepwise extinction events (steps 1-12) transecting the Cretaceous-Tertiary mass extinction interval, with mainly Tropical-Subtropical generic data summarized below the K-T boundary (from many sources) and North Temperate (mainly Danish) species data summarized above the K-T boundary for well studied Bryozoa and Brachiopoda. This pattern reflects ecological grading and change in taxonomic magnitude of extinction events through time from more sensitive (stenotopic) Tropical toward more adaptive (eurytopic) Temperate biotas. Width of graphs below K-T boundary generally scaled to generic diversity in specific data sets; species ranges shown in black lines, terminating in arrows where range boundaries were not precisely defined. Note scale change at K-T boundary; the temporal scale vs. metric thickness of Paleocene strata is based on average rock accumulation rate of 5 cm per thousand years for the Danish chalk.

321

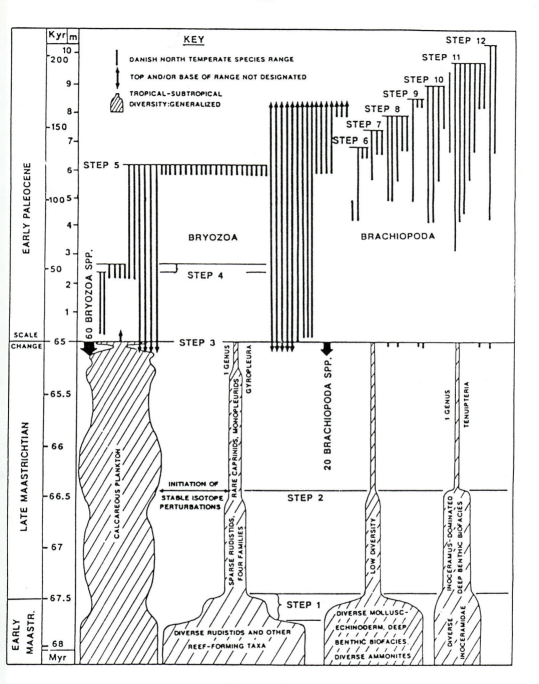

Figure 17

and Mexico (in progress). The paucity of shallow water marine sites reflects extensive eustatic drawdown near the Cretaceous-Tertiary (K-T) boundary and the global development of an erosional unconformity in association with this widespread regression. Kauffman (1984b, and Fig. 17 herein, from Hut et al. 1986, in review) and Alvarez et al. (1984) have summarized evidence for marine extinction patterns based on a decade of careful work by colleagues throughout the world. Whereas there is little argument with the profound effect on extinction of the K-T impact event, detailed biological sampling and stratigraphic inter-pretation has revealed a more complex extinction history than origi-nally portrayed by "catastrophist, one-cause" theory (e.g. the impact event) (Kauffman 1984b, and in Hut et al., in review, 1986).

Environmentally the K-T boundary interval was associated with complex factors describing a general decline in marine conditions for life: (a) a major eustatic drawdown associated with shrinking and loss of prime ecospace in shallow marine and coastal settings; (b) a change from predominantly chalk-pelagic carbonate mud to siliciclastic sedi-mentation ; (d) rapid fluctuations in (first) ^{13}C and Corg values and (at and after the K-T boundary) in ^{18}O values reflecting climatic and oceanic perturbations of unusual magnitude for the Cretaceous (but comparable to those around the C-T boundary); (e) rapid temperature fluctuations (rapid fall, rise, fall events immediately adjacent to and crossing the K-T boundary); and (f) evidence for stagnation in broad, deeper oceanic settings. These changes spanned a 2.5-2.75 Ma interval beginning near the Middle-Upper Maastrichtian boundary and extending into the lowest Paleocene. The K-T boundary meteorite impact caused acceleration of these perturbations (Fig. 17), and of the rate of mass extinction.

Kauffman (1984b; Fig. 17 herein from Hut et al. 1986, in review) attempted to summarize scattered macrofossil and extensive marine microfossil data for the K-T boundary interval and, as at the C-T boundary (Elder 1985, in press, and Fig. 13), found it to be stepwise in nature and ecologically graded through time with the more Tropical, temperature-sensitive, stenotopic taxa becoming extinct earlier and at higher levels of magnitude than those of Temperate areas and/or eury-topic adaptive ranges. Each step seems to mark a regionally or globally correlative, short-term (100 Ka or less) to (at the K-T impact boundary) isochronous biological event. These major events, based on limited data for macro-faunas, are as follows (Fig. 17):

(a) A major loss of Tropical platform biotas associated with reefoid facies, including great decline in diversity and numbers of rudistid bivalves, larger foraminifera, massive oysters (e.g. Exogyra), large echinoids, hermatypic corals, and Tropical nerineid and actaeonellid gastropods characterized a narrow interval of time spanning the Middle-Upper Maastrichtian boundary (base of A. mayaroensis Biozone); only small bioherms and biostromes built of monospecific to paucispecific, generalized rudist are known from younger strata, and no reef-building rudistids are known to reach the K-T boundary; in fact only one rudistid specimen, a small generalized epibiont Gyropleura, has been found at the boundary in Denmark. Diverse mollusc-echinoderm communities of deep benthic settings also seem to be decimated at this time, yielding to low diversity, inoceramid bivalve-dominated communities more tolerant of low oxygen-high stress benthic environments. This narrow interval of widespread extinction occurred about 2.3-2.5 Ma below the K-T boundary, and was associated with onset of the last major pulse of Maastrichtian eustatic drawdown.

(b) Major extinction among remaining generalized rudistids (Radio-litidae, Hippuritidae, Caprinidae) and loss of most core lineages of Inoceramidae were associated with major decline in specialized ammonite lineages about 1.5 Ma below the K-T boundary (Fig. 17) and initiation of the first rapid large-scale fluctuations in temperature and the carbon cycle.

Temperate shelf biotas showed little extinction during (a) and (b), however, if the well documented Danish sequence is typical. Instead, these taxa experienced their main extinction at and shortly after the K-T impact boundary. This probably reflects the greater adaptive range of Temperate organisms in the face of broadly deteriorating global marine environments during the Late Maastrichtian.

(c) The main part of the K-T mass extinction occurred within the few thousand years bounding the Cretaceous-Tertiary boundary, and expecially in association with the asteroid impact event marking the boundary. This boundary comprises an outstanding global event and bio-event marker. The K-T boundary extinction was associated with several major changes in the global environment. Impact-related, rapid temperature decline and reduction in light penetration through a dust-choked atmosphere; peak terminal Cretaceous eustatic drawdown; great enhancement of ^{13}C stable isotope fluctuations marking destabilization of the oceanic carbon cycle; and with initiation of extraordinarily large, frequent ^{18}O stable isotope fluctuations marking rapid, widely varying marine temperature (and possibly salinity) levels over intervals of a few thousand to 100,000 years. Whereas both planktonic foraminifera and nannoplankton showed important changes a few thousand years prior to impact, manifest in decreasing complexity of ecological structure and declining population size, the catastrophic K-T boundary extinction of most of the world's calcareous and siliceous plankton near a peak in their evolutionary history constitutes an extraordinary bio-event, and a prominent horizon of global correlation. Well documented Danish marine faunas further indicate that the final extinction of already decimated Cretaceous group (ammonites, belemnites, inoceramid and other bivalves, including ru- distids), and mass extinction among species of ecologically successful brachiopods and Bryozoa (among many larger invertebrate groups) coincide with the Cretaceous-Tertiary impact boundary. This appears to be the best documented, and most defensible example of a global biotic cata- strophe related to extraterrestrial impact. Papers cited in Kauffman (1984b), Alvarez et al. (1984), and in Hut et al. (1986, in review) summarize this event and provide detailed data for these conclusions (Fig. 17). Whereas it can be argued that decreased light penetration, rapid cooling, and destabilization of ocean chemistry were the primary causes of this biotic catastrophe, the atmospheric dust cloud produced by the K-T impact event, and the already stressed nature of many Late Cretaceous shallow water marine biotas in response to eustatic fall at the time of impact, were clearly the most important factors in deter- mining the global magnitude of this extinction step - the main element of the Cretaceous-Tertiary Mass Extinction interval.

(d) Data on the evolutionary and extinction history of lowest Danian (Paleocene) marine taxa following the K-T impact event are well docu- mented at present only for Denmark (papers in Birkelund & Bromley 1979; Christensen & Birkelund 1979, Alvarez et al. 1984, and associated Scienc articles, Kauffman 1984b, and references therein). An extensive body of data is currently being developed by T.A. Hansen (Hansen et al. 1984) for the Texas Gulf Coast, especially the Brazos River area. Figure 17 summarizes the Danish data for Brachiopoda and Bryozoa, and indicates a period of about 45 Ka following impact which characterized by a few, small ecological generalists that survived the K-T boundary event, or evolved shortly afterward. This interval was followed by three major, short-term, evolutionary bio-events -- rapid radiations of new taxa at about 50 Kyr, 83-100 Kyr, and especially around 120 Kyr after the K-T boundary catastrophe (ages based on average sedimentation rates). These radiations comprise punctuated evolutionary bio-events, but their regional persistence is yet untested. Associated with this recovery interval are nine abrupt, stepwise extinction events among these two groups (the only data for which careful stratigraphic detail is currentl available) at about 55, 125 (major event), 140, 150, 160, 175, 200 and

210 Ta after the K-T boundary event. Inasmuch as these steps occur within similar environments (Bryozoan mound facies of the Danian Chalk) they are considered to be true extinction phenomena; these events extend the stepwise extinction pattern beyond the K-T boundary into the Danian, within the interval characterized by continuing oceanic and atmospheric disruption suggested by rapid, large-scale fluctuations in the earliest Danian stable isotope record.

The Cretaceous-Tertiary boundary interval, therefore, spans about 2.5-2.7 Myr and, like the C-T boundary, is associated with major global perturbations in atmospheric, temperature, and oceanographic parameters that are well beyond background fluctuation. The mass extinction is composed of three major steps of extinction 2.5 and 1.5 Ma below, and at the K-T boundary (a catastrophic bio-event), of several minor events in the latest Cretaceous close to the boundary, and of nine species-level extinction steps among Danish Brachiopoda and Bryozoa in the basal 0.25 Ma of the Paleocene (Fig. 17). Collectively, species-level K-T extinction may exceed 80 percent worldwide, with higher percentages in the Tropics and progressively lower percentages toward the poles. These stepwise extinction bio-events are regional to global in aspect where tested; the K-T boundary extinction event is a biotic catastrophe associated with the many manifestations of a major asteroid impact, superimposed on an already destabilized ocean-climate system and, especially, on the shallow warm water marine biota already stressed by deterioration of prime ecospace associated with major Late Maastrichtian eustatic drawdown. In combination, these factors create a first-order mass extinction on the Raup-Sepkoski scale (1984, 1986) of profound global dimensions -- the most impressive of all well documented global bio-events. Multiple Iridium enrichment "spikes" in the Brazos River sections of Texas (Asaro 1983, Ganapathy et al. 1981), in Austrian boundary sections (D. Herm, personal communication, 1986) and a double boundary clay in south Russia (Naidin, personal communication, 1984), one of which has an Iridium enrichment, suggest the possibility of multiple asteroid/comet impacts associated with the K-T boundary; one impact seems to be conclusively demonstrated, opening the question of extraterrestrial forcing for large, cyclic mass extinction events.

The Eocene-Oligocene mass extinction event. The youngest well documented mass extinction event occurs in Late Eocene and at the Eocene-Oligocene boundary, and has been well studied by Keller (1983, microplankton, tektite horizons) and Hansen (in review, 1986); a summary of their work appears in Hut et al. (1986, in review, Science); with their permission I provide the following brief summary of events for comparison with

Cretaceous mass extinction data.

The Eocene-Oligocene (E-O) mass extinction is clearly stepwise for
both foraminifera and molluscs at both the population and species/genus
level. It spans at least 4 Myr (early Late Eocene to basal Oligocene).
An earlier major extinction event among foraminifera, in the middle Middle
Eocene (top of G. lehneri Biozone) may or may not be related to the
overall E-O mass extinction. Discrete steps within the last 3 Myr of the
E-O mass extinction are contained within an interval characterized by
large-scale stable isotope fluctuations (^{18}O, reflecting temperature)
and by two to three widespread Late Eocene microtektite layers in deep
sea sediments. These represent multiple asteroid and/or comet impacts
associated with the mass extinction. At least two impact crater ages
fall within this interval. In one case, at the top of the Gr. semiinvoluta foraminiferal biozone (middle Late Eocene), massive population
decline and species extinction among foraminifera and molluscs, and
major enhancement of stable isotope fluctuations, is directly correlative
to a microtektite layer inferring impact by a meteorite or comet; like
the K-T boundary, a cause and effect relationship is suggested. Each
stepwise extinction seems to occur within a very narrow stratigraphic
interval, and represents a regionally correlative bio-event. The major
E-O extinction steps are as follows:

(1) A major extinction step characterized by the loss of 89 percent
of Gulf Coast gastropod species, 84 percent of Gulf Coast bivalve
species, and 60-70 percent of planktonic foraminifer species occurred
abruptly within a meter of the Middle-Upper Eocene boundary, about 39.2
Ma ago. (2) A second major extinction step occurred in association with
the lowest Late Eocene microtektite layer (middle Gr. semiinvoluta Biozone, and near top of Chiasmolithus oamaruensis nannofossil Biozone);
it is characterized by extinction of 72 percent of Gulf Coast gastropod
species, 63 percent of the bivalve species, and major decline in warm
water planktonic foraminifer populations (genus Globigeraspis). Molluscan
loss took place over 5-10 m of siliciclastic coastal marine sediments,
and foraminiferal loss within a narrow interval of deep sea core.
(3) A third step of mass extinction involving loss of six major planktonic foraminifer species, and major population decline in others, is
associated with a microtektite layer at or just above the top of the
Gr. semiinvoluta foraminiferal Biozone (Fig. 18); no significant
molluscan extinction is known, but major ^{18}O fluctuations are associated
with this event. (4) The fourth, and final step of extinction comprising
the E-O mass extinction interval, occurs at and just below the Eocene-
Oligocene boundary, 36.25 Ma ago, and follows small steps of foramini-

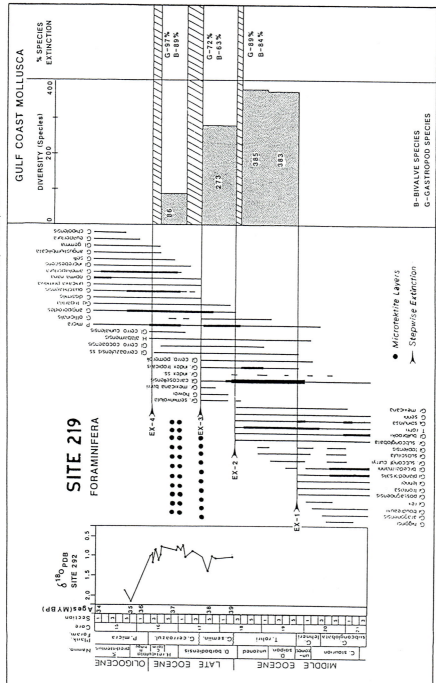

Figure 18. Microtektite layers, species abundance changes, species ranges and oxygen isotope data during the Late Eocene to Early Oligocene in the west equatorial Pacific DSDP Sites 292 and 219. Data at right are Eocene-Oligocene Molluscs extinctions from Gulf Coast of North America (T.A. Hansen, in press).

feral population decline and stable isotope disruption in the latest
Eocene (Gl. cerraozulensis foraminiferal Biozone) associated with two
microtektite layers, implying meteorite and/or comet impacts. At the
E-O Stage Boundary, 97 percent of existing gastropod species and 89
percent of existing bivalve species died out within an interval represented by 10-15 m of siliciclastic strata; the lower part of this
extinction interval may be related to the Late Eocene microtektite layers
and impact, but no precise correlation is yet possible. The molluscan
extinction is correlative with extinction of 2-3 planktonic foraminifer
species, and major reduction in population size of remaining speices
in the open ocean. There is no associated evidence of impact, but the
largest ^{18}O stable isotope excursion of the boundary interval, indicating
rapid temperature change, is coeval with the final E-O extinction step.

A theory for mass extinction

There exist numerous similarities between the Cenomanian-Turonian (C-T),
Cretaceous-Tertiary (K-T) and Eocene-Oligocene (E-O) mass extinction
events: (1) All span 2.5-4 Ma, and the 26 and 30 Ma intervals between
them support, in part, the analyses of Raup and Sepkoski (1984, 1986)
depicting a generally cyclic mass extinction history during the Mesozoic
and Cenozoic; (2) Each mass extinction is composed of a series of
discrete steps of short duration (thousands of years), in some cases
representing geologically "instantaneous" catastrophies, (i.e., the
K-T boundary), between which extinction proceeds at background or
moderately accelerated rates; (3) In a general sense, extinction proceed,
through an ecological-biogeographical gradient; initial steps mainly
affect more Tropical and/or stenotopic groups to a greater extent, and
later steps mainly affect more Temperate and/or eurytopic groups. Overall, mass extinctions are more severe in warm and/or shallow water
marine environments than in cooler and/or deeper environments; (4) All
mass extinctions during the past 250 Ma, including the C-T, K-T, and
E-O events, seem to be partially or completely associated with extraordinarily large, frequent oscillations in climate, temperature, and
ocean chemistry (especially oxygen, salinity, and the carbon cycle), as
mainly depicted from biological and stable isotopic data; individual
extinction steps commonly correlate with individual environmental
oscillations; (5) Several mass extinctions, though not all, are associated with evidence of one or more asteroid and/or comet impacts on
Earth (i.e., Iridium and other rare element concentrations, microtektites or their alteration products, shocked quartz and feldspar, impact

craters, and layers of atmospheric dust and other impact debris). This includes the C-T (craters), K-T (Iridium, spherules, shocked minerals, crater), and E-O events (Iridium, microtektites, craters); in some but not all cases, these impact events are directly correlated to steps in mass extinction history and dramatic stable isotope fluctuations (e.g., the K-T boundary).

A mass extinction theory may be derived from these observations, as follows (Fig. 19): At roughly cyclic 26-30 Ma intervals, the Earth's atmospheric and ocean systems became highly perturbed by some force of probably extraterrestrial origin that is associated with higher than normal frequence of impacts on Earth by comets and meteorites. Extraterrestrial impacts themselves may be directly responsible for some of these large Earthbound perturbations, or at least for the initiation of extraordinarily large, rapid climate, temperature and ocean chemistry fluctuations which then perpetuate themselves for prolonged intervals (1-4 Ma) through dynamic feedback mechanisms. Subsequent impacts, especially in the oceans, would enhance the magnitude of these perturba-

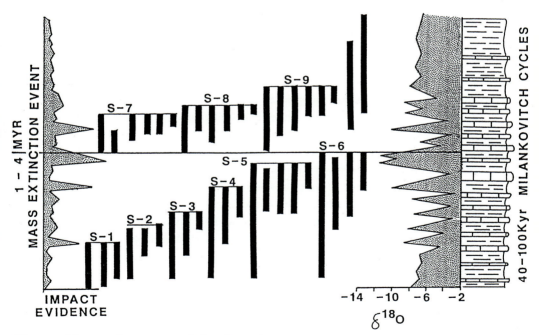

Figure 19.   Summary model of environmental and biological dynamics proposed for the stepwise mass extinction theory. The 1-4 Ma mass extinction is contained within an interval of extraordinary stable isotope disruptions, many in concert with normal Milankovitch cycles (independent catalyst), and steps of extinction. Extraterrestrial impacts may initiate, and enhance the level of destabilization of the marine realm.

tions and reset the feedback process clock; such destabilized climate/ ocean systems could be further driven to unusual fluctuations by normal Earthbound forcing mechanisms, e.g., Milankovitch and larger-scale climatic cycles related to predictable orbital and solar fluctuations. The net effect of these processes would be oscillations of global temperature, climate patterns, and marine chemistry, each of a magnitude that rapidly exceeded the adaptive range of many world species; this would be especially true during warm equable global climate intervals like the Cretaceous, when many taxa evolved narrow thermal and chemical survival limits (stenotopes). Further, the rates of change in these factors could be so rapid (100 Ka or less) that time for evolutionary change to adapt to new conditions would be insufficient for most taxa. In this scenario, the most specialized stenotopic (especially stenothermal) taxa would be affected first, and especially those of shallow marine habitats which provided little protection from such perturbations. Perpetuation of large scale environmental oscillations over millions of years, however, would progressively stress more eurytopic, more temperate, and even deeper water taxa through longer term buildup of detrimental environmental conditions. Short-term acceleration of extinction rates (extinction steps, bio-events) would result from times of extraordinary rate and magnitude of environmental oscillation, and especially from individual impact events and their immediate aftermath. Dust, smoke and other impact-related material thrown into the upper atmosphere and troposphere by a large terrestrial impact (especially if followed by flash fires) would quickly envelope a large portion of all of the earth as a dense smoke and debris cloud, shielding solar radiation, and causing rapid global cooling. Rapid cooling would first affect the continents, and subsequently the sea surface (the time differential potentially causing giant coastal storms along the zone of intense thermal gradients). This would be associated with decreased light intensity for at least a few years, and massive shock to the global food chain as the primary plant base became rapidly depleted. Oceanic impact of a large body might further trigger destratification, overturn, rapid changes in thermal and chemical systems, as well as watermass properties and circulation. Such oceanic impacts, or even rapid tempeature changes resulting from impact debris clouds, could initiate extraordinary climatic and oceanic oscillations. These rapid changes, the real "killers" in mass extinction, would then be perpetrated by dynamic feedback processes for thousands (climate) to even millions of years (oceans), with the initial and many subsequent climate/ocean oscillations, and additional impacts, producing discrete steps (bio-events) in the overall mass extinction.

This theory best accomodates existing detailed data on the C-T, K-T, and E-O mass extinctions, and diverse theories for causal mechanisms being debated today.

Conclusion

High-resolution event-stratigraphic analysis of Cretaceous sequences in the Western Interior of North America, observing stratigraphic data down to the cm or less level of distinction, has demonstrated that biological events of various types and regional extent are common and very useful components of precise regional correlation. Local, regional and inter-continental to global bio-events can be differentiated, the last two normally driven by allocyclic or at least basin-wide forcing mechanisms. Biological events are classified into eight distinct categories: (1) Punctuated evolutionary events; (2) population bursts; (3) productivity events; (4) immigration and emigration events; (5) ecostratigraphic events; (6) regional colonization events; (7) mass mortality events; and (8) extinction events. The Cretaceous of North America, and the Phanerozoic worldwide, provide excellent examples of various types and levels of bio-events, their probable causes, and their importance to high-resolution, regional to global correlation. Many such bio-events can be demonstrated to be very short term or essentially synchronous over broad areas when tested against volcanic ash beds or other iso-chronous deposits/surfaces. Biostratigraphy may draw heavily on bio-events in its construction, although the two systems of stratigraphy have fundamentally different concepts of origin, methodologies, and applications.

Mass extinction events comprise one of the most important bases for global bio-event correlation. Comparison of detailed stratigraphic and biological data from three successive mass extinction intervals at the Cenomanian-Turonian, Cretaceous-Tertiary, and Eocene-Oligocene bound-aries reveal important similarities: (a) stepwise mass extinction patterns (multiple bio-events) spread over 2.5-4 Ma; (b) close corre-lation of extinction steps and entire mass extinction events to short intervals of massive climate, thermal, and oceanic disruption depicted in biological and stable isotope data; (c) ecological grading of mass extinctions from stenotopic and/or Tropical shallow water forms to more eurytopic, Temperate and even deeper water taxa through time and succes-sive extinction steps; (d) diverse evidence for impact of meteorites and/or comets on Earth (craters, Iridium and other rare element enrich-ment, microtektites, shocked mineral grains, etc.) are common to many mass extinction events, but not to all of them, or to all steps within any mass extinction interval.

A new theory of causal mechanisms for mass extinction is proposed which involves roughly cyclic disruption of global climate, thermal, and ocean systems at 26-30 Ma intervals due to some extraterrestrial forcing mechanisms that accelerated the rate of comet and meteorite impact on Earth. Once perturbed directly by extraterrestrial phenomena and/or by impact events, the climate and (especially) the oceans begin, or enhance the effects of, a series of rapid, large scale oscillations in temperature, chemistry, and watermass dynamics that are perpetuated by internal feedback mechanism for prolonged intervals, and at frequencies measured in tens to hundreds of thousands of years. The magnitude and rate of these extraordinary environmental fluctuations exceeds the adaptive range and evolutionary rate among genetically and ecologically diverse organisms worldwide, but in an ecologically and biogeographically graded manner. Mass extinction steps represent the most severe of these environmental fluctuations; they are further enhanced by additional impact events during mass extinction, as well as by the direct effects of impact-related atmospheric dust clouds in causing rapid cooling, giant coastal storms, reduced light penetration, and collapse of the global food chain.

The study of bio-events is in its early, mainly documentary stages so that many additional kinds and stratigraphic levels of biological events should be discovered in coming years, enhancing their role in high-resolution stratigraphic correlation worldwide. This will further provide us with a much better understanding of the dynamics of biological evolution and extinction, including the nature of biological catastrophes.

REFERENCES

ALVAREZ, L.W.; ALVAREZ, W.; ASARO, F. & MICHEL, H.V. (1980): Extraterrestrial cause for the Cretaceous-Tertiary extinction.- Science 208, 1095-1108.

ALVAREZ, W.; KAUFFMAN, E.G.; SURLYK, F.; ALVAREZ, L.; ASARO, F. & MICHEL, H.V. (1984): The impact theory of mass extinctions and the marine and vertebrate fossil record across the Cretaceous-Tertiary boundary.- Science 223, 1135-1141.

-- & MULLER, R.A. (1984): Evidence from crater ages for periodic impacts on the Earth.- Nature 308, 718-720.

ARTHUR, M.A.; SCHLANGER, S.O. & JENKYNS, H.C. (1985): The Cenomanian-Turonian oceanic anoxic event, II. Paleoceanographic controls on organic matter production and preservation.- in: BROOKS, J. & FLEET, A. (eds.): Marine petroleum source rocks. Geol. Soc. London, Spec. Publ. 315, 216-218.

ASARO, F. (1983): A detailed study of geochemical anomalies associated with lithologic changes in a Brazos River section of the Cretaceous-

Tertiary boundary.- Geol. Soc. Amer., Abstr. with Progr. 14, 1, p. 33.
BARRON, E.J. & WASHINGTON, W.M. (1982): Cretaceous climate: a comparison
    of atmospheric simultations with the geologic record.- Palaeogeogr.,
    Palaeoclim., Palaeoecol. 40, 103-133.
-- ARTHUR, M.A. & KAUFFMAN, E.G. (1985): Cretaceous rhythmic bedding
    sequences -- A plausible link between orbital variations and climate.-
    Earth & Planetary Sci. Letters 72, 327-340.
BIRKELUND, T. & BROMLEY, R.G. (eds.) (1979): Cretaceous-Tertiary Boundary
    Events: Symposium v. 1. The Maastrichtian and Danian of Denmark,
    Copenhagen, Univ. of Copenhagen, Inst. of Hist. Geol. & Paleont.
CALDWELL, W.G.E. (1984): Early Cretaceous transgression and regressions
    in the southern Interior Plains.- in: SCOTT, D.F. & GLASS, D.J.
    (eds.): The Mesozoic of Middle North America.- Can. Soc. Petrol.
    Geologists, Mem. 9, 173-203.
CHRISTENSEN, W.K. & BIRKELUND, T. (eds.) (1979): Cretaceous-Tertiary
    Boundary Events: Symposium v. 2.- Proc. Univ. Copenhagen.
EICHER, D.L. (1969): Cenomanian and Turonian planktonic foraminifera from
    the Western Interior of the United States.- in: BRILL, E.J. (ed.):
    1st International Conference on Planktonic Microfossils, Geneva
    Switzerland, 1967, Proc. 2, 163-174, Leiden (Netherlands).
-- & DINER, R. (1985): Foraminifera as indicators of water mass in the
    Cretaceous Greenhorn sea, Western Interior.- in: PRATT, L.M.; KAUFF-
    MAN, E.G. & ZELT, F.G. (eds.): Fine-grained deposits and biofacies
    of the Cretaceous Western Interior Seaway: Evidence of cyclic
    sedimentary processes. Soc. Econ. Paleont. Mineral.,Fieldtrip Guide-
    book 4, 60-71 (Midyear meeting, Golden, CO).
ELDER, W.P. (1985): Biotic patterns across the Cenomanian-Turonian
    extinction boundary near Pueblo, Colorado.- in: PRATT, L.M.; KAUFF-
    MAN, E.G. & ZELT, F.B. (eds.): Fine-grained deposits and biofacies
    of the Cretaceous Western Interior Seaway: Evidence of cyclic
    sedimentary processes. Soc. Econ. Paleont. Mineral., Fieldtrip Guide-
    book 4, 157-169 (Midyear meeting, Golden, CO).
-- (1986): Cenomanian/Turonian stage boundary extinctions in the Western
    Interior of North American.(Unpubl. Ph. D. dissertation), Univ. of
    Colorado.
FISHER, A.G.; HERBERT, T. & PREMOLI SILVA, I. (1985): Carbonate bedding
    cycles in Cretaceous pelagic and hemipelagic sequences.- in: PRATT,
    L.M.; KAUFFMAN, E.G. & ZELT, F.B. (eds.): Fine-grained deposits and
    biofacies of the Cretaceous Western Interior Seaway: Evidence of
    cyclic sedimentary processes. Soc. Econ. Paleont. Mineral., Field-
    trip Guidebook 4, 1-10 (Midyear meeting, Golden, CO).
FRUSH, M.P. & EICHER, D.L. (1975): Cenomanian and Turonian foraminifera
    and palecenvironments in the Big Bend region of Texas and Mexico.-
    in: CALDWELL, W.G.E. (ed.): The Cretaceous system in the Western
    Interior of North America. Geol. Assoc. Can., Spec. Pap. 13, 277-301.
GANAPATHY, R. (1971): A major meteorite impact on the Earth 65 million
    years ago: evidence from the Cretaceous-Tertiary boundary clay.-
    Science (AAAS) 209, 4459, 921-923.
-- , GARTNER, S. & JIANG, M.J. (1981): Iridium anomaly at the Cretaceous-
    Tertiary boundary in Texas. Earth & Planetary Sci. Letters 54, 393-
    396.
GRIEVE, R.A.F. (1982): The record of impact on Earth: Implications for
    a major Cretaceous/Tertiary impact event.- Geol. Soc. Amer., Spec.
    Pap. 190, 25-37.
HANSEN, T.A.; FARRAND, R.; MONTGOMERY & BILLMANN, H. (1984): Sedimen-
    tology and extinction pattern across the Cretaceous-Tertiary boundary
    interval in east Texas.- GSA Field Guidebook, 1984.
HATTIN, D.E. (1971): Widespread, synchronously deposited, burrow-mottled
    limestone beds in Greenhorn Limestone (Upper Cretaceous) of Kansas
    and central Colorado.- Amer. Assoc. Petrol. Geologists, Bull. 55,
    412-431.
-- (1965): Stratigraphy of the Graneros Shale (Upper Cretaceous)in Central

Kansas.- Kansas Geol. Surv., Bull. 178, 83 p.
-- (1985): Distribution and significance of widespread, time-parallel
    pelagic limestone beds in Greenhorn Limestone (Upper Cretaceous) of
    the central Great Plains and southern Rocky Mountains.- in: PRATT,
    L.M.; KAUFFMAN, E.G. & ZELT, F.B. (eds.): Fine-grained deposits and
    biofacies of the Cretaceous Western Interior Seaway: Evidence of
    cyclic sedimentary processes.- Soc. Econ. Paleont. Mineral., Field-
    trip Guidebook 4, 23-37 (Midyear meeting, Golden, CO).
HUT, P.; ALVAREZ, W.; ELDER, W.P.; HANSEN, T.; KAUFFMAN, E.G.; KELLER,
    G.; SHOEMAKER, E.M. & WEISSMAN, P.R. (1986) (in review; Science):
    Comet Showers as a possible cause of stepwise mass extinctions.-
    MS, 48 p.
JABLONSKI, D. (1986): Background and Mass Extinctions: The Alternation
    of Macroevolutionary Regimes.- Science 231, 129-133.
JENKYNS, H.C. (1980): Cretaceous anoxic events -- from continents to
    oceans.- J. Geol. Soc. London 137, 171-188.
KAUFFMAN, E.G. (1977): Geological and biological overview: Western
    Interior Cretaceous Basin.- in: KAUFFMAN, E.G. (ed.): Cretaceous
    facies, faunas and Paleoenvironments across the Western Interior
    Basin.- Mountain Geologist 14, 3/4, 75-99.
-- (1981): Ecological reappraisal of the German Posidonienschiefer.-
    in: GRAY, T.; BOUCOT, A.J. & BERRY, W.B.N. (eds.): Communities of
    the Past.- p. 311-282, Hutchinson Ross Publ. Co., Stroudsburg,
    Pennsylvania.
-- (1984a): Paleobiogeography and evolutionary response dynamic in the
    Cretaceous Western Interior Seaway of North America.- in: WESTERMANN,
    G.E.G. (ed.): Jurassic-Cretaceous biochronology and paleogeography
    of North America.- Geol. Assoc. Can., Spec. Pap. 27, 273-306.
-- (1984b): The fabric of Cretaceous marine extinctions.- in: BERGGREN,
    W.A. & VAN COUVERING, J. (eds.): Catastrophies and earth history --
    the new uniformitarianism.- p. 151-246, Princeton Univ. Press,
    Princeton, N.J.
-- (1984c): Toward a synthetic theory of mass extinction (Abstr.).-
    Abstract with programs, 1984, Geol. Soc. Amer., 97th Ann. Meeting,
    Reno, 16 , 555-556, September 1984.
-- (1985a): Depositional history of the Graneros Shale (Cenomanian), Rock
    Canyon Anticline.- in: PRATT, L.M.; KAUFFMAN, E.G. & ZELT, F.B.
    (eds.): Fine-grained deposits and biofacies of the Cretaceous Western
    Interior Seaway: Evidence of cyclic sedimentary processes. Soc. Econ.
    Paleont. Mineral., Field Guidebook 4, 90-99 (Midyear meeting, Golden,
    CO).
-- (1985b): Cretaceous evolution of the Western Interior Basin of the
    United States.- in: PRATT, L.M.; KAUFFMAN, E.G. & ZELT, F.B. (eds.):
    Fine-grained deposits and biofacies of the Cretaceous Western Interior
    Seaway: Evidence of cyclic sedimentary processes. Soc. Econ. Paleont.
    Mineral.,Field Guidebook 4, IV-XIII (Midyear meeting, Golden, CO).
-- & HANSEN, T.A. (1985): Stepwise mass extinction associated with major
    oceanic and climatic perturbations; Evidence for multiple impacts?-
    Abstr. Amer. Geophys. Union Ann. Meeting, EOS 66, 813-814.
-- & PRATT, L.M. (1985): A field guide to the stratigraphy, geochemistry
    and depositional environments of the Kiowa-Skull Creek, Greenhorn
    and Niobrara marine cycles in the Pueblo-Canon City Area, Colorado.-
    in: PRATT, L.M.; KAUFFMAN, E.G. & ZELT, F.B. (eds.): Fine-grained
    deposits and biofacies of the Cretaceous Western Interior Seaway:
    Evidence of cyclic sedimentary processes. Soc. Econ. Paleont. Mineral.
    Field Guidebook 4, FRS1-FRS26 (Midyear meeting, Golden, CO).
KELLER, G. (1983): Paleoclimatic analyses of Middle Eocene through
    Oligocene planktic foraminiferal faunas.- Paleogeogr., Paleoclim.,
    Paleoecol. 43, 73-94.
KOCH, C.F. (1975): Evolutionary and ecological patterns of Upper Ceno-
    manian (Cretaceous) mollusc distribution in the Western Interior of
    North America.- Geo. Washington Univ., unpubl. Ph.D. thesis, 72 p.

KOCH, C.F. (1980): Bivalve species duration, aerial extent and population
   size in a Cretaceous sea.- Paleobiology 6, 184-192.
LECKIE, R.M. (1985): Foraminifera of the Cenomanian-Turonian boundary
   interval, Greenhorn Formation, Rock Canyon Anticline, Pueblo, Colo-
   rado.- in: PRATT, L.M.; KAUFFMAN, E.G. & ZELT, F.B.: Fine-grained
   deposits and biofacies of the Cretaceous Western Interior Seaway:
   Evidence of cyclic sedimentary processes. Soc. Econ. Paleont.
   Mineral.,Field Trip Guidebook 4, 139-149 (Midyear meeting, Golden,
   CO).
McLAREN, D.J. (1985): Mass extinction and iridium anomaly in the Upper
   Devonian of western Australia: a commentary.- Geology 13, 170-172.
PRATT, L.M. (1985): Isotopic studies of organic matter and carbonate in
   rocks of the Greenhorn marine cycle.- in: PRATT, L.M.; KAUFFMAN, E.
   G. & ZELT, F.B. (eds.): Fine-grained deposits and biofacies of the
   Cretaceous Western Interior Seaway: Evidence of cyclic sedimentary
   processes. Soc. Econ. Paleont. Mineral., Field Trip Guidebook 4, 38-48,
   (Midyear meeting, Golden, CO).
-- , KAUFFMAN, E.G. & ZELT, F.M. (eds.) (1984): Fine-grained deposits
   and biofacies of the Cretaceous Western Interior Seaway: Evidence
   of cyclic sedimentary processes. Soc. Econ. Paleont. Mineral., Field
   Trip Guidebook 4, 288 p. (Midyear meeting, Golden, CO).
RAUP, D.M. & SEPKOSKI, J.J. (1984): Periodicity of extinctions in the
   geologic past.- Proc. Nat. Acad. Sci. U.S.A. 81, 801-805.
-- & -- (1986): Periodic extinctions of families and genera.- Science
   231, 833-836.
RICKEN, W. (1986): Diagenetic bedding; A model for marl-limestone
   alternations.- Lecture Notes in Earth Sci. 6, 210 p., Springer-Verl.,
   Berlin.
SAGEMAN, B.B. (1985): High-resolution stratigraphy and paleobiology of
   the Hartland shale member: Analysis of an oxygen-deficient epicontinental sea.- in: PRATT, L.M.; KAUFFMAN, E.G. & ZELT, F.B. (eds.):
   Fine-grained deposits and biofacies of the Cretaceous Western Interior
   Seaway: Evidence of cyclic sedimentary processes. Soc. Econ. Paleont.
   Mineral., Field Trip Guidebook 4, 110-121 (Midyear meeting, Golden,
   CO).
-- & JOHNSON, C.C. (1985): Stratigraphy and Paleobiology of the Lincoln
   limestone member, Greenhorn Limestone, Rock Canyon Anticline, Colo-
   rado.- in: PRATT, L.M.; KAUFFMAN, E.G. & ZELT, F.B. (Eds.): Fine-
   grained deposits and biofacies of the Cretaceous Western Interior
   Seaway: Evidence of cyclic sedimentary processes.- Soc. Econ. Paleont.
   Mineral.,Field Trip Guidebook 4, 100-109 (Midyear meeting, Golden,
   CO).
SILVER, L.T. & SCHULTZ, P.H. (eds.) (1982): Geological implications of
   impacts of large asteroids and comets on the Earth.- Geol. Soc. Amer.,
   Spec. Pap. 190, 528 p.
VAIL, P.; MITCHUM, R.M. & THOMPSON, S. (1977): Seismic stratigraphy and
   global changes of sea level.- in: PAYTON, C.E. (ed.): Seismic strati-
   graphy-applications to hydrocarbon exploration.- Amer. Assoc. Petrol.
   Geol., Mem. 26, 49-212.
VAIL, P.R. & MITCHUM, R.M. (1978): Global cycles of relative changes of
   sea level from seismic stratigraphy.- Bull. Amer. Assoc. Petrol.
   Geologists 62, 469-472.
WATKINS, D.K. (1985): Biostratigraphy and paleoecology of calcareous
   nannofossils in the Greenhorn marine cycle.- in: PRATT, L.M.; KAUFF-
   MAN, E.G. & ZELT, F.B. (eds.): Fine-grained deposits and biofacies
   of the Western Interior Seaway: Evidence for cyclic sedimentary
   processes. Soc. Econ. Paleont. Mineral., Field Trip Guidebook 4,
   151-156 (Midyear meeting, Golden, CO).

TRACE METAL ACCUMULATION IN BLACK SHALES FROM THE CENOMANIAN/TURONIAN BOUNDARY EVENT

A contribution to Project
GLOBAL BIO-EVENTS

BRUMSACK, Hans-Jürgen *)

Abstract: Black shales from the Cenomanian/Turonian Boundary Event show a remarkable enrichment in trace metals (Cd, Ag, Zn, Sb, Mo, V, Cu, etc.) relative to normal shales, Lower Toarcian black shales from SW Germany or TOC-rich sediments from recent upwelling areas. If these enrichments are authigenic in origin, trace metals may serve as indicators for the paleoenvironmental conditions during deposition of black shales. Low sediment accumulation rates, a comparably low organic primary productivity and enhanced preservation of TOC due to anaerobic conditions in the water column then are indicated for the Cenomanian/Turonian time interval.

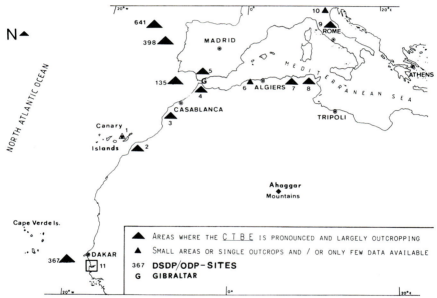

Figure 1. Important outcrops of the Cenomanian/Turonian Boundary Event (CTBE).

1 Fuerteventura (Canary Islands) (not clearly proven)	: Slope to Basin Plain Deposits
2 Tarfaya Basin (SW Morocco)	: Coastal Basin
3 Agadir Basin (SW Morocco)	: Coastal Basin to Outer Shelf Area
4 Rif Mountains (N Morocco) (Gibraltar Arch Area)	: Slope to Basin Plain Deposits
5 Penibetic (S Spain)	: Intramarginal High
6&7 Tell Atlas (N Algeria)	: Slope to Basin Plain Deposits (comparable with Rif)
8 Bahloul Formation (Tunisian Atlas)	: Shelf Sea (? with local depressions)
9 Umbrian Apennines (Central Italy)	: Intramarginal Basin
10 Euganean Hills (Northern Italy)	: Trough margin
11 Casamance Area	

*) Geochemisches Institut, Universität Göttingen, D-3400 Göttingen, F.R.G.

Lecture Notes in Earth Sciences, Vol. 8
Global Bio-Events. Edited by O. Walliser.
© Springer-Verlag Berlin Heidelberg 1986

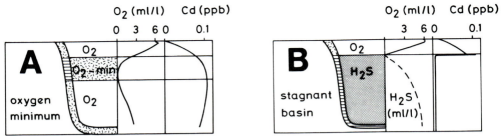

Figure 2. Idealized oxygen (hydrogen sulphide) and cadmium profiles in different paleoenvironmental settings: A = oxygen-minimum zone, B = stagnant basin.

The Cenomanian/Turonian Boundary Event (CTBE) marks a distinct oceanographic event in the Atlantic Ocean as well as in the western Tethys (see Fig. 1) and is characterized by its constancy in time, the deposition of biosiliceous black shales, and a distinct change in the faunal pattern of micro- and macrofossil associations (Thurow & Kuhnt 1986, Wiedmann et al. 1978).

For this investigation 20 black shale samples from outcrops on land from the Gibraltar Arch area (Morocco) and 20 TOC-rich black shales from Site 367 (DSDP Leg 41), besides 25 Lower Toarcian black shales from SW Germany and recent diatomaceous upwelling sediments from the Gulf of California have been analyzed for 5 major and 14 minor elements (see Table 1). Samples from both CTBE settings, outcrops on land and DSDP drill hole 367, show a remarkable enrichment of elements which form stable sulfides (Ag, Cd, Cu, etc.) or are strongly associated with organic matter (U, V) or both (Mo) (see Table 2). The Toarcian shales and Gulf of California oozes, in contrast, are less enriched in these particular elements, even though their TOC and S contents are comparable.

TOC-rich black shales may be formed under two different environmental settings: areas of high biological primary production, like presently occurring in regions of intense upwelling, and restricted, anoxic basins with hydrogen sulfide being present in the water column, like the Black Sea (see Fig. 2). In case of an upwelling environment TOC enters the sediment due to the high production rate of plankton in surface waters, the resulting diminished oxygen content of the intermediate water (oxygen minimum zone), and the high sediment accumulation rate and rapid burial. In contrast to this, anoxic basins accumulate TOC even under comparably low primary production rates. Here preservation of TOC in completely oxygen-free subsurface waters is important.

Trace metals may provide the tool to distinguish between both environments. Reliable, 'oceanographically consistent' analytical data

element	DSDP Site 367 (n=21)		CTBE Morocco (n=20)		Toarcian shales (n=25)		Gulf of California (n=50)	average shale low in TOC
ppm Ag	3.1	(1.9)	3.4	(2.1)	n.d.		0.2[a]	0.07
ppm Ba	771.	(713.)	190.	(164.)	191.	(165.)	566.	580.
ppm Cd	14.	(2.1)	18.2	(4.1)	2.7	(0.58)	2.5	0.13
ppm Co	34.	(29.)	3.3	(1.8)	21.	(17.)	6.6	19.
ppm Cr	263.	(249.)	106.	(92.)	65.	(54.)	44.	90.
ppm Cu	186.	(138.)	250.	(176.)	72.	(62.)	27.	39.
ppm Mn	282.	(201.)	86.	(33.)	908.	(812.)	193.	850.
ppm Mo	64.	(25.)	14.6	(12.)	19.	(10.)	11.9	2.6
ppm Ni	201.	(171.)	58.	(50.)	101.	(80.)	38.	68.
ppm Pb	15.3	(14.6)	11.7	(9.6)	59.	(51.)	17.[a]	22.
ppm Sb	15.	(4.6)	5.9	(4.7)	n.d.		4.[a]	1.
ppm Sr	223.	(208.)	60.	(45.)	1721.	(1248.)	167.	230.
ppm V	1080.	(766.)	499.	(467.)	163.	(138.)	101.	130.
ppm Zn	942.	(414.)	1009.	(383.)	285.	(102.)	88.	115.
% Al	4.32	(4.05)	2.52	(2.29)	4.85	(4.10)	4.72	8.84
% Ckarb.	1.10	(0.34)	0.57.	(0.22)	4.7	(4.2)	0.37	-
% Fe	3.36	(3.24)	1.31	(1.13)	3.14	(3.04)	2.15	4.85
% S	3.00	(2.60)	0.40	(0.19)	2.63	(2.44)	0.49	0.24
% TOC	8.94	(5.67)	4.94	(4.33)	6.73	(4.14)	4.35	0.2

values in brackets = geometric means   n.d. = not determined   a = only determined in composite sample

Table 1. Average chemical composition of CTBE black shales from DSDP Site 367 and outcrops on land, and other TOC-rich sediments in comparison to 'average shale', low in TOC.

for a large number of trace metals in the marine environment have been generated during the past 5 to 10 years by applying new analytical techniques and avoiding contamination during sampling. In seawater the concentration profiles of many elements closely follow those of nutrient like phosphate, nitrate or silica, with surface minima due to the up-take by organisms, and deep water enrichments resulting from regeneration processes within the oxic water column (see Fig. 3). Therefore labile elements, like Cd, should not accumulate in sediments deposited at greater water depth because they are already remineralized in the upper water column.

In areas of high primary productivity, like the Gulf of California, the sediment chemistry reflects the plankton chemistry and regeneration processes taking place in the water column (Brumsack 1986). Even though nutrients and associated trace metals may reach comparably high concen-tration levels in marine plankton (see compilation of plankton data in Brumsack 1986), only a relatively small fraction of these elements is finally buried in the underlying sediments, even in the depth range of 400 to 800 m at continental margins. Under present day oceanographic conditions high trace metal levels, comparable to those found in many CTBE black shales, are never found in TOC-rich upwelling sediments.

Unfortunately, relatively few data exist about the trace metal content of stagnant basins. Sediments from the Black Sea are enriched in Mo, V, Cu, and Ni, and to a lesser degree in Co, Ag, Cd, and Zn. Recent investigations on the behaviour of Cu, Zn, and Cd at the oxygen/hydrogen sulfide interface of an anoxic fjord (Jacobs et al. 1985) showed a dramatic solubility decrease of these elements (see also Fig. 2) and corresponding enrichments in the underlying sediments. The ability of an anoxic water column for accumulating several trace metals may there-fore be documented in the geological record by increased metal contents. The absolute metal enrichment then should be inversely correlated to the sediment accumulation rate.

The extremely high enrichment of Cd and Ag in CTBE black shales relative to 'average shale' by more than a factor of 100, as well as the enrichment of Mo, Sb, Zn, V, and Cu by factors of 10 to 100 (Table 2) indicates periods of complete stagnation for the Cenomanian/Turonian oceans. Sediment accumulation rates must have been very low, preserva-tion of TOC very high. The metal/aluminum ratios of CTBE black shales are much higher than those reported for recent upwelling sediments, element/TOC ratios of these sediments exceed those found in organic-rich particulates from sediment trap experiments (Table 3). The chemical composition of Toarcian black shales from SW Germany and recent Gulf of

enrichment factor	>100	<100 - 10	<10 - 3	<3 - 1	<1
locality					
CTBE DSDP Site 367	Cd Ag	Mo Sb Zn V Cu	Ni Cr Co Ba	Pb (Sr)	Mn
CTBE Morocco	Cd Ag	Zn Sb Mo Cu V	Cr Ni	Pb Ba Co	Mn
Lower Toarcium		Cd Mo (Sr)	Mn Zn Cu Pb Ni	V Co Cr	Ba
Gulf of California		Cd	Mo Ag Sb	Ba Zn V Pb Cu Ni	Cr Co Mn

the enrichment factor was calculated as follows:
element/Al (black shale) / element/Al (average shale); underlined elements are accumulated in comparable levels in CTBE samples; for details see Table 1

Table 2. Comparison of trace metal enrichments in TOC-rich sediments relative to 'average shale', low in TOC.

element ratios	sediment-trap particles	CTBE Morocco & DSDP Site 367	Toarcium SW-Germany	Gulf of California
Ag/TOC $* 10^{-6}$	1.5^a	35.-68.	-	3.7
Ba/TOC $* 10^{-3}$	$18.- 38.^{b,c}$	3.8-8.6	2.8	13.
Cd/TOC $* 10^{-5}$	1.8^d	15.7-36.8	4.0	5.8
Cr/TOC $* 10^{-3}$	6.3^e	2.2-2.9	1.0	1.0
Cu/TOC $* 10^{-3}$	$0.5-2.7^{b,c,d,e,f}$	2.1-5.1	1.1	0.6
Ni/TOC $* 10^{-3}$	$0.9-3.3^{b,c,d,e}$	1.3-2.3	1.5	0.9
Pb/TOC $* 10^{-3}$	$0.23-1.4^{d,e,f}$	0.17-0.25	0.88	0.39
V/TOC $* 10^{-3}$	0.8^d	10.-12.	2.4	2.3
Zn/TOC $* 10^{-3}$	$1.0-9.0^{b,c,d,e}$	11.-21.	4.2	2.0

a=MARTIN et al., 1983 b=COBLER & DYMOND, 1980 c=DYMOND et al., 1981
d=JICKELLS et al., 1984 e=LANDING & FEELY, 1981 f=CHESTER et al., 1978

Table 3. Comparison of average element/TOC ratios in sediment-trap particles and TOC-rich sediments.

California sediments compares quite well with sediment trap particles. The paleoenvironmental conditions during the CTBE must therefore have been completely different from those found in recent oceans.

Preliminary results from ODP Site 641A and DSDP Site 367 seem to indicate that also Au and U are accumulated in the CTBE black shales. These elements, too, are trapped in anoxic basins and may be concentrated to comparably high levels when sediment accumulation rates are low.

This study of trace metal accumulations in CTBE horizons will be extended to other localities (DSDP and ODP drill sites 398 and 641, the Penibetc of S Spain and the former upwelling area of the Tarfaya basin in SW Morocco) to further check the importance of diagenetic metal accumulation and former upwelling centers.

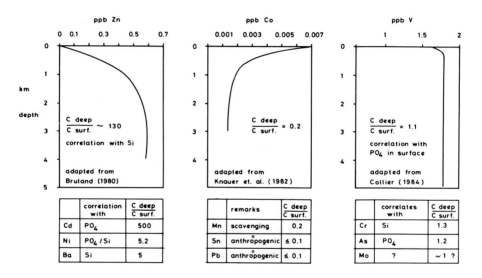

Figure 3. Idealized concentration - depth profiles for trace metals in seawater.

Acknowledgements

I would like to thank J. Thurow (Tübingen) for cooperation and providing the CTBE samples from Morocco.

REFERENCES

BRULAND, K.W. (1980): Oceanographic distributions of cadmium, zinc, nickel, and copper in the North Pacific.- Earth Planet. Sci. Lett. 47, 176-198.

BRUMSACK, H.-J. (1986): The inorganic geochemistry of Cretaceous black shales (DSDP Leg 41) in comparison to modern upwelling sediments from the Gulf of California.- in: SUMMERHAYES, C.P. & SHACKLETON, N.J. (eds.): North Atlantic Paleoceanography. Spec. Publ. Geol. Soc. London, Oxford, Blackwell.

CHESTER,R.A.;GRIFFITHS, A. & STONER, J.H. (1978): Minor metal content of surface seawater particulates and organic-rich shelf sediments.- Nature 275, 308-309.

COBLER, R. & DYMOND, J. (1980): Sediment trap experiment on the Galapagos Spreading Centre, Equatorial Pacific.- Science 209, 801.

COLLIER, R.W. (1984): Particulate and dissolved vanadium in the North Pacific ocean.- Nature 309, 441-444.

DYMOND, J.; FISCHER, K.; CLAUSON, M.; COBLER, W.; GARDNER; W.; RICHARDSON, M.J.; BERGER, W.; SOUTAR, A. & DUNBAR, R. (1981) A sediment trap intercomparison study in the Santa Barbara Basin.- Earth Planet. Sci. Lett. 53, 409.

JACOBS, L.; EMERSON, S. & SKEI, J. (1985): Partitioning and transport of metals across the O_2/H_2S interface in a permanently anoxic basin: Framvaren Fjord, Norway.- Geochim. Cosmochim. Acta 49, 1433-1444.

JICKELLS, T.D.; DEUSER, W.G. & KNAP, A.H. (1984): The sedimentation rates of trace elements in the Sargasso Sea measured by sediment trap.- Deep-Sea Res. 31, 1169-1178.

KNAUER, G.A.; MARTIN, G.A. & GORDON, R.M. (1982): Cobalt in north-east Pacific waters.- Nature 297, 49-51.

LANDING, W.M. & FEELY, R.A. (1981): The chemistry and vertical flux of particles in the northeastern Gulf of Alaska.- Deep-Sea Res. 28A, 19-37.

MARTIN, J.H.; KNAUER, G.A. & GORDON, R.M. (1983): Silver distribution and fluxes in north-east Pacific waters.- Nature 305, 306-309.

THUROW, J. & KUHNT, W. (1986): Mid-Cretaceous of the Gibraltar Arch area.- in: SUMMERHAYES, C.P. & SHACKLETON, N.J. (eds.): North Atlantic Paleoceanography. Spec. Publ. Geol. Soc. London, Oxford, Blackwell.

WIEDMANN, J.; BUTT, A. & EINSELE, G. (1978): Vergleich von marokkanischen Kreide-Aufschlüssen und Tiefseebohrungen (DSDP): Stratigraphie, Paläoenvironment und Subsidenz an einem passiven Kontinentalrand.- Geol. Rdsch. 67, 454-508.

THE CENOMANIAN-TURONIAN BOUNDARY EVENT: SEDIMENTARY, FAUNAL AND GEOCHEMICAL CRITERIA DEVELOPED FROM STRATIGRAPHIC STUDIES IN NW-GERMANY

A contribution to Project GLOBAL BIO-EVENTS

HILBRECHT, Heinz *), ARTHUR, Michael A. **) &
SCHLANGER, Seymour O. ***)

Abstract: "Oceanic Anoxic Events" are time envelopes of increased organic carbon burial, rich in carbon-12 by biogenic fractionation, accompanied by a strong positive shift of the del carbon-13 values of the carbonate in the late Cenomanian - early Turonian. The return to lower values reflects erosion and reoxidation of the carbon-12 rich organic material during times of tectonic activity. The evolution and diversity of planktonic foraminifera is closely bound to the event. The diversity is inversely correlated to del carbon-13 and the occurrence of distinct species in the stratigraphical record provides evidence of changes of the Mid-Water Oxygen Minimum Zone.

In late Cenomanian and early Turonian sequences deposited in globally distributed, diverse basinal settings a marked organic carbon burial event is evident and was termed "Oceanic Anoxic Event" (OAE) by Schlanger & Jenkyns (1976). Their use of this term did not imply global anoxia in the entire world ocean but was used as a descriptive term to denote a period of increased deposition of organic material in marine basins. Black shales may, from basin to basin, differ in details of their stratigraphy, duration of deposition, sedimentary processes and palaeoceanography, that lead to their formation.

The geochemical expression of the Cenomanian - Turonian OAE was found by Scholle & Arthur (1980) who recognized a strong positive carbon-13 excursion in the archaeocretacea foraminiferal zone, which spans a time of about 1 m.y. at the Cenomanian - Turonian boundary transition. This excursion reflects the shift of the oceanic carbon isotope reservoir by biogenic fractionation and intensified burial of carbon-12 enriched organic matter. Based on that assumption a rate of 20 - 40 % more burial of organic matter than in modern times was calculated by Arthur et al. (1986).

The stratigraphy of the isotope excursion was studied by Schlanger et al. (1986) and Hilbrecht & Hoefs (1986). It occurs in the archaeocretacea-Zone of the standard foraminiferal zonation and appears to be isochronous in NW-Europe also in the ammonite and inoceramid zonation.

*) Institut für Paläontologie der Freien Universität Berlin, D-1000 Berlin 33, F.R.G.

**) Graduate School of Oceanography, University of Rhode Island, Narragansett, RI 02882, U.S.A.

***) Department of Geological Sciences, Northwestern University, Evanston, IL 60201, U.S.A.

Table 1: Predicted environmental and sedimentary development based on the carbon isotope record after ARTHUR et al. (1985) and independent observations from NW-Germany by HILBRECHT (1986) and DAHMER & HILBRECHT (1984, in prep.). For explanations and interpretations see text. Note the times of low tectonical activity in the basal Upper Cenomanian and the plenus bed

Stratigr.	predictions	sedimentology	palaeontology
Middle Turonian	*(vertical text spanning: increasingly efficient erosion and oxydation of 13C-rich rocks, decreasing preservation potential of organic matter)*	slumping, sliding, mass flows common. No black shales. Sandy turbidites include Cenomanian fossils in Westphalia.	Start of the Inoceramus lamarcki/cuvierii-lineage. Increasing diversity of Marginotruncana and abundant other planktonic foraminifera, incl. keeled (deeper water) species.
mytiloides-event		lensoidal tempestites common, partly mass flows, abundant bioclastic wacke--and pack-stones, shallow water sediments also in basins (ramp facies).	highest diversity of Mytiloides-type inoceramids and planktonic foraminifera, abundant brachiopods and occasionally solitary corals and benthos-related fishes (Ptychodus), high morphological variability of inoceramids, echinoids at swells.
		thinner, partly bioturbated black shales (1-2 % org. C), proximal mud turbidites in basins increasingly present, lensoidal tempestites become abundant in swell facies.	abundant Mytiloides-type inoceramids with comparable morphotypes in basinal and swell facies, occasionally brachiopods, increasing diversity of planktonic foraminifera including first twin-keeled (deeper water) Marginotruncana and abundant dinoflagellates in basinal facies.
Lower Turonian = range of Mytiloides		thick black shales (1-3% org.C) only in central basins, distal tempestites and sheet flows, minor mass flows at swells, slides in basins occ. large.	Low diversity of Mytiloides and late Inoceramus with "black shale"morphotypes in central basins, intermediate diversity of planktonic foraminifera (mainly globular, shallow water forms), fish scales and bioturbation abundant in black shales.
Upper Cenomanian	*(vertical text spanning: Transgression, maximum of org. C burial, black shales abundant, MWOMZ expansion and oxygen depletion at a maximum)*	thick black shales (2-3% org.C) and distal mud turbidites in basins, marls and partly laminated "fish shale" in tectonically active swell areas (trap).	rare to common "black shale" morphotype of Inoceramus in basins and occasionally in the "fish shale" at swells, very low diversity of planktonic foraminifera (nearly all globular, shallow water forms), rare dinoflagellates.
		erosional relicts of black shales (2-5% org. C), slides common in basins, swell sediments eroded in tectonically passive areas (no traps) during latest Cenom.-earliest Turonian.	no macrofossils, except for rare inoceramid debris (redeposited), rare planktonic foraminifera and marine phytoplankton, mud turbidites are bioturbated at the top.
plenus limestone bed (ca. 1m)		upper and lower limits are erosion surfaces, little thickness variation and internal reworking, tectonically quiet.	lowest diversity of planktonic foraminifera, abundant low diverse macrofauna of inoceramids brachiopods, echinoids, ichnofossils include common Thalassinoides
extinction of Rotalipora	*(vertical text spanning: increasing rate of burial and preservation of org. C)*	thin, mostly bioturbated black shales (1-2%%org.C) and mud turbidites in basins, thin mass flows at swells, rich ichnofauna	sudden drop of diversity of planktonic foraminifera after interval of intermediate diversity below extinction of Rotalipora, small inoceramids and oysters common, abundant Chondrites.
facies change		sudden drop of carbonate contents above an erosion surface	low plankton diversity (foraminifera), omission suite burrows highly diverse.
		limestones with intercalated marl seams, little lateral differentiation of facies and thickness, basal sequence of mud turbidites and intercalated marls in basins (little lateral extend).	low diversity of macrofauna (inoceramids, echinoids, oysters) partly abundant in distinct beds, rich ichnofauna, Lingula and diverse benthonic foraminifera in grey marls of basal basinal facies (they resemble bioturbated black shales).

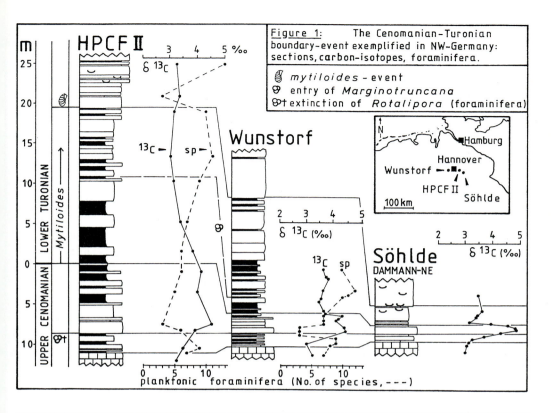

Figure 1: The Cenomanian-Turonian boundary-event exemplified in NW-Germany: sections, carbon-isotopes, foraminifera.

mytiloides - event
entry of *Marginotruncana*
†extinction of *Rotalipora* (foraminifera)

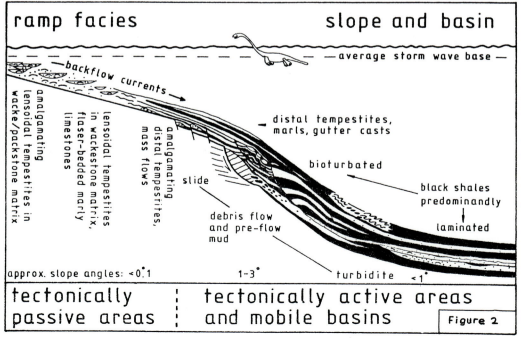

Figure 2

Less carbon-13 enriched carbonates may be present above and below that
interval. The long-term isotopic signal is always interrupted by a marked
peak at its maximum (Fig. 1). The sharp definition of the peak (e.g.
at Söhlde and Wunstorf) indicates hiati which are common in the upper
archaeocretacea-Zone.

The application of the foraminiferal zonation in NW-Germany is problematic in the light of general rarity of the critical species Praeglobotruncana helvetica, that defines the overlying zone. Other species
enter at different stratigraphical levels in basinal and swell facies
(e.g. the genus Marginotruncana occurs earlier at swells. Hilbrecht
1986). It is evident from the relationship of the carbon-13 excursion
to the diversities of planktonic foraminifera (Fig. 1) that the isotope
signal is more or less inversely correlated to diversity and may therefore indicate ecological restrictions of the ranges of foraminifera and
their stratigraphy.

Although the water depth increased in the course of the late Cenomanian - early Turonian transgression, the deeper water morphotypes of
planktonic foraminifera become extraordinarily rare or absent in times
of the carbon-13 peak . The fauna consists of globular forms of Hedbergella, Dicarinella and occasionally Praeglobotruncana. Below the peak
occurs the extinction of the keeled genus Rotalipora, while above the
peak the first fully twin-keeled forms, i.e. the genus Marginotruncana,
enters at stratigraphically different levels in the different facies.
According to Douglas & Savin (1978) the latter keeled morphotypes would
represent the middle to deep water planktonic assemblage which is apparently absent during the carbon isotope anomaly.

Based on the occurrence of black shales in mid-water depth, the
development of an expanded mid-water oxygen minimum zone (MWOMZ) was
suggested by Schlanger & Jenkyns (1976). This agrees well with the record
of planktonic foraminifera. In their review of black shale occurrences
Schlanger et al. (1986) came to the conclusion that otherwise important
mechanisms of black shale formation (e.g. upwelling, barred basins) could
not explain the widespread occurrences at continental slopes, oceanic
rises and in shelf- and epicontinental seas. That model is also supported
by considerations on the formation of deep water masses by thermohaline
processes. In the Cretaceous the existence of warm saline bottom waters
(WSBW), generated in areas of evaporation at shelves, controlled the potential oxygen depletion in the MWOMZ.

An example of the sedimentary processes during the OAE is provided
by the NW-German pelagic sequences, which were studied by Dahmer & Hilbrecht (1984 & in prep.). Their analysis was based on sections in differ-

ent palaeogeographical, structural,and tectonical positions near saltdomes, faulted blocks,and at the Rhenish Massif.

The pelagic Cenomanian - Turonian boundary successions are developed
in Rotpläner facies near swells and the Black Shale facies of the local
basins (Ernst, Schmid & Seibertz 1983). The Rotpläner comprise reddish-
brown, white and greenish sequences of in-situ reworked, mass flow,
sheet flow and storm deposits (tempestites). The Black Shale facies is
built up by mud turbidites, occasionally mud flows and intercalated laminated and bioturbated black shales. The laminae were deposited from episodic suspensions, as is equivalent from graded bedding, imbrication and
sole marks. Where inoceramids (pelecypods are the only benthic larger
organisms) occur they were mainly buried in living position. These sediments were secondarily bioturbated by burrowing organisms (ichnofauna:
Chondrites, Zoophycos, Planolites et aut.) which started their activity
from a distinct level after the deposition of a laminated sequence.

The lateral facies relationships between both facies are shown in
Fig. 2 after Dahmer & Hilbrecht (in prep.). In Table 1 we have summarized
the sedimentological and palaeontological record and compared to predic-
tions by Arthur et al. (1985) which were based on the assumption that
the carbon isotope excursion reflects the global increase in preservation of organic carbon in the sediments and a return to normal isotopic
compositions by erosion and oxidation of organic matter after the carbon-
13 peak , because otherwise the return to less positive isotope values
would not have occurred.

It is therefore important to recognize that the laminated black shales become increasingly younger towards the basin-centres. The "fish
shale" of expanded Rotpläner sections is only preserved in tectonic traps
in areas of active faulting. In the marginal Black Shale facies of the
Teutoburger Wald, thin erosional relicts of black shales are preserved
as lenses between successive mud turbidites, while in more basinal sections the infill of burrows consists of black shale material (erosion),
in the limestone interval above the extinction of Rotalipora. The uppermost Cenomanian black shales are devoid of benthonic inoceramids in the
few preserved deeper beds, while this group becomes increasingly abundant
towards the Cenomanian - Turonian boundary, in the late phase of the
carbon-13 peak . The return to normal carbon-13 values of the carbonate
is also a phase of redeposition by slides, related to syndepositional
faults in the basins. It correlates with the "fish shale" occurrences
in the Rotpläner facies.

From the diachroneity of the black shales, the decreasing organic
carbon contents and proportion of terrigenous derived organic material

until the higher early Turonian it is evident that most of the preserved black shale beds are redeposits which were transported successively from more marginal areas towards the basins. The decrease of organic carbon contents is even more marked if the diagenetic enrichment by carbonate dissolution is taken into account (Hilbrecht & Hoefs 1986).

The "stepwise" redeposition of organic matter towards the central basins corresponds to tectonical activity which is increasingly active until a maximum in the Middle Turonian. This activity precedes the well known "subhercynian phase", which starts in the uppermost Turonian. The early and Middle Turonian tectonic activity was not as marked as the later movements, but can be recognized in a comparable geographical range and is found independently of the local structural conditions (Hilbrecht & Kaplan, in prep.).

Whereas the introduction of the OAE/carbon-13 peak is clearly bound to the latest Cenomanian transgression (Schlanger et al. 1986), its decay was controlled by increasingly active tectonical movements and we expect that future studies of their relationships in other areas of the world will greatly improve our understanding of such events.

REFERENCES

ARTHUR, M.A.; SCHLANGER, S.O. & JENKYNS, H.C. (1986): The Cenomanian-Turonian oceanic anoxic event, II. Paleooceanographic control on organic matter production and preservation.- in: BROOKS, J. & FLEET, A. (eds.): Marine Petroleum Source Rocks. Geol. Soc. London, Spec. Publ. 24.

DAHMER, D.D. & HILBRECHT, H. (1984): Beziehungen zwischen Stratigraphie und Fazies im Cenoman/Turon-Grenzbereich in NW-Deutschland.- Geotagung 1984, Hamburg (Abstr.), 20-22.

-- & -- (in prep.): The pelagic basin and swell facies of the late Cenomanian and early Turonian of NW-Germany.

DOUGLAS, R.G. & SAVIN, S.M. (1978): Oxygen isotopic evidence for the depth stratification of Tertiary and Cretaceous planktic foraminifera. - Mar. Micropal. 3, 175-196.

ERNST, G.; SCHMID, F. & SEIBERTZ, E. (1983): Event-Stratigraphie im Cenoman und Turon von NW-Deutschland.- Zitteliana 10, 531-554.

-- ; WOOD, C.J. & HILBRECHT, H. (1984): The Cenomanian-Turonian boundary problem in NW-Germany with comments on the north-south correlation to the Regensburg area.- Bull. Geol. Soc. Denmark 33, 103-113.

HILBRECHT, H. (1986): On the correlation of the Upper Cenomanian and Lower Turonian of England and Germany (Boreal and N-Tethys).- Newsl. Stratigr. 15, 115-138.

-- & HOEFS, J. (1986): Geochemical and palaeontological studies of the d carbon-13 anomaly in boreal and North-Tethyan Cenomanian-Turonian sediments in Germany and adjacent areas.- Palaeoecol., Palaeoclim., Palaeogeogr. 53, 169-189.

-- & KAPLAN, U. (in prep.): Slumping, Rutschungen und debris flows im Unter- und Mittel-Turon NW-Deutschlands als Folge regionaler tektonischer Bewegungen.

SCHLANGER, S.O. & JENKYNS, H.C. (1976): Cretaceous oceanic anoxic events: causes and consequences.- Geol. Mijnbouw 55, 179-184.

SCHLANGER, S.O.; ARTHUR, M.A.; JENKYNS, H.C. & SCHOLLE, P.A. (1986):
    The Cenomanian-Turonian oceanic anoxic event, I. Stratigraphy and
    distribution of organic carbon-rich beds and the marine d carbon-13
    excursion.- in: BROOKS, J. & FLEET, A. (eds.): Marine Petroleum
    Source Rocks. Geol. Soc. London, Spec. Publ. <u>24</u>, 342-375.
SCHOLLE, P.A. & ARTHUR, M.A. (1980): Carbon isotope fluctuations in Cre-
    taceous pelagic limestones: potential stratigraphic and petroleum
    exploration tool.- Amer. Assoc. Petr. Geol. (AAPG) Bull. <u>64</u>, 67-87.

UPPER CRETACEOUS EVENT-STRATIGRAPHY IN EUROPE

DAHMER, Dirk-Daniel & ERNST, Gundolf *)

A contribution
to Project
GLOBAL
BIO-
EVENTS

IUGS
UNESCO

Abstract: Within the framework of the IGCP Major Project "Mid-Cretaceous Events" an innovative correlation scheme was established. Until now 40 - 50 stratigraphical events of different nature have been identified in the stages of the Cenomanian to the Campanian, providing correlation within NW-Germany and to England. The method and selected events, such as the <u>primus</u> Event (Middle Cenomanian), the long-ranged Mid-Cenomanian Event and the complex <u>Hyphantoceras</u> Event (Upper Turonian) were described.

General aspects

It becomes increasingly apparent to stratigraphers that it is impossible to construct a sufficient functional stratigraphic framework based on single methods only (e.g. bio- or lithostratigraphy, cycle geochronometry etc.). In order to solve this problem for the Upper Cretaceous of NW-Germany a combination of methods was introduced by Ernst, Schmid & Klischies (1979) as the concept of "Multistratigraphy". Further refinement of the stratigraphic subdivision was attained by the establishment of stratigraphical events (marker horizons), which subdivide poorly defined biozones. This concept, "Event-stratigraphy", was developed by Ernst, Schmid & Seibertz (1983) in NW-Germany, in co-operation with the Berlin working group (e.g. Badaye, Farman, Hilbrecht, Kott, Rasemann).

Isochronism of events

The question of isochronism of particular horizons has to be discussed in comparison with the biostratigraphic framework, which provides the best time resolution until yet.

It cannot be presumed that the lower or upper boundaries of biozones, when traced throughout a local basin, are isochronous (e.g. Matthews 1984). The actual range (local range) of index species in the outcrop often extends over only a part of the hypothetical range and should therefore be regarded to represent a "local range zone" or "Teilzone".

In the Upper Cretaceous of Europe zonal boundaries are often coupled with lithological changes or hiatuses (e.g. Birkelund et al. 1984). Where the occurrence of index species is restricted to single horizons (e.g. <u>Acanthoceras jukesbrownei</u> of the upper Middle Cenomanian), zonal

*) Institut für Paläontologie der Freien Universität, D-1000 Berlin 33, F.R.G.

Lecture Notes in Earth Sciences, Vol. 8
Global Bio-Events. Edited by O. Walliser
© Springer-Verlag Berlin Heidelberg 1986

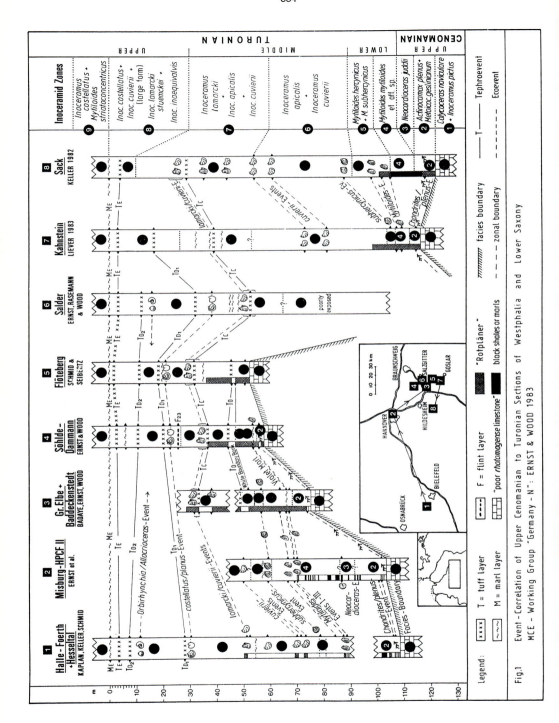

Fig.1 Event-Correlation of Upper Cenomanian to Turonian Sections of Westphalia and Lower Saxony

MCE - Working Group "Germany - N": ERNST & WOOD 1983

boundaries are arbitrarily defined. The record of ammonites in particular is determined by preservational conditions. In addition, the ecological factors regulating faunal distribution in a basin must be taken into account, but for most Upper Cretaceous index species these factors have not been sufficiently studied.

The strict isochroneity of events has thus to be proven within a rough framework of biozones, but the time resolution afforded by Event-stratigraphy is in many cases much higher than that of the biostrati-graphical "yardstick". We regard events to be of short duration and to be isochronous since the constant sequence of discrete marker beds allows us to subdivide biozones.

The consequent application of Event-stratigraphy can therefore be used to provide solutions for the aforementioned problems of biostrati-graphy. The concept of Event-stratigraphy was developed in conjunction with the IGCP Major Project 58 ("Mid-Cretaceous Events"). The application of this concept led to a remarkable refinement of the Upper Cretaceous correlation scheme of the Cenomanian to Coniacian and Campanian according to Ernst, Schmidt & Seibertz (1983), Ernst, Wood & Hilbrecht (1984), Wood, Ernst & Rasemann (1984).

Short-term phases of synsedimentary tectonic activity (lateral thick-ness variation) can now be studied in detail (Dahmer & Hilbrecht, in prep.) thanks to the refined stratigraphical subdivision. It became also possible to calculate the extent of stratigraphical gaps.

The use of Event-stratigraphy as a prospecting aid was first tested in a project to correlate the Cenomanian to Coniacian sequence between Lower Saxony and Westfalia in collaboration with the stratigraphers U. Kaplan and E. Seibertz. After this successful application of Event-stratigraphy, the correlation of the Cenomanian to Coniacian and Campa-nian sequence between NW-Germany and England was undertaken in co-opera-tion with C.J. Wood from the British Geological Survey (Nottingham). The correlation of Upper Turonian sequences between both regions is based chiefly on the correlation of the marl seams T_{D2}, T_E and M_E (T: tuff, i.e. tephro-event) and flint maxima in the sequences. Figures 1 and 3 indicate the correlation of German sequences according to these tephro-events.

It is difficult to determine the causes of peak occurrences of single taxa or foreign faunal elements. Some of these eco-events may have been caused by the influx of waters from adjacent oceanic basins due to the opening of seaways and the formation of new ecological niches in NW-Germany (e.g. primus Event of the Middle Cenomanian, Fig. 2). However, regression is probably one of the chief causes of bio-events (e.g. Pyc-

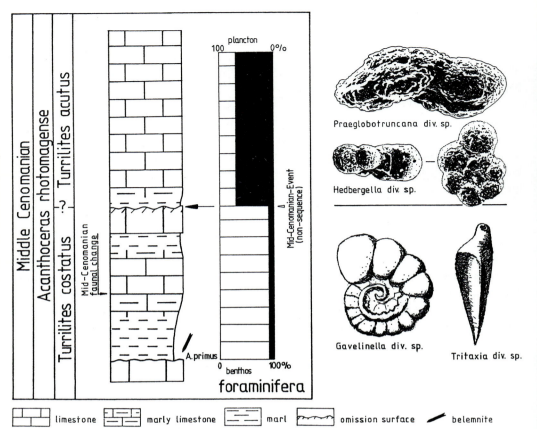

Figure 2. Schematic section of the lower Middle Cenomanian of NW-Germany. The traditional biostratigraphy is refined by a sequence of well defined events in the sequence upwards: primus Event, Mid-Cenomanian faunal change (MCFC), Mid-Cenomanian event. On the right hand side some typical foraminifera from below and above the Mid-Cenomanian Event are pictured.

nodonte/jukesbrownei Event, upper Middle Cenomanian; Mytiloides Event, Lower Turonian). The upper of two Didymotis Events is of particular interest in defining the Turonian-Coniacian boundary as proposed by Wood, Ernst & Rasemann (1984). These events indicate that this typical delicate mollusc invaded western Europe only during short time intervals.

The following examples are intended to illustrate the methods of Event-stratigraphy and the internal structure of some events.

Primus Event (Fig. 2) *)

This event belongs stratigraphically into the lower Middle Cenomanian
ammonite zone of Turrilites costatus (Lamarck).Above an omission surface
the facies usually passes from limestones into dark marls, which yield
the species-rich Actinocamax primus-fauna (Ernst, Schmid & Seibertz
1983). This primus-fauna comprises the belemnites A. primus and A.
boveri, small solitary corals, thin-shelled inoceramids, pectinids, ser-
pulids, crustaceans, and an association of small brachiopods. The occurrence of these faunal elements is confined to an interval approx. 2 m
thick. They are usually concentrated in thin bedded mass flows (sheet
flows). These layers are interstratified with background sediments yielding poorly preserved ammonites.

Mid-Cenomanian faunal change (Fig. 2)

The primus-fauna disappears coincidentally with a change to higher carbonate contents. Preliminary data indicate the presence of coeval but
locally differentiated ammonite associations above the Mid-Cenomanian
faunal change (MCFC). The division of these ammonite associations into
four groups is based on analysis of the acme occurrences. Single scattered specimens probably do not reflect paleoecologically interpretable
distribution patterns.

These four ammonite associations are distinguishable in NW-Germany
and recur at certain levels in the interval from directly below the
primus Event to the base of the Mid-Cenomanian Event. The associations
are:

1. Austiniceras austeni, Sciponoceras baculoide (either both or one of
   them predominates)
2. Acanthoceras rhotomagense (dominant)
3. Anisoceras plicatile, Schloenbachia coupei (both abundant)
4. Turrilites div. sp. (predominant)

Within a given horizon the predominant association varies from locality
to locality within the Turrilites costatus Zone, in other words these
associations are mutually substitutable. The presence of Association 1
excludes the abundant presence of forms from Associations 2 - 4. This
pattern could be fortuitous due to infelicities in collection, but the
occurrence of single anomalous forms in a clearly dominated association
cannot render this grouping implausible. The geographical distribution
of the ammonite associations appears to be the result of primary i.e.
paleoecological factors.

*) Concerning this chapter we are deeply indepted to our co-operator
   C.J. Wood.

Mid-Cenomanian Event (Fig. 2)

The term Mid-Cenomanian Event was established by Ernst, Schmid & Seibertz (1983) for a glauconitized nodular limestone horizon in the HPCF II quarry in Misburg (Hannover). The coeval horizon in England was termed "Mid-Cenomanian non-sequence" by Carter & Hart (1977). Based on the drastic increase in the planktonic/benthonic ratio of foraminifera (Fig. 2) above this Mid-Cenomanian Event horizon, this event of non-sequence is traceable throughout NW Europe and correlates with a horizon in the Granerous Shale of Colorado, USA (Carter & Hart 1977, Fig. 11). In NW-Germany this change in foraminifera was first observed and researched in detail by one of the authors (D.D.); this observation was corroborated in the Wunstorf quarry (Lower Saxony) by T. Meyer (pers. comm.). In Söhlde (Lower Saxony) the change in the planktonic/benthonic ratio results from a marked change in preservation of foraminifera. Here the benthonic foraminifera below the Mid-Cenomanian Event are poorly preserved, whereas keeled and globular plankton is extremely rare and etched. In other sections Hedbergella planispira is the dominant plank-tonic foram below and directly above the Mid-Cenomanian Event. We assume that this species inhabited the uppermost part of the water column and that its predominance indicates relatively shallow water. Taxa which indicate deeper water, according to Hart & Bailey (1979), occur only higher up in the sequence above the Mid-Cenomanian Event.

The data from NW-Germany tally well with the interpretation of Carter & Hart (1977), that a major transgression caused the change in the planktonic/benthonic ratio above the Mid-Cenomanian Event.

Present data suggest , that the environmental changes affecting faunal compositions in the sequence from the primus- to the Mid-Ceno-manian Event were caused by a rise in sea-level. This transgressive trend is evidenced by upward fining of sediments and a general increase in carbonate content. The primus-fauna was apparently a shallow water fauna as is indicated by the coarse sediment. This fauna was vertically replaced (MCFC) by other faunal associations (one of the ammonite associations, 1 - 4), which reflect the entry of locally differentiated ecological conditions. The European-wide change in the planktonic/benthonic ratio of foraminifera marks the entry of the more or less uniform conditions of a pelagic environment in the course of transgression.

Hyphantoceras Event (Fig. 3) *)

The Upper Turonian Hyphantoceras Event identified by Ernst, Schmidt &

─────────────────

*) A detailed paper on the Hyphantoceras Event is in preparation by
   U. Kaplan, F. Schmid and the authors.

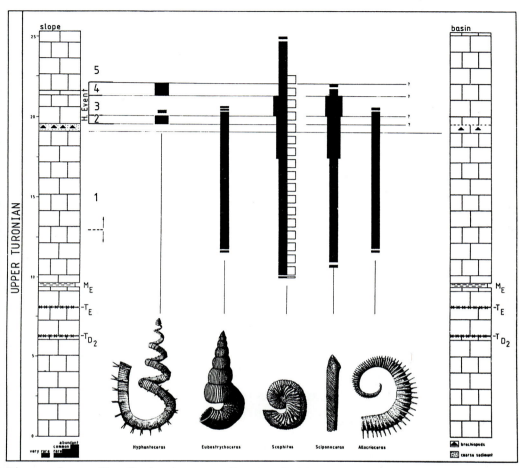

Figure 3. The <u>Hyphantoceras</u> Event of NW-Germany and its vertical and lateral development. T_{D2}, T_E and M_E are tephro-events (tuff layers), which provide correlation throughout NW-Germany and to England (M_E). The presence and ranges of typical ammonites are indicated in black for the slope facies and in white for the basinal facies (off swell). The ammonite associations and sedimentological characteristics of the environment are discussed in the text.

Seibertz (1983) is defined by the occurrence of the heteromorph ammonite <u>Hyphantoceras reussianum</u> (d'Orbigny). Our studies revealed an internal zonation of this 2 - 3 m thick horizon. The concomitant fauna includes heteromorph ammonites, normally coiled ammonites, brachiopods, inocera-mids, pectinids, small solitary corals, bryozoans, irregular echinoids, crinoids, sponges and occasionally snails. This extremely diverse "<u>reussi-anum</u>-fauna" is concentrated in the vicinity of two acme occurrences of <u>Hyphantoceras</u>, which form the upper and lower boundary of the event. Five distinct ammonite associations (1 - 5 in Fig. 3) can be discerned in the vertical sequence; the Associations 2 - 4 encompass the <u>Hyphan-</u>

toceras Event.

Association 1 consists of _Scaphites_, _Sciponoceras_, _Allocrioceras_, and _Eubostrychoceras_, as well as _Subprionocyclus neptuni_ (Geinitz), the index species of cephalopod zonation. The occurrence of ammonites and the accompanying fauna is restricted to thin layers, approximately 5 cm thick, which are interbedded with strata of poorly fossiliferous sediments up to 10 cm thick. These ammonite assemblages are dominated either by _Scaphites_ or _Sciponoceras_.

The bed just below the first _Hyphantoceras_ flood (Association 2) is characterized either by a coarse sediment (casts of ammonite debris, other bioclasts), or by acme occurrences of brachiopods. In other locations this horizon is a highly bioturbated marly limestone poor in fauna and bioclasts.

Association 2 contains the first occurrence of _Hyphantoceras_ in the designated sequence. _Hyphantoceras_ is generally preserved in small fragments. The top of Association 2 is not sharply defined, as sparse _Hyphantoceras_ fragments range into the basal Association 3 (Fig. 3). In one location (Gustedt, Salzgitter area) _Hyphantoceras_ is rare in the horizon of Association 2. Here the accompanying fauna and _Hyphantoceras_ are restricted to thin layers, whereas in other sections _Hyphantoceras_ and the _reussianum_-fauna are abound at this level.

Association 3 is dominated either by peak occurrences of _Scaphites_ or _Sciponoceras_, whereby _Scaphites_ floods are associated to common sponges. _Allocrioceras_ and _Eubostrychoceras_ are still present in the lower half of this interval, with _Eubostrychoceras_ ranging somewhat higher up in the sequence than _Allocrioceras_. As for the echinoid fauna it is remarkable that _Micraster bourchardi_ (Quenstedt), otherwise found only in the sequence below M_E (Figs. 1 and 3), recurs in this association.

Association 4 is defined by the second flood occurrence of _Hyphantoceras_, which is usually more pronounced than the first _Hyphantoceras_ flood (Ass. 2). _Hyphantoceras_ in Association 4 is more completely preserved and larger than specimens of Association 2. Single _Hyphantoceras_ fragments in Association 4 obviously derive from large specimens whose existance is attested by the frequent occurrence of huge U-shaped living chambers. _Lewesiceras mantelli_ is generally more frequent than in the underlying associations, although in the Groß Flöte quarry (Oderwald) this ammonite is found throughout the section. Typical echinoids in Association 4 are _Sternotaxis ananchytoides_ (Elbert) and the huge _S. latissimus_ (Roemer), which is restricted to the Teutoburger Wald range (Westfalia).

Association 5 is usually dominated by _Scaphites_, but the fauna in general is depleted. _Hyphantoceras_ is entirely missing. This drop in

diversity and absence of Hyphantoceras define the upper limit of the
Hyphantoceras Event.

The sediments of the designated sequence can be described as an
intercalation of poorly fossiliferous mudstones and bioclastic wackestones yielding the ammonites. Between these limestones thin marl seams
(2 - 3 cm) are interstratified. Traces, which obviously stem from these
marl seams, piped marly material down into the limestones, where the
diagenetic alteration by pressure solution led to a wavy flaser bedding.
Planolites/Thalassinoides burrows are often the inhomogeneity at which
pressure solution originated. The ichnofossil assemblages comprise
several types of Chondrites and Muensteria sp. In some of the bioclastic
wackestones fossils are concentrated at the base of beds, giving an
intimation of graded bedding, but usually the fauna is matrix supported
without any gradation. As a rule the limestone beds are composite beds,
i.e. an intercalation of layers rich and poor in fossils. We interpret
the matrix supported fossiliferous layers as debris flows and regard the
poorly fossiliferous mudstones as the background sediment.

An important aspect of the Hyphantoceras Event is the distinct
lateral variation of the faunal composition. The occurrence of Hyphantoceras and the species-rich reussianum-fauna is restricted to slopes of
swell areas, constituted by saltdomes, in Lower Saxony and to structural
highs in Westfalia. In the more basinward settings the fauna is depleted,
and Hyphantoceras is extremely rare or absent. This is the case for instance in the Söhlde quarries (Salzgitter area, Lower Saxony), where rare
Scaphites geinitzi occur above a horizon yielding Orbirhynchia sp. and
other brachiopods. We regard this brachiopod occurrence in basinal sections as correlative with the brachiopod horizon in swell areas, as is
indicated in Figure 3.

The lateral and vertical variation of the faunal composition is obviously due to ecological factors, since the preservational conditions
cannot be regarded to generate such distinct distribution patterns.

One of these ecological factors could have been breeding habits. This
possibility is suggested by the observation that the flood occurrences of
Hyphantoceras, Scaphites and Sciponoceras contain predominantly mature
specimens. It is well known that certain recent cephalopods form breeding
swarms. Squids for example travel over 1000 km without eating, in order
to reach their breeding ground, whereupon they die of consumption, after
copulating and depositing their eggs.

The missing of immature Hyphantoceras, Scaphites and Sciponoceras
suggests that the juvenile forms preferred another environment. However,
the preservation of ammonitellas is bound to fortuitous preservational

362

conditions. It could be assumed that the adults had to travel from
pelagic environments to reach their breeding ground, where they died
after copulation and deposition of eggs, due to consumption like the
squids.If so, they preferred swell regions for breeding from where the
shells were transported downslope by gravity flow (debris flows) after
death. The evolution of suitable conditions for breeding, such as the
formation of swells due to local tectonic uplift or a regression in
general could have been the chief cause for the development of the
Hyphantoceras Event and the occurrence of the accompanying fauna.

REFERENCES

BIRKELUND, T.; HANCOCK, M.J.; HART, M.B.; RAWSON, P.F.; REMANE, J.; ROBASZYNSKI, F.; SCHMID, F. & SURLYK, F. (1984): Cretaceous stage boundaries - Proposals.- Bull. geol. Soc. Denmark 33, 3-20.

CARTER, D.J. & HART, M.B. (1977): Aspects of Mid-Cretaceous stratigraphical micropaleontology.- Bull. Br. Mus. nat. Hist. (Geol.) 29, 1-135.

ERNST, G.; SCHMID, F. & KLISCHIES, G. (1979): Multistratigraphische Untersuchungen in der Oberkreide des Raumes Braunschweig-Hannover.- Aspekte der Kreide Europas, IUGS Ser. A 6, 11-46.

-- ; SCHMID, F. & SEIBERTZ, E. (1983): Event-Stratigraphie im Cenoman und Turon von NW-Deutschland.- Zitteliana 10, 531-554.

-- WOOD, C.J. & HILBRECHT, H. (1984): The Cenomanian-Turonian boundary problem in NW-Germany with comments on the north-south correlation to the Regensburg Area.- Bull. geol. Soc. Denmark 33, 103-113.

HART, M.B. & BAILEY, H.W. (1979): The distribution of planktonic Foraminiferida in the mid-Cretaceous of N.W. Europe.- Aspekte der Kreide Europas, IUGS Ser. A 6, 527-542.

MATTHEWS, R.K. (1984): Dynamic Stratigraphy.- Prentice Hall, New Jersey.

WOOD, C.J.; ERNST, G. & RASEMANN, G. (1984): The Turonian-Coniacian Stage Boundary in Lower Saxony (Germany) and adjacent areas: the Salzgitter-Salder Quarry as a proposed international standard section.- Bull. geol. Soc. Denmark 33, 225-238.

CRETACEOUS/TERTIARY BOUNDARY

THE DECCAN TRAPPS (INDIA) AND CRETACEOUS-TERTIARY BOUNDARY EVENTS

A contribution to Project GLOBAL BIO-EVENTS

BESSE, Jean *), BUFFETAUT, Eric **), CAPPETTA, Henri ***), COURTILLOT, Vincent *), JAEGER, Jean-Jacques **), MONTIGNY, Raymond ****), RANA, Rajendra, S. *****), SAHNI, Ashok *****), VANDAMME, Didier *) & VIANEY-LIAUD, Monique ***)

Abstract: It has been suggested that the eruption of the Deccan Trapps has contributed to the events of the Cretaceous-Tertiary transition as a possible cause of Iridium enrichment and/or other physicochemical disturbances. However, no precise chronological framework was available for the emplacement of these Trapps, which form one of the largest known occurrences of continental basalt flows. A joint Indian-French project, still under way, is providing more accurate informations about the chronology of the Deccan Trapps and its possible implications for Cretaceous-Tertiary boundary events.

Up to now, the available evidence could be interpreted according to two main scenarios:

-- a long one, from supposedly Turonian dinosaur-bearing strata to Eocene levels, mainly supported by K/Ar dating, some palaeomagnetic data, and some palaeontological evidence;

-- a short one, supported by palaeomagnetic evidence, suggesting durations ranging from 5 Myr (McElhinny 1968) to less than 1 Myr (Officer & Drake 1985).

The classical succession of the Deccan Trapps consists of:

-- the Lameta Beds, containing a dinosaur fauna of supposedly Turonian age;

-- the basalt flows of the lower Trapps;

-- very thin lacustrine Intertrappean sediments, locally intercalated between the basalt flows;

-- the basalt flows of the upper Trapps.

In one place only, marine sediments belonging to the P2 zone of the Palaeocene directly overlie the upper Trapps. In the same borehole,

*)   Laboratoire de Paléomagnétisme et Géodynamique, Institut de Physique du Globe, 75005 Paris, France.

**)   Laboratoire de Paléontologie des Vertébrés, Université Paris VI, 75252 Paris, France.

***)   Laboratoire de Paléontologie, Institut des Sciences de l'Evolution, Université des Sciences et Techniques du Languedoc, 34060 Montpellier, France.

****)   Laboratoire de Géochimie, Institut de Physique du Globe, 67084 Strasbourg, France.

*****)   Centre of Advanced Study in Geology, Panjab University, Chandigarh, 160014 India.

Lecture Notes in Earth Sciences, Vol. 8
Global Bio-Events. Edited by O. Walliser
© Springer-Verlag Berlin Heidelberg 1986

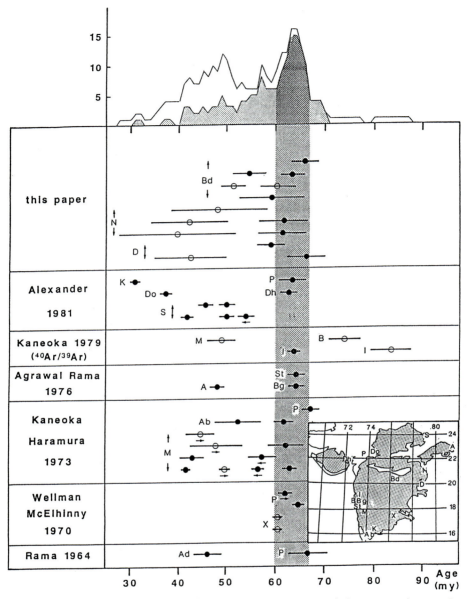

Figure 1. K-Ar ages for the Deccan Trapps, with approximate location (inset), from this study (see also Fig. 2), and also Alexander (1981), Kaneoka (1979, $^{39}Ar-^{40}Ar$), Agrawal and Rama (1976), Kaneoka and Haramura (1973), Wellman and McElhinny (1970) and Rama (1964). Open symbols are for altered samples according to the original papers; 1 σ error bars are indicated. On the top of the Fig. are two histograms for the age distributions (shaded for non altered samples only, unshaded for all samples). Arrows in the Kaneoka and Haramura (1973) study indicate K/Ar ages which are in severe contradiction with stratigraphical position in a given lava pile (arrows toward right, stratigraphically older, arrows toward left, stratigraphically younger).

Intertrappean and Infratrappean sediments have yielded Maastrichtian Foraminifera (Govindan 1981).

New researches in the eastern, central and western part of the Deccan Trapps have yielded new data in the following three fields:

K/Ar dating (Fig 1):

Eight new ages for the less altered sample from the eastern part of the Deccan Trapps range from 55 to 66 Myr, with 1σ error bars between 2 and 4 Myr.

A statistical review of the previously published K/Ar ages confirms a peak in the number of data between 60 and 67 Myr. Younger ages may correspond to Argon loss due to alteration, as already suggested by several authors.

Magnetostratigraphy:

Twenty-one sites have been sampled on a traverse from Nagpur to Bombay, corresponding to nine distinct geographical locations. The mean virtual geomagnetic pole deduced from these sites is located at 290.5°E and 32.5°N, with a confidence interval of 5° at the 95% probability level. The resulting palaeolatitude reduced at Nagpur is 29°S ($\pm 4°$) at the time of lava emplacement. Eighteen out of the twenty-one sites have a reversed magnetisation. This completes and confirms previous findings which indicate that the remanent magnetisation is predominantly reversed. Interpretation of new data in the light of numerous previously published results supports the hypothesis that Deccan Trapps volcanism may have covered no more than three polarity intervals, one reversed sequence between two normal one.

Palaeontology:

The Lameta Formation has often been referred to the Turonian because of the similarity of its dinosaur fauna with faunas from other southern continents which are now referred to the Campanian or Maastrichtian. New studies of Lameta dinosaurs (Chatterjee 1978, Berman & Jain 1982) also suggest a Maastrichtian age. This is confirmed by the recent discovery of a typical Maastrichtian selachian (Igdabatis) in the Lameta Formation just below the lower Trapps at Jabbalpur. Large scale washing and screening of Intertrappean sediments at several localities has yielded the same diversified fish fauna (Gayet et al. 1984), associated with a Laurasian pelobatid frog (Sahni et al. 1982) and dinosaur teeth and egg-

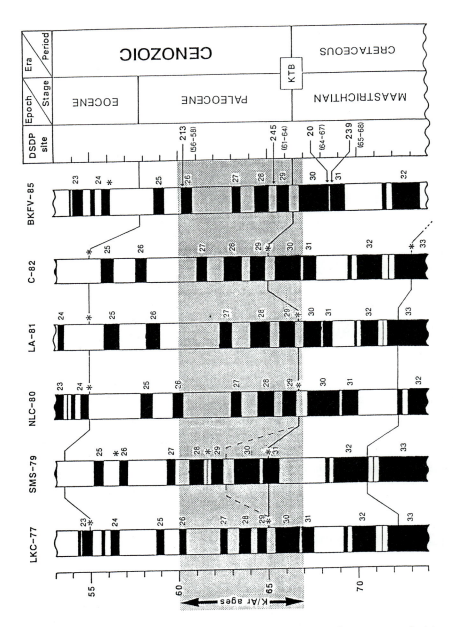

Figure 2. A comparison of six recent magnetic reversal time scales (LKC 77: Labrecque et al. 1977. SMS 79: Schlich et al. 1979. NLC 80: Ness et al. 1980. LA 81: Lowrie and Alvarez 1981. C 82: Cox 1982. BKFV 85: Berggren et al. 1985). Normal polarities are shown in black, reversed ones in white. Chron 29R is at the Cretaceous-Tertiary boundary. The range of K-Ar ages of Deccan basalts from this study is shaded. Biostratigraphic tie-points used in the various time scales are shown as stars (most of them are revisions of the ages given by Hardenbol and Berggren 1978), and DSDP sites are listed with their biostratigraphic age ranges.

shells (Sahni & Gupta 1982). All these Intertrappean dinosaur levels fall within the reversed palaeomagnetic interval and overlie basalts dated from 66.2 $\pm$ 3.9 Myr (Dongargaon) and 61.7 $\pm$ 5.0 (Nagpur). Evidence based on fossil vertebrates is thus in agreement with the very few marine micropalaeontological data mentioned above.

In conclusion, palaeontological evidence indicates that the latest fossiliferous Intertrappean sediments in the eastern part of the Deccan Trapps are still Maastrichtian in age -- if dinosaurs did not survive into the Palaeocene in India, as suggested by Van Valen and Sloan (1970).

Conclusion

Constraints from palaeontological data indicate that the main part of the Deccan Trapps was emplaced before the Cretaceous-Tertiary boundary. Palaeomagnetic data allow several possible correlations with the geomagnetic polarity time scale. However, if K/Ar constraints are also taken into account, the most likely correlation of the main Deccan Trapps geomagnetic reversed interval is with anomaly 29 R. It is generally accepted that the Cretaceous-Tertiary boundary falls within this interval. Therefore, it is suggested that the Deccan Tapps eruptions were contemporaneous with the events of the Cretaceous-Tertiary transition, and that volcanic activity was of comparatively short duration (less than 1 Myr) (Fig. 2).

It therefore appears that hypotheses linking terminal Cretaceous events with the eruptions of the Deccan Trapps are not unlikely from a chronological point of view. However, much remains to be done to gain a better understanding of possible causal links between Deccan Trapps eruptions and terminal Cretaceous events. For instance, the search for an Iridium-enriched level in the Deccan Trapps has so far yielded only negative results.

REFERENCES

BERMAN, D.S. & JAIN, S.L. (1982): The braincase of a small sauropod dinosaur (Reptilia: Saurischia) from the Upper Cretaceous Lameta Group, Central India, with review of Lameta Group localities.- Ann. Carnegie Mus. 51, 405-422.

CHATTERJEE; S. (1978): Indosuchus and Indosaurus, Cretaceous carnosaurs from India.- J. Paleont. 52, 570-580.

GAYET, M.; RAGE, J.C. & RANA, R.S. (1984): Nouvelles ichthyofaune et herpétofaune de Gitte Khadan, le plus ancien gisement connu du Deccan (Crétacé/Paléocène) à microvertébrés. Implications paléogéographiques. - Mém. Soc. Géol. France, n.s. 147, 55-65.

GOVINDAN, A.(1981): Foraminifera from the Infra and Intertrappean subsurface sediments of Narsapur Well-I and age of the Deccan Trapp flows.-

Proc. IX Ind. Coll. Micropaleont. Strat., 81-93.

McELHINNY, M.W. (1968): Northward drift of India - Examination of recent paleomagnetic results.- Nature 217, 342-344.

OFFICER, C.B. & DRAKE, C.L. (1985): Terminal Cretaceous environmental events.- Science 227, 1161-1167.

SAHNI, A. & GUPTA; V.J. (1982): Cretaceous egg-shell fragments from the Lameta Formation, Jabalpur, India.- Bull. Ind. Geol. Assoc. 6, 85-88.

-- ; KUMAR, K.; HARTENBERGER, J.L.; JAEGER, J.J.; RAGE, J.C.; SUDRE, J. & VIANEY-LIAUD, M. (1982): Microvertébrés nouveaux des Trapps du Deccan (Inde): mise en évidence d'une voie de communication terrestre probable entre la Laurasie et l'Inde à la limite Crétacé-Tertiaire.- Bull. Soc. géol. France 24, 1093-1099.

PALAEOFLORISTIC AND PALAEOCLIMATIC CHANGES IN THE CRETACEOUS AND TERTIARY PERIODS (FACTS, PROBLEMS AND TASKS)

A contribution to Project GLOBAL BIO-EVENTS

KNOBLOCH, Ervín *)

Cretaceous

In the recent twenty years the author and D.H. Mai (Berlin) have studied in detail Cretaceous sediments in Central Europe, using the palaeocarpological method. Washing of sediments has yielded a large amount of carbonized seeds and fruits belonging to 85 genera and 27 families. These fossils have proved to provide the most reliable evidence, thus far obtainable, of the development of vegetation in the sense of the natural plant system. The morphological features of Upper Cretaceous fruits and seeds demonstrate that the angiosperms had then attained the organizational level of the contemporary ones. The assemblage corresponds to the evergreen, subtropical and temperate vegetations of the northern hemisphere.

From the stratigraphical point of view, the fossils document three basic evolutinary stages: 1. Cenomanian, 2. late Turonian-Campanian and 3. Maastrichtian. The Maastrichtian is dominated by genera that also occur in recent floras, whereas in earlier stages most species belong to extinct genera. Under the existing conditions it is advisable to expand this research method to further areas and to attempt to obtain some finds from the Turonian, whence very few are known.

The abundance of these small-sized plant remains (0.5 - 2 mm), found in the Carpathian Flysch and in the Gosau Formation of Austria was surprising. They are relics of a terrestrical arboreal vegetation, and the author does not think it probable that turbidity currents are responsible for their presence in sediments which are regarded as abyssal deposits.

The Bohemian Massif has yielded the richest floras (chiefly leaf impressions) from the Cenomanian (the Peruc Member) of Europe, representing one of the richest Cretaceous floras throughout the world. In the coming years we intend to study in detail the old finds and new collections from this area.

Besides fossil angiosperms, whose sudden world-wide development in the Cenomanian was connected with a fundamental palaeogeographic change, and also resulted in the evolution of a new terrestrial fauna, there exist other plant groups and organs that can be used for correlation

---

*) Ústřední ústav geologický, 11821 Praha, Č.S.S.R.

studies or whose paths of migration should be clarified. This group
involves numerous ferns (also arboreal), conifers, pteridophyte megaspores or the genus Costatheca of uncertain systematic position.

With respect to the changes in the configuration of continents and
seas during the Cretaceous, the Arctic region plays a very relevant
role in the propagation of plants. It seems probable that some identical
or at least very similar plant species were growing in North America and
Central Europe at approximately the same time. All these questions
demand our attention.

Tertiary

From the palaeofloristic view, the least recognized series is the Palaeocene of Europe, and yet the knowledge of Palaeocene floras is, in consideration of the floras of the uppermost Cretaceous, of special importance. The investigation of carbonized seeds and fruits in the Palaeocen
of Europe has shown that the genera occurring in the Palaeocene have
appeared as early as the late Maastrichtian. This implies that between
the Cretaceous and the Tertiary, as it were between the Permian and
Triassic, the evolution of vegetation had outrun the evolution of some
significant faunal groups, on the basis of which these boundaries of
primary importance are drawn. In the following years the Palaeocene floras of England and the German Democratic Republic will be elaborated.

In the Tertiary of Europe migration and interrelationship of the
two following components of the vegetation cover are of fundamental
significance: 1. the palaeotropical geoflora and 2. the acrotertiary
geoflora. The palaeotropical geoflora involves the evergreen species,
whose analogues occur nowadys, for example, in subtropical to tropical
forests of south-eastern and southern Asia. The acrotertiary geoflora i
represented by deciduous, summergreen trees and shrubs analogous to
those growing at the present time in temperate forests of the northern
hemisphere, such as beech, alder, birch, maple and other genera. Wherea
in Eocene times the arctotertiary geoflora dominated in some Arctic
regions, the palaeotropical geoflora existed in Central Europe. The
immigration of the arctotertiary geoflora into Central Europe started
in the Lower Oligocene. From the Lower Oligocene until the Upper Miocen
either arctotertiary or palaeotropical geoflora predominated, in Europe
in particular, which was caused by climatic changes (subtropical vs.
temperate climate). This suggests a phasic development of climate
showing a general tendency to cooling, but with at least one warm maximum between the Lower and Middle Miocene (Ottnangian/Karpatian). This
phenomenon can well be used for stratigraphic purposes (whence the term

climatic stratigraphy). In essential, it is associated with palaeogeographic, geophysical and astronomical changes. The change in the position
of the Earth's axis resulted in a different distribution of climatic
and thus also vegetation zones, and the altered palaeogeographic conditions made the migration of plants between transitory land masses either
easier or more difficult, even impossible.

An important requirement of the climatic-stratigraphic study is to
use only rich floras described on the basis of the same organs (leaves,
fruits or seeds) and woody plants, because aquatic plants are not so
sensitive to climatic changes. Otherwise, there is a risk of interpreting
local differences and peculiarities as due to climatic oscillations.
Geochronological data may be used with advantage as a controlling factor.

The construction of temperature curves for the Tertiary period is not
simple; it cannot be unambiguous and is invariably very schematic because the climate at a certain site is not characterized only by mean
annual temperature but also by a number of other factors. The present-
day phytogeographic provinces with which the fossil plants are correlated
are extensive and are characterized by many local climatic diagrams,
differing one from another.

On account of climatic and palaeogeographic changes and of the
general evolution of flora, another significant bio-event in the
Tertiary flora occurred in Europe during the Upper Miocene. It is
distinguished by the disappearance of the elements of the palaeotropical
geoflora, by the development of herbs and, in result of the obliteration
of Tethys, by the continentalization of climate and thus by the propagation of several xerothermal elements.

Another question that should be settled, that is when and where the
true recent plant species began to appear, concerns the bio-event from
the end of the Tertiary. Our task will be to establish and prove which
of the contemporary European plant species have developed from the
Tertiary European species and which had immigrated into Europe from Asia
in the Pliocene or Quaternary.

The problems mentiond above will be studied either by re-evaluation
of earlier data or on the basis of new finds. Our endeavours should be
aimed at elaborating the succession of plant biocoenoses in individual
regions (floristic zones!) and their correlation. Only in this way it
will be possible to comprehend the intricate mechanism of migration of
plant species and communities, which differed from one region to another,
and to outline the picture of climatic changes in Tertiary times.

TWO EXAMPLES OF EVOLUTION CONTROLLED BY LARGE SCALE ABIOTIC PROCESSES: EOCENE NUMMULITIDS OF THE SOUTH-PYRENEAN BASIN AND CRETACEOUS CHAROPHYTA OF WESTERN EUROPE

A contribution to Project GLOBAL BIO-EVENTS

MARTÍN-CLOSAS, Carles & SERRA-KIEL, Josep *)

In the last few years, geological if not cosmical events have been proposed as the ultimate cause for controlling global environmental changes, and as a result of this, they also control the evolutionary rate (Walliser 1984). In this work we present two different cases in which this scheme can be recognized.

The evolution of Eocene larger foraminifera in the South-Pyrenean Basin

Larger foraminifera are useful biostratigraphical and palaeoenvironmental tools in the study of the Palaeocene marine facies. In the Eocene South-Pyrenean Basin, the evolution of the group demonstrates long-lasting stable periods truncated by short ruptures than can be referred to the main palaeogeographical changes within the basin. In this work we propose that the evolution of Eocene larger foraminifera was controlled by large-scale tectonic events, related to the emplacement of Pyrenean thrust sheets (Fig. 1).

The palaeontological data are taken from Serra-Kiel (1984) and tectonic data from Puigdefabregas et al. (1986). During Ilerdian, Cuisian and Lower Lutetian times, larger foraminifera were composed of basin-wide homogenous faunas that evolved with gradualistic trends. This was related to the existance of a widespread shallow carbonate platform, little influenced by the initial stages of Pyrenean thrusting.

During Middle and Upper Lutetian gradualism continued to be the main evolutionary trend and the nummulitid fauna experienced a progressive provincialism: in the Catalan Basin a typical off-shore fauna evolved, mainly composed of <u>Assilina</u> and <u>Nummulites</u>, whereas in part of the South-Pyrenean Basin, which was connected with the Atlantic, this fauna was formed by both, <u>Assilina</u> and <u>Nummulites</u> gr. <u>distans-millecaput</u>. Provincialism can be related to the emplacement of the important submarine South Pyrenean nappes, which began in the Cuisian and lasted to the Lower Lutetian time. As a consequence of this, there was a progressive shallowing and basin restriction; Catalan and Aquitanian Basins became paleogeographically separated and Atlantic nummulitids could not enter the Catalan Basin.

*) Departament de Paleontologia, Facultat de Geologia, Universitat de Barcelona, 08007 Barcelona, Spain.

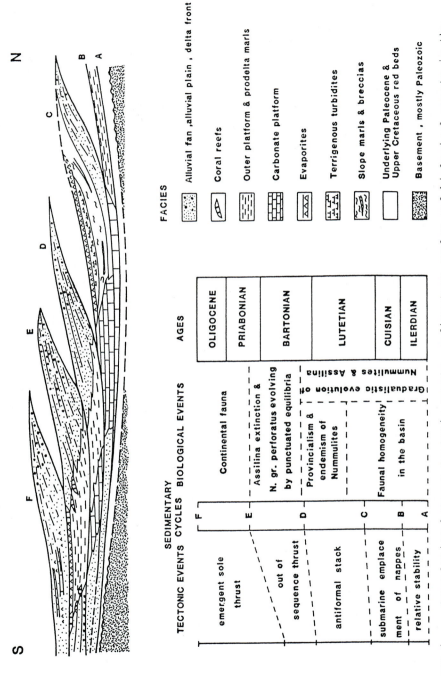

Figure 1. Correlation between tectonic events, sedimentary cycles and biological events in the Catalan South-Pyrenean Basin. Note the reduction in the width of the carbonate platforms between D and E, the migration through time of these platforms southwards and the presence of evaporitic events below C and D.

It was at the Lutetian-Bartonian boundary where larger foraminiferid faunas suffered the main change in their evolution. Assilina and also the large flat Lutetian Nummulites disappeared completely from the basin. After this boundary crisis, Bartonian faunas were characterized by morphologically, highly variable Nummulites (N. gr. perforatus) that do not show phylogenetic connection with the earlier faunas. These nummulitids evolved by punctuated equilibria in the Bartonian widespread prodelta facies, as it did N. gr. fabianii in the Priabonian reef environments. The Lutetian-Bartonian boundary crisis followed a strong reduction of the platform environments which was produced by the progradation into the basin of deltaic systems. This deltaic progradation is related to the relief created as a result of thrust-sheets piling up (antiformal stack) in the northern margin of the basin.

In this example the control of evolutionary mechanisms presents a hierarchical structure, natural selection and evolutionary rate being controlled by environmental change (sedimentary sequences and ecological changes) that are a function of regional tectonic processes. This structure is referred in the heterochronism of boundaries defined by tectonic, environmental, and biological changes. Final biological changes occur following environmental changes produced by regional tectonic events.

These conclusions observed in a relatively small basin can also probably be recognized in a more global scale.

Our second example demonstrate this in a completely different setting: the Cretaceous Charophyta of the Atlantic passive margins and Tethyan domain.

The evolution of Cretaceous Charophyta in Western Europe

Charophytes are in freshwater living, biostratigraphically useful algae. Their calcified oogonium, called a gyrogonite, is the structure that characterizes fossil species and from which the evolutionary trends of the group have been deduced. The apical pore is an important feature of the gyrogonite that allows the fertilisation of the egg-cell and the germination of the zygote. This pore seems to have played major role in the Cretaceous evolution of the Charophyta.

At the Lower Cretaceous-Upper Cretaceous boundary, worldwide important changes took place in the floral composition of charophyte associations and in their evolution (Grambast 1974) (Fig. 2). Lower Cretaceous dominant families (Clavatoraceae, Porocharaceae), that have an apical pore in their gyrogonite, suffered a strong regression and had almost disappeared in the Upper Cretaceous.

After the Mid-Cretaceous events, which includes a 13-15 ma gap (Turon-

378

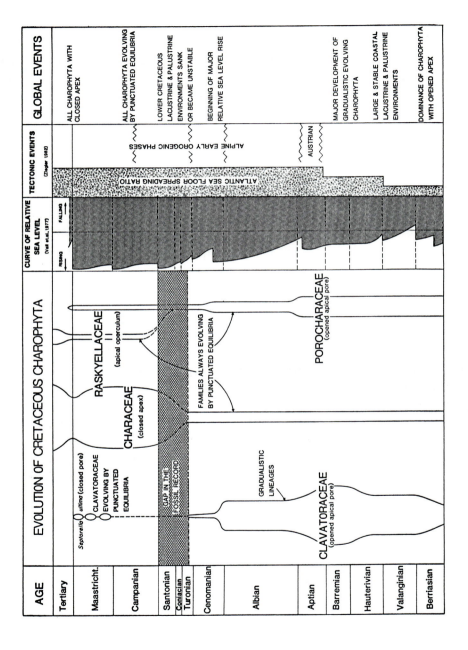

Figure 2. Evolution and evolutionary trends of Cretaceous Charophyta related with the tectonic events and the relative sea level rise occurred during Mid Cretaceous time. Both are proposed here as the main control of evolutionary patterns of the group.

ian-Santonian) in the charophyte's fossil record, these families attempted to close the apical pore. So, by Campanian, Raskyellaceae -- a family with an apical operculum -- evolved from porocharaceae. Clavotoraceae only achieved the closure by Upper Maastrichtian (_Septorella ultimata_), just before the final extinction of the family at the Cretaceous-Tertiary boundary. On the other hand, Characeae, a family with a closed apex which were a minor component of the Jurassic and Lower Cretaceous associations, became dominant after the Mid-Cretaceous event.

These floral substitutions were accompanied by a dramatical change in the mode of evolution, that passed from being mainly gradualistic in the Lower Cretaceous to showing punctuated equilibria after the Mid-Cretaceous crisis.

We propose that the changes described above in the charophyte flora were mainly produced by large-scale geological events that occurred at the Lower Cretaceous - Upper Cretaceous boundary. In Western Europe these events are characterized by an important sea level rise, that was related with an increase of the Atlantic sea floor spreading ratio (Ziegler 1982). This sea level rise produced the flooding of the stable and extensive Purbeck-Wealden marshy environments. The remaining lacustrine and palustrine environments of the still emergent areas became isolated from each other and were probably much smaller than the Lower Cretaceous ones.

During Campanian and Maastrichtian, the early phases of alpine orogeny intensified the stress situation of the Upper Cretaceous freshwater environments. The palaeogeographic heterogeneity, resulting from the tectonic activity, produced small, short-living lakes in which gradualistic evolution of charophytes became impossible. On the other hand, frequent subaerial exposure by lake desiccation made the gyrogonite's apical pore lethal for the survival of the fertilised egg-cell, and nature selected the species with closed apex.

In this example, that is referred to a global biological event, we have also proposed a large scale geodynamical process as the final factor controlling evolution and evolutionary rates.

In both examples, stable geological settings and palaeogeographic homogeneity allow the development of gradualism. Geodynamically caused stress situations produce palaeogeographic diversification and environmental instability, which force the organisms to accelerate their evolutionary rate (punctuated equilibria) or cause their extinction.

REFERENCES

GRAMBAST, L. (1974): Phylogeny of the Charophyta.- Taxon 23, 463-481.

PUIGDEFABREGAS, C.; MUNOZ, J.A. & MARZO, M. (in press): Thrust belt development in the Eastern Pyrenees and related depositional sequences in the Southern foreland basin.- I.A.S. Spec. Publ. 8.

SERRA-KIEL, J. (1984): Estudi dels Nummulites del grup N. perforatus (Montfort).- Inst. Cat. Hist. Nat. Treballs 11, 1-244.

VAIL, P.R.; MITCHUM, R.M., Jr. & THOMPSON, S. (1977): Global Cycles of Relative Changes of Sea Level.- A.A.P.G. Mem. 26, 83-97.

WALLISER, O.H. (1984): Global Events and Evolution.- Proc. 27th Internat. Geol. Congr. 2, 183-192.

ZIEGLER, P.A. (1982): Geological Atlas of Western and Central Europe.- Shell Internat. Petrol., 130 p.

THE DIACHRONOUS C/T PLANKTON EXTINCTION IN THE DANISH BASIN

A contribution to Project GLOBAL BIO-EVENTS

HANSEN, Hans J., GWOZDZ, R., HANSEN, Jens M., BROMLEY, Richard G. & RASMUSSEN, K.L. *)

At Stevns Klint, Denmark (see Fig. 1), an un-sheared Cretaceous/Tertiary boundary section has been located.

Figure 1. Map of Denmark with localities mentioned in the text.

Its Ir content is the highest so far recorded (185 ppb). Carbonate and non-carbonate carbon fractions are inversely correlated, the lower part of the fish clay being almost free of carbonate (see Fig. 2). We there-fore studied the non-carbonate dinocyst composition across this profile. The black fish clay is dominated by a Maastrichtian species that does not occur in the white chalk. The first Danian dinocysts occur 4.5 cm above the base of the fish clay at this locality. Thus, if the C/T bound-ary were to be fixed on the basis of positive biostratigraphical evi-dence, it would be placed within the fish clay at the level of first occurrence of Danian dinocysts. Placing the boundary at the level of disappearance of carbonate-shelled organisms such as planktonic forami-nifera and coccoliths is inadvisable, since carbonate organisms tempo-rarily disappeared from the stratigraphic column at Stevns Klint (and other places) at the time of fish clay deposition.

The evidence from the dinocysts further implies that Cretaceous non-carbonate sedimentation continued in fish clay time and that the fish clay is not a residuum following dissolution of an earlier chalk deposit. If the latter were the case, the fish clay would contain a white chalk dinocyst flora.

Previous investigators have been unable to trace the world-wide

*) Geological Central Institute, University of Copenhagen, 1350 Copenhagen K, Denmark.

Lecture Notes in Earth Sciences, Vol. 8
Global Bio-Events. Edited by O. Walliser
© Springer-Verlag Berlin Heidelberg 1986

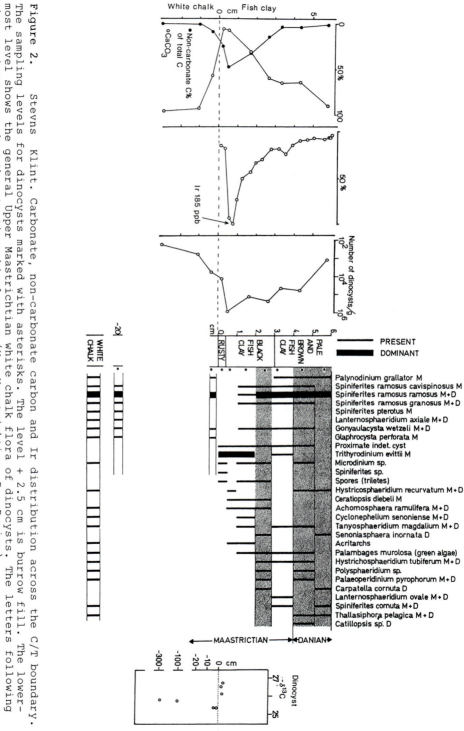

Figure 2. Stevns Klint. Carbonate, non-carbonate carbon and Ir distribution across the C/T boundary. The sampling levels for dinocysts marked with asterisks. The level + 2.5 cm is burrow fill. The lower-most level shows the general Upper Maastrichtian white chalk flora of dinocysts. The letters following the species names show the stratigraphical range (M - Maastrichtian, D - Danian).

¹³C anomaly at the C/T boundary at Stevns Klint; their measurements were based on carbonate organisms. Since, however, carbonate was not deposited in fish clay time at Stevns Klint, we made measurements for ¹³C from dinocysts, as these were deposited in the relevant time period. We found a very clear ¹³C anomaly in the basal fish clay. This anomaly occurs shortly after the disappearance of the planktonic foraminifera and coccoliths.

Stevns Klint is generally accepted as representing rather shallow water deposits, whereas the Nye Kløv section (Fig. 3) in Jylland is considered to have been deposited in rather deeper water. We have found a small Ir anomaly in the fish clay at this locality. For comparison we have made measurements of ¹³C from dinocysts from the uppermost Maastrichtian at this locality. We found the ¹³C anomaly to occur 15 cm below the base of the fish clay.

If the ¹³C anomaly were linked with the decline of coccolith carbonate production, we would expect a drastic drop in coccolith abundance at or slightly below the ¹³C anomaly. We examined in the SEM a series of

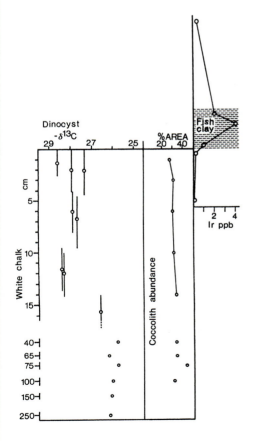

Figure 3. Nye Kløv. Carbon isotopes and coccolith abundance in the Upper Maastrichtian white chalk. Ir values by INAA.

closely spaced samples and made counts of whole and fragmentary coccoliths. The number of coccolith-containing fields, each 10 x 10 m, were
counted. No correlation was found to exist between coccolith abundance
and the ^{13}C anomaly. At Nye Kløv, the disappearance of planktonic foraminifera and coccoliths occurs 0.5 mm below the fish clay.

These investigations indicate the existence of two distinct, presumably isochronous, geochemical horizons in the Danish Basin at the C/T
boundary: a ^{13}C anomaly and an Ir anomaly. At Nye Kløv, planktonic foraminifera and coccoliths continued to be deposited for longer than at
Stevns Klint, corresponding to at least 15 cm of white chalk (Fig. 4).
Relative to the isochrons, therefore, the local extinctions observed at
the two localities are diachronous. The hypothesis of meteoritic impact
as a cause for the C/T extinction event requires that extinctions are
mutually simultaneous and geologically instantaneous. This requirement
is not upheld by evidence from the Danish Basin.

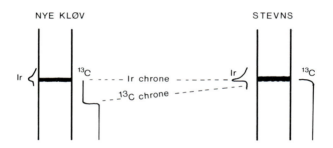

Figure 4.   Diagram (not to scale) showing relative position of the Ir
and carbonisotope chrones.

SOME BIOSTRATIGRAPHIC AND PALEOGEOGRAPHIC OBSERVATIONS ON THE CRETACEOUS/TERTIARY BOUNDARY IN THE HAYMANA POLATLI REGION (CENTRAL TURKEY)

A contribution to Project GLOBAL BIO-EVENTS
IUGS UNESCO

SİREL, Ercüment, DAĞER, Zeki & SÖZERİ, Biler *)

Abstract: This study aims to define the biostratigraphic and litho-stratigraphic characteristics of Maastrichtian and Lower Paleocene units, cropping out in the Haymana-Polatlı region (SW of Ankara).

As a consequence of the tectonic movements in the oldest Paleocene (Danian), the non-marine sediments (as, e.g., the Kartal Formation) are most widespread in the studied area and various parts of the territory of Turkey. Principal changes of benthic Foraminifera occurred at the end of Maastrichtian or at the beginning of Danian stages. The extinction of some benthic Foraminifera (Assemblage I) and appearance of endemic (?) foraminiferal taxa (Assemblage II) have been well recognized in the measured sections from the Beyobası Fm. (in the shelf environments) and Çaldağ Fm. (in the reef environment). Consequently, from this bio-stratigraphic point of view, the Danian Stage must be attributed to the Tertiary.

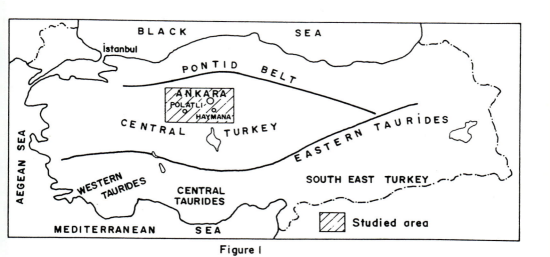

Figure I

Introduction

This study aims also to define the biostratigraphic and lithostratigraphic characteristics of Maastrichtian-Lower Tertiary (Danian-Thanetian) sediments, cropping out in the Haymana-Polatlı region S and SW of Ankara and in Central Turkey (Fig. 1).

Up to now non-marine sediments of Maastrichtian time have not been

---

*) MTA General Directory, Jeologji Etütleri Dairesi, Ankara, Turkey.

observed in the territory of Turkey. But marine Maastrichtian sediments occur most widespread in the studied area and various parts of Turkey. In the western and southern parts of the Haymana-Polatlı area Lower Paleocene sediments are represented, in which at the beginning of the Paleocene important paleogeographic, sedimentary and faunal changes occurred. They may be connected with tectonic movements. These changes were reflected in different ways in the area, causing different facial developments of sediments and different developments of life.

The oldest Paleocene (Danian) of Haymana-Polatlı area is characterized by tectonic movements which have resulted in sedimentary diversi fication; thus during Lower Paleocene time in the Haymana-Polatlı region different sedimentation environments with associated foraminiferal assem blages are recognized in three formations:Kartal Fm. (fluviatile at the base, supratidal or supralittoral with shallow marine foraminifers, such as Fabulariidae, Miliolidae, Discorbidae and Elphididae), Çaldağ Fm. (reefal environment with newly identified benthic foraminiferal assemblage II; Fig. 7) and Yesilyurt Fm. (deep sea environment with planktoni Foraminifera and Nannoplankton, Fig. 7). These three formations are interfingering in some localities (Fig. 8).

Principal changes of the benthic larger Foraminifera occurred at the end of Maastrichtian or at the beginning of Lower Paleocene (Danian) Stages). Disappearance of some benthic foraminiferal taxa (Assemblage I) at the end of Maastrichtian and first occurrence of new foraminiferal taxa (Assemblage II) have been well recognized in the measured sections from the Çaldağ Fm. (Figs. 3A, B), Bahçecik (Figs. 4A, B), Erif (Figs. 5A, B), and Kayabası (Figs. 6A, B). Consequently, from the biostratigraphic point of view, the Danian Stage must be attributed to Tertiary.

The geology and paleontology of this basin have been studied by many authors such as: Rigo & Cortesini (1959), Schimidt (1960), Yüksel (1970), Akarsu (1970), Çapan & Buket (1975), Sirel (1975, 1976a, b), Sirel & Gündüz (1976), Ünalan et al. (1976), Dizer (1964), Gökçen (1976) and Toker (1979). Unfortunately, the biostratigraphy of the Upper Cretaceous/Lower Tertiary boundary has not been investigated in detail, except Toker (1980) in the Haymana-Polatlı region. This latter author studied on planktonic Foraminifera and Nannoplankton.

Lithostratigraphy

In the Haymana-Polatli region, various lithostratigraphic units of Triassic, Upper Jurassic/Lower Cretaceous, Upper Cretaceous, Lower Tertiary (Danian-Lutetian) and Neogene age crop out (Fig. 2). In this study, Maastrichtian aged Beyobası Fm. (in the shelf environment) and

SYSTEM	SERIES	STAGE	FORMATION	THICKNESS Meters	LITHOLOGY EXPLANATIONS	FORAMINIFERA
NEOGENE	MIO.-PLIO		AGASİVRİ		ALLUVIUM	
					Lacustrine limestone and conglomerate	Ostracoda, Gastropoda
PALEOGENE	EOCENE	CUISIAN-LUTETIAN	a.Yamak b.Çayraz c.Beldede		a - Red conglomerate, sandstone and marl	Assilina spira, Nummulites pinfoldi
						Assilina exponens, Nummulites laevigatus
					b - Sandy and clayey limestone and marl	
					c - Conglomerate, sandstone and marl	Alveolina canavarii, Alveolina bayburtensis
						Nummulites planulatus, Nummulites partschi
			Eskipolatlı		Gray sandstone, marl, limestone	Nummulites planulatus, N. partschi
						Alveolina oblonga
		ILERDIAN	İlginlık dere		Gray sandstone, conglomerate and sandy limestone	Nummulites exilis, Nummulites fraasi
						Alveolina cucumiformis, Assilina pustulosa
	PALEOCENE	THANETIAN	Kırkkavak	1363	Algal limestone	Alveolina (Glomalveolina) primaeva
					Marl and sandstone	Discocyclina seunesi, Vania sp.
					Algal limestone	
			Yeşilyurt		a - Red conglomerate, sandstone, marl and argilleous limestone	Bolkarina aksarayı, Cuvillierina n. sp.
		DANIAN-MONTIAN	Çaldağ			Miscellanea sp.
					b - Algal limestone	Laffitteina mengaudi, Mississipina
						Miliolidae
			Kartal		c - Marl with limestone blocks	New foraminiferal genera, Planorbulina, Miliolidae
CRETACEOUS	UPPER CRETACEOUS	MAASTRICHTIAN	b Haymana	1842	a - Alternation of sanstone and conglomerate	a) Fallotia n.sp., Sulcoperculina n. sp.
						Sirtina orbitoidiformis, Smoutina n. sp.?
						Hellenocyclina beotica, Siderolites calcitrapoides,
						Cuvillierina sözerii, Omphalocyclus macroporus
			a Beyobası			b) Globotruncana mayorensis
						Glt. gansseri.
						Glt. havanensis.
					b - Light and dark gray shale with sandstone and conglomerate	Glt. elevata.
	Campa-nian	Dereköy			Ophiolites, limestone, radiolarite volcanics	
UPPER JURASSIC / LOWER CRETACEOUS		Temirözü	Makresul		Massive limestone	Trocholina alpina, Protopeneroplis striata
TRIASSIC						Labyrinthina, Deberina, Clypelina jurassica
					Limestone blocks within the metagreywacke	

Figure 2.   Generalized columnar section of Haymana-Polatlı area.

ÇALDAĞ COLUMNAR SECTION
(EAST FLANK OF ÇALDAĞ ANTICLINE)
(Sirel, Dağer, Sözeri 1985)

SYSTEM	SERIES	STAGE	FORMATION	THICKNESS	SAMPLE NO	BIOZONE	LITHOLOGY	FORAMINIFERA
T E R T I A R Y	P A L E O C E N E	MONTIAN	Ç A L D A Ğ	~1187 m.	17.	Miscellanea n. sp.		Miliolidae, Rotaliidae
					16. 15. 14. 13. 12. 11. 10. 9.	Miscellanea aff. mi- nuta		Miliolidae, Rotaliidae
					8.	Cuvillerina n. sp.		Small Rotaloid forms, Miliolidae
					7. 6. 5.	Heterillina n. sp. Miscellanea n. sp.	Algal limestone	Rotaliidae, Miliolidae
					25.	Laffitteina mengaudi Scandonea aff. samnitica		Ataxophragmiidae, Mississippina sp.
					24. 23. 22. 21. 20. 19. 18. 17. 16. 15. 14. 13. 12. 11.	Laffitteina mengaudi Orduna aff. erki conica	Assemblage II	Mississippina sp. fragment of the new genus I., Miliolidae.
		DANIAN			10. 9.			New genus (1,2,3,4) Planorbulina sp. Mississippina sp. Ataxophragmiidae Miliolidae.
CRETACEOUS UPPER CRETACEOUS		UPPER MAASTRICHTIAN	BEYOBASI	100 m.	8. 7. 6. 5. 4. 3. 2. Çl.1	Assemblage I	Sandstone argilleous and sandy limestone	Hellenocylina baotica, Orbitoides medius Omphalocylus macroparus Cuvillerina sözeri Siderolites calcitrapoides, Lepidorbitoides Pseudomphalocyclus cf. blumenthali Sulcoperculina sp, Loftusia sp. Goupillaudina sp. Sirtina sp. Nummofallotia sp. Globotruncana sp. Fallotia n. sp.

Figure 3 A

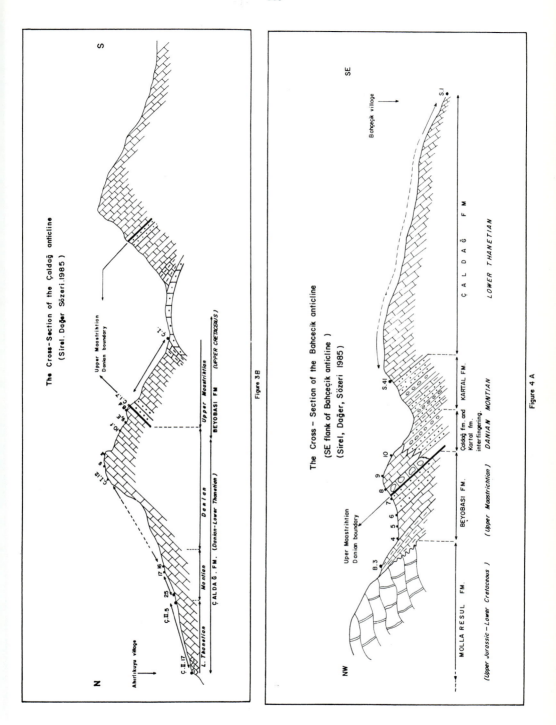

389

The Cross-Section of the Çaldağ anticline
(Sirel. Dağer Sözeri.1985)

Figure 3B

The Cross – Section of the Bahcecik anticline
(SE flank of Bahçeçik anticline)
(Sirel, Dağer, Sözeri 1985)

Figure 4 A

BAHÇECİK SECTİON

(Sirel, Dağer, Sözeri, 1985)

Erathem	SERIES	STAGE	FORMATION	THICKNESS	SAMPLE NO	BİOZONE	LİTHOLOGY	FORAMİNİFERA
Mesozoic	SENOZOIC — PALEOCENE	Upper Tha.	KIRKKAVAK		I		Marl and argillaceous limestone	
		DANIAN-MONTIAN — Lower Thanetian	ÇALDAĞ	820 m.	1514	Assemblage III	Algal limestone	Bolkarina aksarayi, Pseudolacazina oeztemueri Pfendericonus sp., Lacazina? Miscellanea globularis, Hottingerina sp. Kathina sp., Planorbulinidae Daviesina sp., Periloculina? sp. Rotaliidae, Miliolidae, Miscellanea sp.
						Miscellanea aff. minuta		Mississippina sp. Planorbulina sp. Miliolidae Rotaliidae Agglutinated forms.
					3029	Heterillina n.sp. / Miscellanean.sp.		Miscellanea sp. Cuvillierina? sp. Danian's form [new genus (I)]
	U.CRET.	Upper Maastrih.	CALDAĞ-KARTAL	200 m.	41	Assemblage II	Red, sandstone, conglomerate and argillaceous limestone / Algal limestone	Ostrocoda, Characea
	U.Juras-L.Cret.	BEYO SUL BASI — MOLLARE-		30 m.		Assemblage I	Sandy and argillaceous limestone / Massive limestone	Hellenocyclina beotica, Orbitoides sp. Omphalocyclus macroporus Cuvillerina sözerii, Sulcoperculina sp. Siderolites calcitrapoides Deberina sp. Labyrinthina sp. Protopeneroplis sp.

Figure 4 B

E R İ F   S E C T I O N
(Sirel, Dağer, Sözeri 1985)

SYSTEM	SERIES	STAGE	FORMATION	THICKNESS	SAMPLE NO	BIOZONE	L I T H O L O G Y	P A L E O N T O L O G Y
T E R T I A R Y	P A L E O C E N E	LOWER THANETIAN			1	Laffitteina mengaudi - Scandonea sp.		
					2		Algal limestone	*Mississippina sp.* *Miliolidae*
					3			
					4			
		MONTIAN	Ç A L D A G	350 m.	5	Laffitteina mengaudi-Orduina aff.erki conica	Algal limestone	*Mississippina sp.* *Rotaliidae* *Miliolidae*
					6			
					7			*Danian's form n. gen.( )*
		DANIAN			8	Assemblage II	Algal limestone	*Assemblage II (N. Gen, 1,2,3,4)* *Planorbulina sp. Mississippina sp.* *Ataxophragmiidae*
					9 1-3		Nodular limestone	*Miliolidae*
CRETACEOUS	UPPER CRETACEOUS	UPPER MAASTRICHTIAN	BEYOBASI	~ 300 m.	10 11 12 S 12 Y 13	Assemblage I	Sandstone Alternations of the sandstone, sandy and argilleous limestone	*Orbitoides medius* *Siderolites calcitrapoides* *Hellenocyclina beotica* *Omphalocyclus macroporus* *Sulcoperculina obesa* *Nummofallotia sp.,Fallotia sp.* *Lepidorbitoides sp.* *Globotruncana sp.* *Goupillaudina sp.* *Cuvillerina sozerii*

Figure 5 A

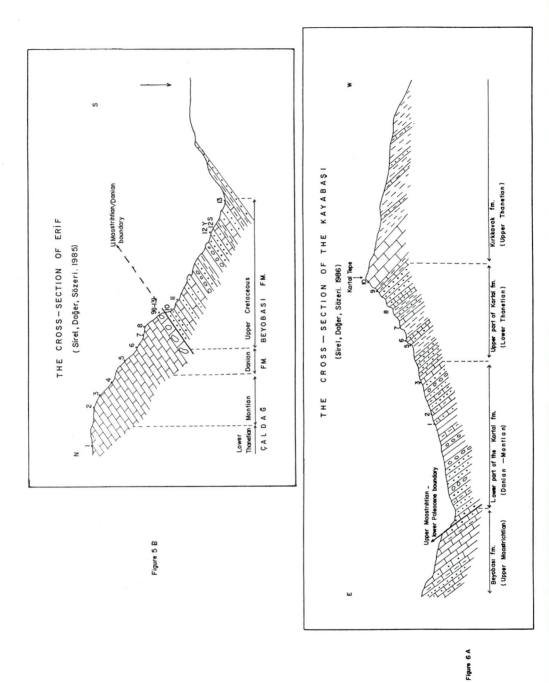

THE CROSS—SECTION OF ERIF

(Sirel, Dağer, Sözeri. 1985)

Figure 5 B

THE CROSS—SECTION OF THE KAYABAŞI

(Sirel, Dağer, Sözeri. 1986)

Figure 6 A

KAYA BAŞI SECTION

SYSTEM	SERIES	STAGE	FORMATION	THICKNESS	SAMPLE NO	ZONES	LITHOLOGY	FORAMINIFERA
TERTIARY	PALEOCENE	DANIAN – MONTIAN ? THANETIAN	KIRKKAVAK / KARTAL	450 → / 1267 m		Laffitteina Mengaudi	Gray marl	
							Algal limestone	Alveolina (Glamalveolina) primaeva, Vania sp.
							Red sandstone / Argillaceous limestone, marl and sandy limestone.	Pseudolacazina oeztemueri Laffitteina? Miliolidae
								Periloculina sp. Miliolidae Rotaliidae
							Algal limestone	Laffitteina mengaudi
							Algal limestone	Small benthic forms. Miliclidae Rotaliidae
							Coal	
							Red sandstone argillaceous limestone and conglomerate	
							Coal	
							Red sandstone and Conglomerate	
CRETACEOUS	UPPER CRETACEOUS	UPPER MAASTRICHTIAN	BEYOBASI(60)m	100 m			Sandy and argillaceous Limestone	Omphalocyculus macroporus Orbitoides medius Siderolites calcitrapoides Pseudomphalocylus sp. Hellenocyclina sp. Loftusia aff. ketini Loftusia minor

Figure 6 B

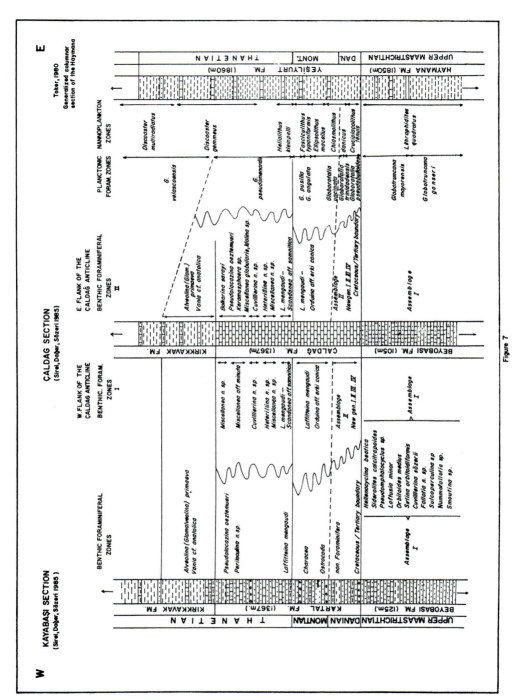

Figure 7. Correlation table; showing the correlation between the biozones of benthic Foraminifera and the biozones of planktonic Foraminifera and Nannoplankton of the Maastrichtian-Lower Thanetian.

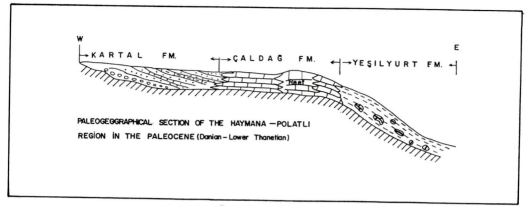

W
├→KARTAL FM.   ←├→ÇALDAĞ FM.   ←├→YEŞILYURT FM. ←├
E

PALEOGEOGRAPHICAL SECTION OF THE HAYMANA—POLATLI
REGION IN THE PALEOCENE (Danian–Lower Thanetian)

Figure 8

Paleocene (Danian-Lower Thanetian) aged Çaldag Fm. (in reefal environment) are examined in detail.

Biostratigraphy

The Maastrichtian/Danian boundary has been examined in four measured
sections and realized in various places in the Haymana-Polatlı region.
All the foraminiferal biozones are shown in the Figs. 3A, 4B, 5A, 6B,
and 7.

Conclusions

(1) In the studied area (Haymana-Polatlı/Central Tureky) there occurred
at the end of the Maastrichtian and at the beginning of the Paleocene
(Danian) important faunal, paleogeographic and sedimentary changes. These
may be connected with some orogenic movements. During the Lower Paleocene (Danian) time, three different developments of sediments and different developments of life are documented in the investigated area (Fig. 8).
(2) The Çaldağ limestone of Danian-Lower Thanetian age, conformably overlies the Beyobası Fm. of Maastrichtian age (Figs. 3A, B, 4A, B, 5A, B,
6A, B). Many genera of benthic Foraminifera (Assemblage I) disappeared
at the end of the Maastrichtian Stage and the new Assemblage II occurred
at the beginning of the Lower Paleocene (Danian).
(3) The boundary Maastrichtian/Danian is accurately defined by biostratigraphical and lithological parameters.
(4) The lower part of the Kartal Fm., which is deposited in a fluviatile
environment, occurs most widespread in the studied area and in various
parts of Turkey. This paleogeographic pattern indicates that the Danian

is a time of some tectonic movements, which continued from the uppermost
Maastrichtian into the Paleocene.

REFERENCES

AKARSU, I. (1971): II. Bölge AR/TPO/747 no lu sahanın terk raporu.-
  Pet. İs. Gen. Md., Ankara (unpubl.).
ÇAPAN, U.Z. & BUKET, E. (1975): Geology of Aktepe-Gökdere region and
  Ophiolitic Melange.- Bull. Geol. Soc. Tureky 18, 11-16.
DİZER, A. (1964): Sur quelques Alveolines de l'Eocene de Turquie.-
  Rev. de Micropaleont. 7, 265-279.
GÖKÇEN, S.L. (1976): Haymana güneyinin sedimanolojik incelenmesi (SW
  Ankara).- Doç. tezi H.Ü. Yebilimleri Enstitüsü, Ankara (unpubl.).
RIGO de RIGHI, M. & CORTESİNİ, A. (1959): Regional studies Central Ana-
  tolian basin, Progress report 1, Turkish Gulf Oil Comp.- Pet. Is.-
  Gen. Md., Ankara (unpubl.).
SCHIMIDT, G.C. (1960): AR/Mem/365-366-367. Sahalarının nihai terk
  raporu.- Pet. İs. Gen. Md., Ankara (unpubl.).
SİREL, E. (1975): Stratigraphy of the South of Polatlı (SW Ankara).-
  Bull. Geol. Soc. Turkey 18, 181-192.
-- (1976a): Description of six new species of the Alveolina found in the
  South of Polatlı (SW Ankara) region.- Bull. Geol. Soc. Turkey 19, 19-
  22.
-- (1976b): Systematic study of some species of the Genera Alveolina,
  Nummulites, Ranikothalia and Assilina in the South of Polatlı (SW
  Ankara).-
-- & GÜNDÜZ, H. (1976): Description and stratigraphical distribution of
  some species of the genera Nummulites, Assilina and Alveolina from
  the Ilerdian, Cuisian and Lutetian of Haymana region (S Ankara).-
  Bull. Geol. Soc. Tureky 19, 31-44.
TOKER, V. (1979): Haymana yöresi (GB Ankara) Üst Kretase Planktonik Foraminifera'ları ve biostratigrafi incelemesi.- Türk. Jeol. Kur. Bült.
  22, 121-134.
-- (1980): Nannoplankton biostratigraphy of the Haymana region (SW Ankara).- Bull. Geol. Soc. Turkey 23, 165-177.
ÜNALAN, G.; YÜKSEL, V.; TEKELİ, T.; GÖNENÇ, D.; SEYIRT, Z. & HÜSEYİN, S.
  (1976): The stratigraphy and paleogeographical evolution of the
  Upper Cretaceous-Lower Tertiary sediments in the Haymana-Polatlı
  region (SW of Ankara).- Bull. Geol. Soc. Turkey 19, 159-176.
YÜKSEL, S. (1970): Etude géologique de la région d'Haymana (Turquie
  Centrale).- Thèse Fac. Sci. Univ., Nancy, France (unpubl.).

MACRO-INVERTEBRATES AND THE CRETACEOUS-TERTIARY BOUNDARY

A contribution
to Project

GLOBAL
BIO-
EVENTS

WIEDMANN, Jost *)

Abstract: Most of Cretaceous macro-invertebrate groups such as ammonites, inoceramids, belemnites, and rudists whow a gradual decline towards the C/T boundary, and some of them disappear long before the boundary level itself. As in the terrestrial vertebrates, their disappearance is unrelated to an extraterrestrial impact as suggested by the widespread iridium anomaly occurring at the boundary. It is also unrelated to fluctuations in temperature, which have been recognized near the boundary level but which can probably be better correlated with such an event. The decline of ammonites is gradual and at the same time periodic in nature. Periodic events occurring through the Upper Cretaceous and the Phanerozoic as a whole are sea level changes. Indeed, the pattern of global transgressions and regressions shows a striking similarity with increasing and decreasing ammonite diversity.

In contrast, the turnover in calcareous plankton of oceanic surface waters as well as in angiosperms (Aquilapollenites Province) is a later and "instantaneous" event which can be related with observed fluctuations in temperature, the iridium anomaly, and presumed impact at the boundary level.

Unlike those groups of organisms affected by the terminal Cretaceous extinctions, a few others flourished at this time. From the boundary beds of Jylland, Denmark, bryozoans and specialized crinoids were reported as having their peak density at the time of mass extinction. Since both groups are very sensitive to changes in salinity or oxic conditions, botl factors can be ruled out as having caused the C/T boundary mass mortality. From available observations it can be concluded that the great faunal and floral break at the C/T boundary was not a monocausal phenomenon, but was the result of a complex scenario of both periodically fluctuating and instantaneous catastrophic factors.

In memory of O.H. Schindewolf (1896-1971)

Introduction

As at all "critical" boundaries (Precambrian/Cambrian, Silurian/Devonian, Permian/Triassic), the initial problem at the C/T boundary is the scarcity of undisturbed marine transitional sections. The available sections in Europe and northwestern Africa are reproduced in Fig. 1; their number is not overwhelming. The next problem becomes immediately obvious: All localities of the epicontinental seas in the Temperate Realm (most of the Danish sections, those of The Vistula River and in the Maastricht area) and even some in the Tethyan Realm (most of the Alpine Gosau sections, the Nierental Beds of Lattengebirge, some of the Appennine sections) are inappropriate for the present purpose since they exhibit reduced thicknesses, hardgrounds, pebble beds, reworking, or condensation

---

*) Institut und Museum für Geologie und Paläontologie, D-7400 Tübingen, F.R.G.

Lecture Notes in Earth Sciences, Vol. 8
Global Bio-Events. Edited by O. Walliser
© Springer-Verlag Berlin Heidelberg 1986

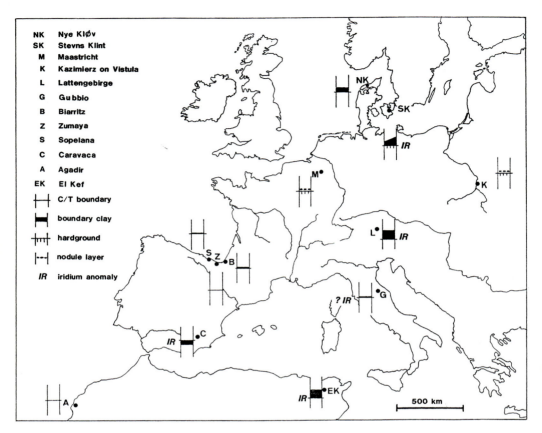

Figure 1. Location map of European and northwest African C/T boundary sections and the presence of boundary clay, iridium anomaly, hardgrounds or nodule layers (revised from Smit & ten Kate 1982).

at the boundary level. Even worse is the fact that the incomplete Temperate sections are for the time being the only ones from which confident macro-invertebrate data are available (Birkelund 1979, Błaszkiewicz 1979). The generally complete and comparatively thick Tethyan and oceanic (Atlantic, Indic, Pacific) C/T boundary sections, however, are much more tectonized, often include turbidite series, and can be characterized by a total lack of macro-invertebrates (the Appennines, Lattengebirge, Caravaca/Spain, all DSDP-IPOD sites) or their extreme scarcity (El Kef, Agadir). Outside of the area considered in Fig. 1 the conditions are even worse.

Yet another problem is related with the scarcity of complete marine and ammonite-bearing sections near the boundary: the impossibility of defining and subdividing the Maastrichtian Stage on the basis of ammonites (Kennedy 1984).

Luckily, the nearly perfect sections on the coast of the Gulf of Biscay between Biarritz and Bilbao provide much necessary information on the nature of the C/T boundary.

The Zumaya section, northern Spain

A series of C/T boundary sections exists on the Gulf of Biscay coastline between Biarritz (France) and Bilbao (Spain). The Biarritz section was described and defined on the basis of calcareous nannoplankton by Perch-Nielsen (1979). Lamolda et al. (1983) described the turnover in plank-tonic foraminifera in the Sopelana section, north of Bilbao. At both sections macro-invertebrates are rare or not yet known. Sopelana became "famous" as a locality with Tertiary inoceramids (Kopp 1959); but the dating was found to be erroneous and based on a series of faults parallel to the bedding plane at the boundary level (Bijvank 1967).

The most important coastline section is that of Zumaya, Guipuzcoa Province, Spain (Fig. 2, 3). First described by Gómez de Llarena (1954, 1956), this section seems -- for the time being -- to be the only one

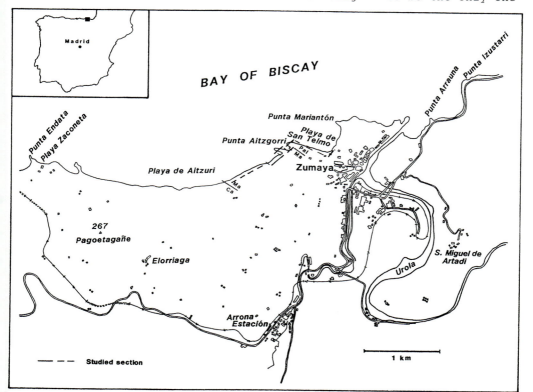

Figure 2. Location map of the Zumaya C/T boundary section, northern Spain.

in the northern hemisphere which has abundant macro-invertebrates within the Maastrichtian and which is tectónically undisturbed at the boundary itself.

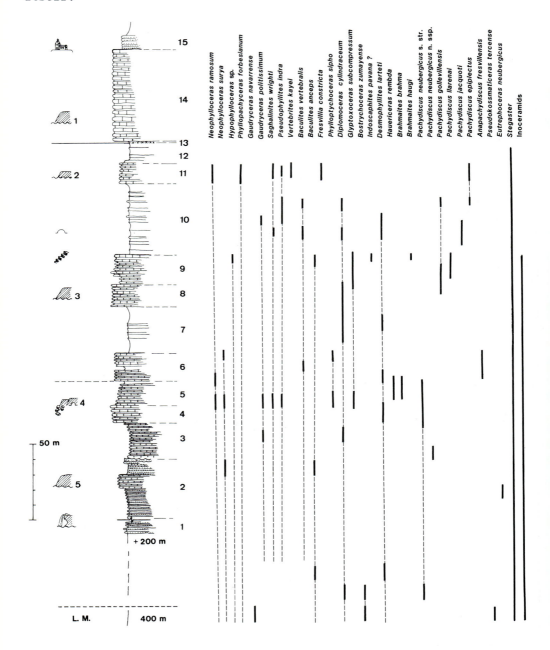

Figure 3. C/T boundary section at Zumaya and macro-invertebrate ranges.

The importance of the sections for dating the Maastrichtian was first stressed by Wiedmann (1960, 1962, 1981). At another occasion (Wiedmann 1969) it was demonstrated that the final decline of ammonites was
(1) unrelated with shell heteromorphy,
(2) accompanied by small-scale stress associations in the late Maastrichtian, and
(3) gradual through the whole of the Upper Cretaceous.

Herm (1965) and v. Hillebrandt (1965) described the Cretaceous and Tertiary part of the boundary section at Zumaya in detail and investigated the planktonic foraminifera. Herm (op. cit.) was likewise able to show stress communities (small size, uncoiling) in these planktonics just below the boundary. Both authors erroneously placed the boundary at the contact between purple marls and overlying red limestones (Fig. 3, units 12/13 and 14, respectively); this is 0.40 m above the true boundary which was drawn by Percival & Fischer (1977) on the basis of calcareous nannoplankton.

New investigations (Ward & Wiedmann 1983; Wiedmann, in press) allow the following conclusions:
(a) The Zumaya section offers several advantages: great sedimentary thickness of the Maastrichtian (about 800 m $\approx$ 100 m/My); decrease in turbidite input towards the boundary (i.e., Middle, Upper Maastrichtian, Lower Paleocene); continuity across the boundary; no extensive tectonics; relative abundance in macro-, microfaunas, and nannofloras.
(b) The turnover in planktonic foraminifera and calcareous nannoplankton is placed 0.40 m below the facies change from purple marls to red limestones (Percival & Fischer 1977, Smit & ten Kate 1982); sedimentation is, thus, continuous across the boundary.
(c) Ammonites are present in high diversity and with good stratigraphic resolution; for the first time a Maastrichtian ammonite zonation (based on pachydiscids) can be proposed (Wiedmann, in press).
(d) Ammonite diversity decreases continuously through the Maastrichtian; ammonites practically disappear 13 m below the boundary ($\approx$ 130,000 years);all four Cretaceous ammonite groups (phylloceratids, lytoceratids, ammonitids, ancyloceratids) are present in comparable percentages up to the final extinction.
(e) No innovations (new families or new genera) can be recognized in late Maastrichtian time.
(f) Layers with small ammonites are concentrated in the late Maastrichtian (Wiedmann 1969). Whether this is the result of mass mortality of juveniles or true dwarfed forms cannot be ascertained due to

Figure 4. Ammonite
decline through the
Upper Cretaceous and
global sea level changes
(after Wiedmann 1969,
Sliter 1976).

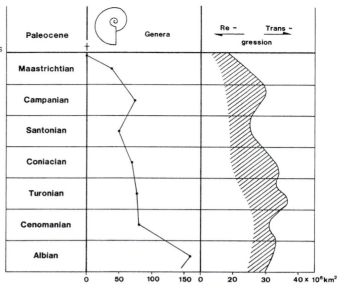

Figure 5. Late Cretaceous
decline of inoceramids (after
Dhondt 1983) and belemnites
(after Christensen 1976).

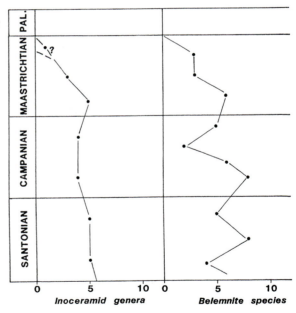

poor preservation; but in any case these forms can be regarded as
proof of stress environments existing long before the boundary itself.

(g) Typical inoceramids disappear even earlier (unit 9 of Fig. 3); only
the doubtful (inoceramid ?) genus _Tenuipteria_ Stephenson persists

up to the close of the Cretaceous; in accordance with Dhondt (1983), the inoceramid decline is also gradual.
(h) Echinoids persist with one or two of the genus Stegaster Pomel up to the boundary.
(i) No boundary clay and no iridium anomaly can be recognized in the Zumaya section (Smit & ten Kate 1982).

From these observations, it can be concluded that -- in contrast to the surface planktonics -- ammonite and inoceramid extinctions are unrelated with the boundary event. The decline of both groups (Figs. 4, 5) is gradual in nature and precedes the iridium anomaly so much that even the argument of a collapsing food chain does not fit.

The boundary scenario

A number of questions relating with the boundary scenario should be treated again:
1. Are the boundary clay and the iridium layer synchronous and global in distrbution?
2. Are there any faunas flourishing at the C/T boundary and if so, what are their relevance?
3. What are the patterns of organic evolution towards the C/T boundary?
4. Which conclusions can be drawn for the C/T boundary scenario?

ad 1. At most localities, the iridium layer seems to be placed within the boundary clay (Smit & ten Kate 1982); but the boundary clay seems neither to have the same age everywhere nor does it exist everywhere. It is not present (Fig. 1) in the Zumaya section, in Maastricht, nor in the Vistula River sections. In addition, it has an uppermost Maastrichtian foraminifera fauna at Caravaca (Smit & ten Kate 1982), while the fauna is of lowermost Paleocene age in most other sections. However, what is common in all these sections is the dwarfed nature of these planktonic associations. The solution could be that there are two or more iridium layers at or near the boundary as recently found in the Lattengebirge, Northern Calcareous Alps (D. Herm, pers. comm.).
Another problem is the fact that a number of localitites with boundary clay do not yield iridium, i.e. Biarritz and even the famous Gubbio section where the iridium anomaly was originally detected (Smit & ten Kate 1982, Fig. 5). The Zumaya section is also void of iridium.
Nevertheless, these restrictions probably do not devaluate the existence of a widespread iridium anomaly (generally associated

with concentrations of sanidine spheres and a depletion of rare
earths) which might be of cosmic origin. Localities in which the
iridium anomaly is undisputed are most of the DSDP-IPOD sites,
Lattengebirge, the Caravaca section, El Kef, and Stevns Klint.

ad 2. In a particularly interesting paper, Birkelund & Hakansson (1982)
were able to demonstrate that during the maximum collapse of Cretaceous faunas and floras, in the inner shelf environment of Nye
Kløv, Denmark, a peculiar, low diverse, benthic, soft bottom stress
community "flourished", consisting of bryozoans and the crinoid
Bourgueticrinus d'Orbigny. The special character of this stress-
related pioneer community of early Danian age has been treated in
detail. One of the most important results is the conclusion that
echinoderms as well as bryozoans were fairly sensitive to water
quality and were therefore unable to tolerate pronounced changes in
salinity or anoxic conditions. In consequence, these factors can be
ruled out from being effective for the C/T boundary scenario.

ad 3. The resulting pattern of organic evolution towards the close of
the Cretaceous is, therefore, much more complex than generally
stated.
Hence we have groups in which a gradual decline can be registered
through the whole or part of the Upper Cretaceous. These are primarily the ammonites; but inoceramids (Dhondt 1983), belemnites
(Christensen 1976), and terrestrial vertebrates (Clemens & Archibald
1980, T.J.M. Schopf 1982) can be added to the same pattern of
extinction. Nearly all these groups (except belemnites) as well
as the reef-building rudists died out before the end of the Cretaceous.
It should thus be stressed that the gradually declining groups do not
exhibit any environmental restriction: nearly all terrestrial and
marine environments and differing climatic zones (North Temperate,
Tethyan, Austral) were affected. Obviously, all these extinctions
cannot be related with the iridium anomaly.
Oceanic calcareous plankton (foraminifera as well as coccoliths),
however, became "instantaneously" extinct coeval with the iridium
anomaly. Contemporaneous -- within the scope mentioned above --
might also have been the drastic turnover in the North Temperate
Aquilapollenites Province of North America and northeastern Asia
(Clemens et al. 1981, Hickey 1981), but the pattern of global
angiosperm turnover at this boundary does not match the picture of
a global impact: The severity of extinction decreases rapidly towards
the equator where we have an increase of diversity at that time.

While the majority of the diverse shelf benthos vanished during the C/T transition, an increase in density is reported from the low diverse, benthic stress community of bryozoans and crinoids from northern Jylland, Denmark (Birkelund & Hakansson 1982). The appearance of this stress community can be correlated with the impact and can exclude the existence of other physico-chemical factors at the same time, i.e., drastic changes in salinity or anoxic conditions. The occasional existence of organic black shales at the boundary level, e.g., the Fish Clay at Stevns Klint, Denmark, can therefore not be overestimated or generalized, nor the possibility of an "Arctic Ocean injection" (Thierstein & Berger 1978).

ad 4. As has been demonstrated by Hsü et al. (1982), Smit & ten Kate (1982), and others, changing temperature might have played an additional role at the C/T boundary. But interpretation of $\delta^{18}O$ and $\delta^{13}C$ data is still highly controversal. As can be seen from Fig. 6, there is some indication of fluctuation in temperature at the boundary; while we have decreasing temperature in the Brazil Basin (DSDP

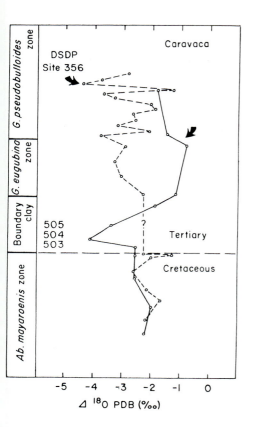

Figure 6.   Fluctuating temperature at the C/T boundary (from Smit & ten Kate 1982).

Site 356), a rapid increase is observed in the boundary clay level
at Caravaca. In any case, this is a short-termed variation of more
than 10°C which would have affected terrestrial angiosperms and
oceanic plankton, and -- if the proposed correlation is correct --
could even be related with an extraterrestrial impact. The cooling
might be the direct consequence of the impact and global atmospheric
darkening by clouds (Milne & McKay 1982), which could develop later
into a greenhouse situation, as is documented by the warming trend
in early Paleocene.

But in any case, these fluctuations occurred more or less synchronously with the impact and are thus irrelevant for the decline of
ammonites and related groups. In addition, the often-used argument
of a collapsing food chain is inappropriate since the extinction of
plankton feeders precedes the extinction of plankton; also the disappearance of terrestrial reptiles anticipates the break-down of the
Aquilapollenites Province.

Figure 7. Course of ammonite
diversity (from House 1985) and
global sea level changes (from
Yanshin 1973 and Silter 1976)
through the Paleozoic.

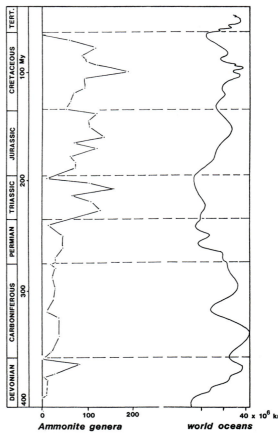

The continuous decline of ammonites, belemnites, inoceramids, and reptiles has to be related with factors causing continuous stress through the whole or most of the Upper Cretaceous and exhibiting a certain periodicity, especially if the earlier "critical" boundaries are likewise considered. If salinity, food chain collapse, oxygen depletion and perhaps even temperature can be ruled out, not too many factors are available. The relevance of magnetic reversals for organic life is still uncertain; it is not applicable to the Upper Cretaceous with its comparatively long lasting quiet zones (Harland et al. 1982).

There is, however, one parameter running nearly parallel to the course of ammonite diversity, which is the pattern of global sea level changes (Figs. 4, 7). Sliter's (1976) sea level curve of the Cretaceous coincides nearly perfectly with the peaks in ammonite diversity during the Cenomanian and Campanian transgressions. But, for the present purpose is more important the congruency between ammonite decline and global regressions, not only in the late Cretaceous but in all previous crises in ammonoid evolution (Fig. 7). If we admit that ammonoid life flourished in the inner shelf seas and was much more bottom-related than generally believed (Wiedmann 1973), this congruency is not too surprising. It is reasonable to imagine that even terrestrial reptiles were strongly affected by late Cretaceous regressions, since inland water bodies would have dried up with the sinking water table (Wiedmann 1969, T.J.M. Schopf 1982).

From our observations on evolutionary patterns of various organic groups during the late Cretaceous we can conclude that a complex scenario has to be postulated for having caused the diverse late Cretaceous extinctions. In addition to instantaneous, catastrophic events at the C/T boundary itself (cosmic impact, fluctuating temperature) long lasting and periodically fluctuating processes (sea level changes) have also to be considered which started to affect organic life much time before. More investigations of the congruency between sea level patterns and the evolution of organic groups and biotas are necessary.

When discussing earlier ideas on the interdependency between tectonic, climatic and eustatic pulses and organic evolution (Umbgrove 1942, 1947; v. Bubnoff 1949), O.H. Schindewolf presented in a little-known talk given at Tübingen University (1950, Fig. 17) an impressive diagram in which the course of organic evolution was shown to parallel that of eustatic sea level changes, but without paying particular attention to it.

Acknowledgements

I greatly enjoyed field work at the Zumaya section conducted with Peter Ward (Seattle) and Marcos Lamolda (Bilbao). Special thanks are due to Linda Hobert for having improved my English. The drawings were prepared by Mr. H. Vollmer.

REFERENCES

BIJVANK, G.J. (1967): Über das Vorkommen von Inoceramen im Tertiär und das Alter der Schichtenfolge in einem Küstenprofil bei Guecho (Vizcaya, Nord-Spanien).- N. Jb. Geol. Paläont. Mh. 1967, 385-397.
BIRKELUND, T. (1979): The last Maastrichtian ammonites.- in: BIRKELUND, T. & BROMLEY, R.G. (eds.): Cretaceous-Tertiary Boundary Events. Symposium, Copenhagen 1979, 1, 51-57.
-- & HAKANSSON, E. (1982): The terminal Cretaceous extinction in Boreal shelf seas - A multicausal event.- Geol. Soc. Amer., Spec. Pap. 190, 373-384.
BŁASZKIEWICZ, A. (1979): Stratigraphie du Campanien et du Maastrichtien de la vallée de la Vistule Moyenne à l'aide d'ammonites et de belemnites.- in: WIEDMANN, J. (ed.): Aspekte der Kreide Europas. I.U.G.S. (A) 6, 473-485.
BUBNOFF, S. v. (1949): Einführung in die Erdgeschichte, II. Teil.- Mitteldt. Druck- & Verl.-Anst., 345-771, Halle, Saale.
CHRISTENSEN, W.K. (1976): Palaeobiography of Late Cretaceous belemnites of Europe.- Paläont. Z. 50, 113-129.
CLEMENS, W. & ARCHIBALD, D. (1980): Evolution of terrestrial faunas during the Cretaceous-Tertiary transition.- Mém. Soc. géol. France, N.S. 139, 67-74.
-- ; ARCHIBALD, J.D. & HICKEY, L.J. (1981): Out with a wimper not a bang.- Paleobiology 7, 293-298.
DHONDT, A.V. (1983): Campanian and Maastrichtian inoceramids: A review.- Zitteliana 10, 689-701.
GÓMEZ DE LLARENA, J. (1954): Observaciones geológicas en el flysch cretácico-numulítico de Guipúzcoa. I.- Mon. Inst. "L.Mallada" Invest. geol. 13, 1-98.
-- (1956): Observaciones geológicas en el flysch cretácico-numulítico de Guipúzcoa. II.- ibid. 15, 1-47.
HARLAND, W.B. et al. (1982): A geologic time scale.- Cambridge Earth Sci Ser. (Univ. Press), 131 p.
HERM, D. (1965): Mikropaläontologisch-stratigraphische Untersuchungen im Kreideflysch zwischen Deva und Zumaya (Prov. Guizpuzcoa, Nord-spanien).- Z. dt. geol. Ges. 115, 277-348.
HICKEY, L.J. (1981): Land plant evidence compatible with gradual, not catastrophic, change at the end of the Cretaceous.- Nature 292, 529-531.
HILLEBRANDT, A. v. (1965): Foraminiferen-Stratigraphie im Alttertiär von Zumaya (Provinz Guipúzcoa, NW-Spanien) und ein Vergleich mit anderen Tethys-Gebieten.- Abh. Bayer. Akad. Wiss., Math.-naturwiss. Kl., N.F. 123, 1-62.
HOUSE, M.R. (1985): The ammonoid time-scale and ammonoid evolution.- in: SNEILLING, E.J. (ed.): The Chronology of the Geological Record. Geol. Soc. Mem. 10, 273-283.
HSÜ, K.J. et al. (1982): Mass mortality and its environmental and evolutionary consequences.- Science 216, 249-256.
KENNEDY, W.J. (1984): Ammonite faunas and the "standard zones" of the Cenomanian to Maastrichtian stages in their type areas, with some proposals for the definition of the stage boundaries by ammonites.-

Bull. geol. Soc. Denmark 33, 147-161.
KOPP, K.-O. (1959): Inoceramen im Tertiär des Mittelmeerraumes.- N. Jb.
  Geol. Paläont., Mh. 1959, 481-492.
LAMOLDA, M.A. et al. (1983): The Cretaceous-Tertiary boundary in Sopelana
  (Biscay, Basque Country).- Zitteliana 10, 663-670.
MILNE, D.H. & MCKAY, C.P. (1982): Response of marine plankton communi-
  ties to a global atmospheric darkening.- Geol. Soc. Amer., Spec. Pap.
  190, 297-303.
PERCH-NIELSEN, K. (1979): Calcareous nannofossils at the Cretaceous/
  Tertiary boundary near Biarritz, France.- in: CHRISTENSEN, W.K. &
  BIRKELUND, T. (eds.): Cretaceous-Tertiary Boundary Events. 2, 151-
  155, Copenhagen Univ.
PERCIVAL, S.F. & FISCHER, A.G. (1977): Changes in the calcareous nanno-
  plankton in the Cretaceous-Tertiary biotic crisis at Zumaya, Spain.-
  Evol. Theory 2, 1-35.
SCHINDEWOLF, O.H. (1950): Der Zeitfaktor in Geologie und Paläontologie.-
  Akad. Antrittsvorlesung, 1-114, Stuttgart (Schweizerbart-Verl.).
SCHOPF, T.J.M. (1982): Extinction of the dinosaurs: A 1982 understanding.
  - Geol. Soc. Amer., Spec. Pap. 190, 415-422.
SLITER, W.V. (1976): Cretaceous foraminifers from the southwestern Atlan-
  tic Ocean, Leg 36, Deep Sea Drilling Project.- in: BARKER, P.F.;
  DALZIEL, I.W.D. et al. (eds.): Init. Reps. DSDP 36, 519-537.
SMIT, J. & ten KATE, W.G.H.Z. (1982): Trace-element patterns at the
  Cretaceous-Tertiary boundary - Consequences of a large impact.-
  Cretaceous Res. 3, 307-332.
THIERSTEIN, H.R. & BERGER, W.H. (1978): Injection events in ocean history.
  - Nature 276, 461-566.
UMBGROVE, J.H.F. (1942): The Pulse of the Earth.- 1st. ed., 179 p., The
  Hague (M. Nijhoff).
-- (1947): idem.- 2nd. ed., 358 p., The Hague (M. Nijhoff).
WARD, P.D. & WIEDMANN, J. (1983): The Maastrichtian ammonite succession
  at Zumaya, Spain.- Abstracts, Symposium on Cretaceous Stage Bounda-
  ries, Copenhagen 1983, 205-208.
WIEDMANN, J. (1960): Le Crétacé supérieur de l'Espagne et du Portugal
  et ses Céphalopodes.- C.R. 84e Congrès Soc. Savantes Paris et Dépt.,
  Dijon, 1959, Sect. Sci., Sous-Sect. Géol., 709-764.
-- (1962): Ammoniten aus der vascogotischen Kreide (Nordspanien).- I.
  Phylloceratina, Lytoceratina.- Palaeontographica (A) 118, 119-237.
-- (1969): The heteromorphs and ammonoid extinction.- Biol. Rev. 44,
  563-602.
-- (1970): Über den Ursprung der Neoammonoideen - Das Problem einer Typo-
  genese.- Eclog. geol. Helv. 63, 923-1020.
-- (1973): Evolution or revolution of ammonoids at Mesozoic System
  boundaries.- Biol. Rev. 48, 159-194.
-- (1981): in: LAMOLDA, M.; RODRÍGUEZ-LÁZARO, J. & WIEDMANN, J.:
  Field Guide: Excursions to Coniacian - Maastrichtian of Basque-Canta-
  bric Basin.- Publ. Geol. Univ. autóm. Barcelona 14, 53 p.
-- (in press): Ammonid extinction and the "Cretaceous-Tertiary Boundary
  Event".-
YANSHIN, A.L. (1973): On so-called world transgression and regressions
  (in russ.).- Bjul. moskovsk. obč. ispyt. prirody, n. ser., otd. geol.
  48, 9-44.

BIO-EVENTS IN THE CONTINENTAL REALM DURING THE CRETACEOUS/TERTIARY TRANSITION: A MULTIDISCIPLINARY APPROACH

A contribution to Project GLOBAL BIO-EVENTS

FEIST, Monique (Coordinator) *)

Abstract: This study is a short progress report of a multidisciplinary cooperation. It deals with animal and plant biotas found in five non-marine Maastrichtian-Paleocene sequences in southern Europe (southern France, north-east Spain). In addition, contributions on North America charophyte floras are taken into account.

The aims of the study are two-fold:

1. stratigraphical

This is essential, since the main problems relating to continental sequences is that of dating, owing to the difficulties of correlation with the marine standard stages. At present, charophytes provide the best means of sub-division and correlation of southern European terminal Cretaceous and basal Paleocene successions; an attempt to correlate between continental and marine sequences using charophytes was presented at the Colloquium on the Coniacian to Maastrichtian Stages in Marseille (1983). For example, the non-marine Tremp Formation in northern Spain can be subdivided into Maastrichtian and Paleocene portions (Table). In the present state of knowledge, distant correlations with America and Asia can be made only at the generic level, with the exception of a few cosmopolitan species such as <u>Platychara</u> <u>compressa</u>.

Current investigations deal with new and precise sampling of fossils (charophytes, pollen, spores, molluscs, ostracods, vertebrates) carried out together with palaeomagnetic and geochemical studies with a view a) to compare the various biozonations; b) to establish, as far as possible, direct correlations with the marine realm by means of palynofloras; c) to determine the position of the Cretaceous/Tertiary (C/T) boundary by palaeomagnetic zonation and geochemical analysis.

2. palaeobiological

A comparative study of the faunas and floras is expected to supply new data on the C/T boundary event: will non-marine biotas give a clearer indication of an abrupt event or of progressive changes in the environment? The studies carried out in other areas, such as N America (Clemens 1982, Smit & Van de Kaars 1984) are far from exhibiting common views on this problem.

---

*) Laboratoire de Paléobotanique, Université des Sciences et Techniques, 34060 Montpellier, France.

412

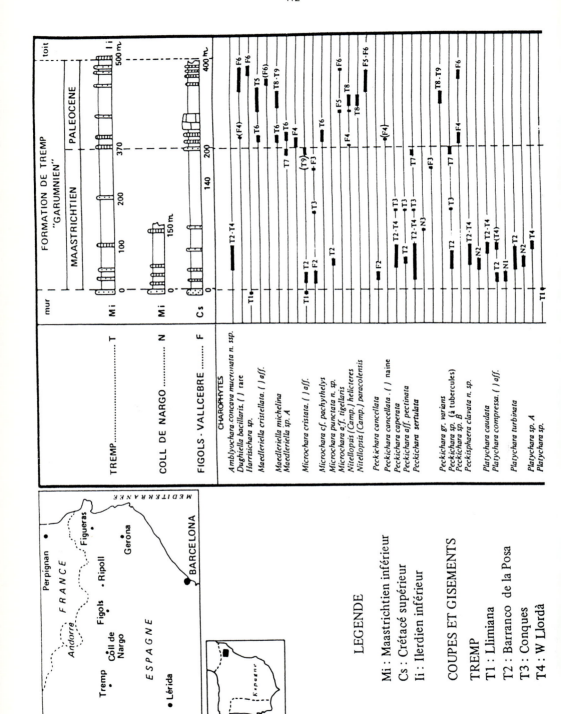

LEGENDE

Mi : Maastrichtien inférieur
Cs : Crétacé supérieur
Ii : Ilerdien inférieur

COUPES ET GISEMENTS

TREMP
T1 : Llimiana
T2 : Barranco de la Posa
T3 : Conques
T4 : W Llordà

Porochara malladae

Porochara oblonga
Pseudoharrisichara isonae n. sp.
Septorella brachycera
Septorella ultima
Sphaerochara edda

FORAMINIFERES

Discorbidae

RUDISTES

Agriopleura moroi
Biradiolites ara
Hippuritella castroi
Praeradiolites boucheroni
Praeradiolites echenensis
Praeradiolites leymerici
Radiolatella pulchella
Radiolites sellesi

GASTEROPODES

? Carychiopsis sp.
Indeterminae generae
Melanopsis sp.
? Palaeocyclophorus
Physa sp.
Planorbis sp.
? Pupilla
Rissoina (? Zebinella) sp.
Syrnola sp.
Valvata cf. indecisa
Valvata sp.

OSTRACODES

Bairdia sp.
Cytherella sp.
Dolocytheridea (= cf. Parakrithe *)
Dordonilla sp.
Frambocythere gr. tumiensis
(=Bisakocypris tumiensis **)
Limburgina sp. (=aff. Hermanites *)
Neocyprideis sp. (=aff. Cyprideis *)
Pterygocythere sp.
? Sphaerolebebris sp.
* sensu Liebau (1971)
** sensu Helmdach (1978)

T5 : Claret
T6 : Sant Salvador de Tolo
T7 : Tossal d'Obà
T8 : La Baronia de Sant Oïsme
T9 : Beniure

COLL DE NARGO
N1 à N3 : N Río Sellent

FIGOLS-VALLCEBRE
F1, F2 : Figols (carrière)
F3 : W Vallcebre
F4 à F6 : N Vallcebre

Distribution des organismes à la limite du Crétacé et du Tertiaire dans le nord-est de l'Espagne (d'après M. Feist et F. Colombo 1983).

Various evolutionary patterns are being investigated in southern Europe, as noted below:

a) charophytes. 50 % of the families became extinct around the C/T boundary, yielding to structural uniformity in fructifications. However, the crisis occurred within the first major phase of expansion of the family Characeae; thus a great diversity of floras offers valuable information on the evolutionary processes during the period under consideration. Until now, C/T boundary charophytes have been studied mainly in Europe and N America. The main research problem deals with the differences observed in the successive floras in these two areas: whereas in Europe the boundary is marked by a floral break, in North America the floral transition across the boundary is progressive, without any abrupt change. Another major problem is the abrupt extinction of the Clavotoraceae at the end of the Cretaceous; this longliving family is represented in terminal Cretaceous solely by the genus Septorella, which not only occurred in flood abundance, but also reached an acme in size of fructification; was this plant so specialised that only a small change in the environment was required to eliminate it, or can its disappearance be attributed to the postulated C/T Event (M.F.).

b) palynofloras. Palynostratigraphy of NW European Maastrichtian and Paleocene is in progress. Preliminary indications are that sections in north-west Spain may provide. material to investigate palynological events at the C/T boundary and afford floras comparisons with North America and Africa. (J.M.)

c) molluscs. Studies of molluscan faunas already published or in progress allow the interpretation of their migration in time and space as a function of shift in life habit, especially the gastropods (such as Lychnus) from the terminal Cretaceous. Information about basal Tertiary faunas, however, remains very limited. However, new data have been obtained from investigations on north Pyrennean faunas: in passing from the Maastrichtian to the Dano-Montian, no new genera appear, there is slow evolution of limnic taxa and some speciation in terrestrial faunas. (J.V.).

d) ostracods. The investigations concern the phyletic evolution of limnic species during the C/T transition: (1) Species temporarily grouped under the name Paracandona and (2) species assigned to Frambocythere. The latter genus had its apogee during the terminal Cretaceous, and in the Tertiary is represented only by parthenogenetic species. An attempt will be made to distinguish the evolutionary species from ecological variants and to determine the effect of sedimentation and climate on the size and ornamentation of the shells. (Y.T.).

e). mammals. The C/T transition corresponded to a period of radiation and dispersal of the modern groups which lie at the origin of the present day population in the different continents. The information is particularly sparse for Europe; it would be of the highest interest to determine whether this mammal province was linked to NW America or to the continental Asia or, alternatively, whether it exhibits an endemic character. (B.S.).

Participants

J.F. BABINOT, ostracods, Marseille; D. BATTEN, megaspores, Aberdeen; M. BILOTTE, stratigraphy, Toulouse; P. BONTE, geochemistry, Gif-sur-Yvette; E. BUFFETAUT, crocodiles, Paris; J. CHIMENT, charophytes, New York; F. COLOMBO, stratigraphy, Barcelona; J.P. DURAND, sedimentology, Marseille; M. FEIST, charophytes, Montpellier; M. FLOQUET, sedimentology, Dijon; B. GALBRUN, magnetrostratigraphy, Paris; B. LEPICARD, sedimentology, Toulouse; J. MEDUS, palynology, Marseille; M. RENARD, geochemistry, Paris; R. ROCCHIA, astrophysics, Gif-sur-Yvette; B. SIGE, mammals, Montpellier; Y. TAMBAREAU, ostracods, Toulouse; P. TAQUET, dinosaurs, Paris; J. VILLATTE, molluscs, Toulouse.
Further co-operation is requested and any new contributions would be welcome.

TERTIARY

GLOBAL TERTIARY CLIMATIC CHANGES, PALEOPHYTOGEOGRAPHY AND PHYTOSTRATIGRAPHY

PANTIĆ, Nikola K. *)

A contribution to Project GLOBAL BIO-EVENTS

Abstract: The most notable mega longtime event in the Tertiary is the global transformation of the climate: the globally warm and equable Mesozoic climate on Earth changes into a globally colder climate. The causes of these climatic changes are extremely complex (various abiotic and biological events in constant mutual interaction).
  Paleophytogeographical changes in the Tertiary are caused by the global paleoclimatic events of that time, primarily by the narrowing of the tropical and sub-tropical climatic belts and migration of land floras.
  Stratigraphic correlations, founded on the study of remnants of land flora, evidently depends basically on climatic events. The classic, long used method of stratigraphic marker must therefore be abandonned. By the use of the event stratigraphy new bases are obtained for refining existing stratigraphic scales and methods of correlation.

Introduction

The present global quasistable state of geological, geographical, climatic and biological systems on earth is the result of the constant interaction of all events and processes in the past geological period: abiotic (geological and extra-telluric) and biological. Every other concrete period in the geological past has its pre-history and its specific abiotic and biological events and processes in the overall action on which depends each concrete state of the natural environments on Earth. The period from the late Cretaceous to the Plio-Quaternary is in this respect exceptionally important, for these events and processes of that time change the natural environments with great intensity.

The most notable mega longtime event in Tertiary is the global transformation of the climate: the globally warm and equable Mesozoic climate on Earth changes into a globally colder climate with cyclic glaciations in the Plio-Quaternary.

Climatic changes in the Tertiary and their causes

Through the Mesozoic right up to the late Cretaceous the climate on Earth was, globally spoken, far warmer than it is today. The warm climatic belts (tropical and sub-tropical) were considerably broader than today (Figure 1) and the temperate climatic belts were limited to relatively narrow areas around the poles (Pantić & Stefanović 1984, Hallam 1981, Vakhrameev et al. 1978).

*) Institute of Geology and Palaeontology, University of Beograd, 11000 Beograd, Yugoslavia.

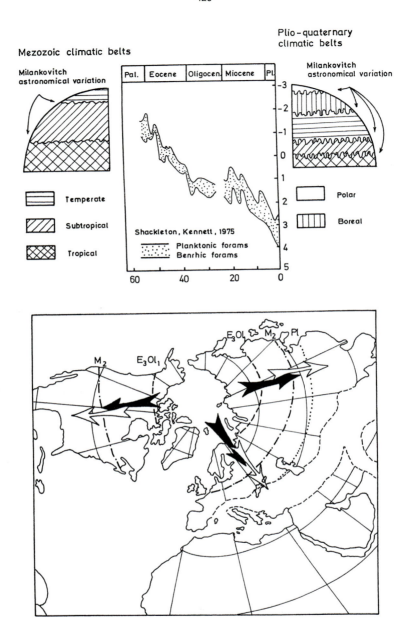

Figure 1. Above: Distribution of climatic belts and phytogeographical realms in Middle Cretaceous. Below: Trajectory (possible projection) of extraterrestrial body - about the boundary Cr/Pg (from Abramović 1984). I. Impact places (1. Karska, 2. Ist-Karska, 3. Kamenskaia, 4. Gusevskaia, 5. Oazis, 6. B.P.). II. Projection of trajectory. III. Possible impact places of largest body. IV. Phytogeographical provinces in which 75 % of land plants have extended (boundary Cr/Pg). V. Locality in which are identified layers enriched with siderophiles.

Since the late Cretaceous the climate became cooler on earth, at
first extremely slowly but since the Eocene with increased intensity.
This cooling process has not been an even one, especially at the beginning, but an irregular one in which the climate on earth became
colder and colder step by step. The causes of these climatic changes
are extremely complex: there are various abiotic and biological events
and processes which are in constant mutual interaction.

I Abiotic events

1) Extra-telluric events - impacts of celestial bodies on the Earth
(asteroids or comets) (Alvarez et al. 1980, O'Keefe 1980 et al.),
effects due to the Earth's place in the galaxy (Delsemme 1985) and
astronomical variations (Milanković).
2) Geological events - a) geodynamic events - horizontal and vertical
movement inside the Earth's crust (especially the position of the land
mass in the pole region), volcanism (Ramping et al. 1979, Campste 1984),
variable heat flow intensity on the Earth (Rozanov 1985), volume of the
ocean basins etc. b) interaction of the Earth's hydrosphere (including
cryosphere) and the Earth's atmosphere (especially sea level changes and
ocean currents).

II Global biological events - quantity of biomass, cyclic variations of
CO_2 concentrations in the atmosphere etc.

Complex coincidental action and interaction (always in a different combination of some geological, extra-telluric and biological events *)
caused the following climatic changes from the late Cretaceous to the
Plio-Quaternary (Figure 2): a) tropical and sub-tropical climatic belts
narrowed intensively. At the same time broad, cold (polar and boreal)
and temperate climatic belts were formed. b) the cryosphere was formed:
first the southern ice cap (from the Oligocene) and then the northern
cap. These caps, depending on changes in the Earth's orbit, increase or
decrease their masses of ice and thereby contribute to the formation
of the climate of the anthropogeny.

---

*) Special mention should be made of the following in this period of
   relevant events: frequent impacts (for our observations impacts
   65 (Fig. 1), 38,35, 28, 15 and 1 million years ago are significant),
   intensive volcanic activity during the formation of the Alpine-
   Himalayan and other orogenic belts, horizontal and vertical movement of parts of the Earth's crust, the formation of new continental parts, especially in areas around the poles, and in this connection the increasingly intensive effects of astronomical variations
   (Milanović), the directions of ocean currents etc.

Paleophytogeographical changes in the Tertiary

Paleophytogeographical changes in the Tertiary are caused by the global paleoclimatic events of that time, primarily by the narrowing of the tropical and sub-tropical climatic belts. We shall mention only a number of the characteristic state or changes in the composition of land vegetation in the northern hemisphere (mainly on the continent of Europe and in the area of Tethys) which demonstrate the dependence of basic paleophytogeographical changes in the Tertiary (migration of land floras on the narrowing of the warm climatic belts (Fig. 2).

1. During the Eocene subtropical and tropical vegetation was spread over an even greater area than is the case today. Temperate vegetation was limited to the relatively narrow zones around the poles. Northern Europe was still in the zone of sub-tropical vegetation-geographical realm. Flora from the Geiseltal are wellknown: palms, wet swamp ferns, a multitude of Normapollis types, specific types of conifers etc. South of the northern land at that time, in the upper Eocene and the lower Oligocene,there were many archipelagoes and larger islands (like the Malayan archipelago) in the scope of Tethys -- the Tethian vegetation-geographical area or the "southern-European province". On the sea coasts (in the zone of tide activity)there was mangrove vegetation . Nipa-palms and other characteristic vegetation exist (Pantić 1983). In the interior of the extensive islands vegetation is drier in nature (Zizyphus, small-leaved leguminous plants, some xerophilous myrica of the type <u>Myri ca bankstiaefolia</u> etc.).

2. The early Oligocene - late Oligocene is the period of the most rapid drop in temperature (and of the most intensive narrowing of the warm climatic belts) as well as of the most rapid migration of paleo-tropic vegetation to the south and the push of the vegetation of arcto-tertiary-type also to the south. This is best demonstrated by the major movement of the northern border of the palm to the south in the period $O_1 - O_3$ (Fig. 2).

3. A parallel study of Oligocene and Miocene flora from the northern parts of Europe and the "fixed" flora of corresponding age from geographically distant regions of southern Europe shows, even at the present level of research, the tempo of southward migration of all types of land vegetation. We note that, inter alia:

-- the Normapollis group remains longer in the southern parts of Europe (Balkans, France etc.) (see Fig. 3),

-- the genus <u>Bombacidites</u> exists in the south of France (Bessedick 1981, Pantić 1983) and on the Balkan peninsula up to the end of the Oligocene,

-- palms and the genus <u>Engelhardtia</u> disappear from south-eastern Europe

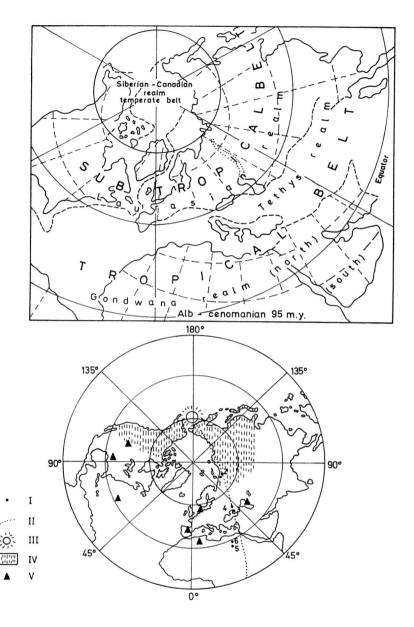

Figure 2. Global Tertiary climatic changes (above) and global paleophytogeography (below) with migration of the Northern boundary of palms during the Cenozoic.

only in the Pliocene,

-- the genus Cinnamomum is to be found in the lower Pliocene sediments of south-eastern Europe, while it retreated from the more northerly parts of Europe in the mid-Miocene. The same is the case with the genus Zizyphus (Pantić 1956, Bessedick & Suc 1983, Hochuli 1984),

-- parallel to this, the more frequent occurrence of some arctotertiary types of the genus Quercus (Q. roburoides-type) and of the genus Fagus (F. Pliocene type) in southern and south-eastern Europe comes at a considerably later date than their occurrence in northern Europe (see Fig. 3).

4. One of the major characteristics in the development of Tertiary land vegetation (quite well studied in European fossil flora) is the emergence and migration of dry vegetation of the savanna-steppe type in the late Oligocene-Mio-Pliocene. Two important phases are noted: one still insufficiently studied -- in the late Miocene (Ottnangian-Carpathian) and another occurrence of dry vegetation in Central Europe, at the Badenian-early Sarmatian interval, which migrates towards the south "crossing across" the then "Balkan" land during the Pannonian. It migrates even farther towards the south encompassing at the end of the late Miocene the Mediterranean area ("Messinian salinity crisis" -- see Pantić & Mihajlović 1979/80). The further study of this event (the migration of the dry climatic belt from the area of northern Europe towards the south) could promote inter-regional stratigraphic correlations between Paratethys and Tethys.

5. After this dry climatic phase, there occurred in south-eastern Europe a period of markedly humid climate which caused the emergence of extensive sub-tropical swamps on the northerly edge of the "Balkan land" where plant mass accumulated, from which many coal deposits arose. This humid climatic belt also migrated gradually towards the south. This can be shown on the basis of an analysis of the age of coal deposits. On the northerly edge of the "Balkan land" these deposits are lower Pliocene (Kolubara, Kostolac, Kreka etc.) while the known coal deposits in Greece are middle Pliocene in age (Ptolemais and others, Weerd 1983).

Phytostratigraphy of the Tertiary

Stratigraphic correlations, founded on the study of remnants of land flora (leaves, seeds, palynomorphs), evidently depend basically on climatic events. The classical, long used method of stratigraphic markers must therefore be abandoned as it is obviously no longer suitable for successful phytostratigraphic correlations. It is, however, also evident that through the use of the event stratigraphy new bases are obtained

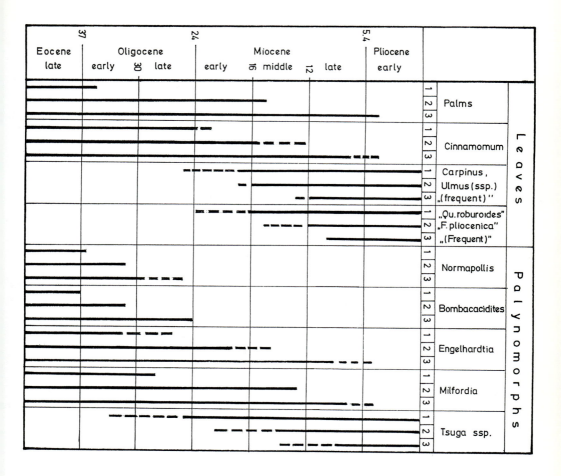

Figure 3. Tertiary phytostratigraphy depends basically on global climatic events (1 N-Europe, 2 Central Europe, 3 S-Europe).

for refining the existing stratigraphic scales and methods of correlation (Steininger et al. 1976). Knowledge to date (Fig. 3) enables the first major corrections of stratigraphic scales and methods of phytostratigraphic correlations.

For the further work on the refining of stratigraphic correlations there should also be the use of the so-called "fixed floras" ("repére" floras, Pantić 1956, or "Leitfloren", Pantić 1967). I have pointed several times (1956-1984) to the need to apply this method in the refining of paleobotanical scales and correlations. The fundamental idea of all my research has been based on the fact that paleobotanical correlations basically depend on climatic events, mainly on climatic changes and, in this context, on the migration of land flora over areas with the passage of time.

Far more is known today of the dependence of paleobotanical corre-lations on climatic events. Fig. 3 shows only a small part of the new knowledge which demonstrates in an extremely convincing manner the depen-dence of phytostratigraphy on climatic events.

The long use of stratigraphic markers in phytostratigraphy is past, it is evident that the so-called "golden pins" have started to loose their shine. The time of event stratigraphy is beginning (maybe, "with multi-stratigraphical methods, we should approach to a kind of holo-strati-graphy", Walliser 1985). It is certain that this is valid to the maximum use of fossil plant remnants in stratigraphy. It should, however, be expected that specific refining will also be achieved by the use of event stratigraphy in the case of sea and land fauna. Some papers are already showing the initial results in this domain of stratigraphic research (Bolli & Krasheninikov 1977).

The refining of phytostratigraphic correlations will also contri-bute to the better understanding of some events in the Tertiary which have not been sufficiently clarified to date, such as for example the case of the Messinian salinity crisis: "The problem of the Messinian salinity crisis is not entirely solved, however, several facts of the model are still poorly understood" (Cita et al. 1978) or only partially justified with proposals for the introduction of a new term for the final parts of the Tertiary ("Ultimogen" - Traverse 1982).

REFERENCES

ALVAREZ, et al. (1980): Extraterrestrial cause for the Cretaceous-Tertiary extinction.- Science 200, 1095-1108.
ABRAMOVIC, I. et al. (1984): Sovremennie idei teoretićeskoi geologii.-NEDRA, Leningrad, 1-280.
AXELROD, D. (1984): An interpretation of Cretaceous and Tertiary biota in Polar Regions.- Palaeogeogr., Palaeoclim., Palaeoecol. 45, 105-147.
BESSEDIK, M. (1981): Recherches palynologiques sur quelques sites du Burdigalien du midi de la France.- Thèse 3-ème cycle, Univ. Sci. et Techn. du Languedoc, Montpellier, 1-43.
-- & SUC, P. (1983): Les caractères du climat du Néogène en Mèditerranèe Nord-Occidentale d'après l'analyse pollinique.- Mediterranean Neogene continental palaeoenvironments and palaeoclimatic evolution. R.C.M. N.S. Interim Colloqium, Montpellier, 33-37.
BOLLI, H. & KRASHENINIKOV, V. (1977): Problems in Paleogene and Neogene correlations based on planctonic foraminifera.-Micropaleont. 23, 436-452.
CAMPSTE, J. et al. (1984): Episodic Volcanism and Evolutionary Crises.-Eos 65, 796-800.
DELSEMME, A.H. (1985): Un explication astronomique à la disparation de

Dinosaures.- L'Astronomie Rev. mem. C. Flammarion, 3-14.
HALLAM, A. (1981): Biogeographic relations between the northern and southern continents during the Mesozoic and Cenozoic.- Geol. Rdsch. 70, 583-595.
HOCHULI, P. (1985): Correlation of Middle and late Tertiary Sporomorph assemblages.- Paléobiologie continentale XIV, 301-314.
KNOBLOCH, E. (1975): Paläobotanische Daten zur Entwicklung des Klimas im Neogen der Zentralen Paratethys und der angrenzenden Gebiete.- Proc. VIIIth Congr. R.C.M.N.S. Bratislava.
O'KEEFE, I.A. (1980): The terminal Eocene event; formation of a ring-system around the Earth.- Nature 285, 309-311.
PANTIĆ, N. (1956): Biostratigraphie des Flores Tertiaires de Serbie.- Ann. Geol. Peninsule Balkanique, Beograd, 199-321.
-- (1967): Die Jungtertiären Floren und der Klimawechsel im Balkanraum.- Abh. Zentr. geol. Inst. Berlin 10.
-- (1983): The Problem of Paleogene formations in Serbia and Palaeobotanical Research Methods.- Glas Acad. Serbe Sci. et Arts, C.S. nat. et mat. 49, 7-22.
-- (1984): Refinement of Palynostratigraphic correlation from studies of the development of Mesozoic and Cenozoic floral realms.- 27th Internat. Geol. Cong. Moscow, Abstracts I, 03.
-- & MIHAJLOVIC, D. (1977): Neogene floras of the Balkan land areas and their bearing on the study of paleoclimatology, paleobiogeography and biostratigraphy (Partz).- Ann. Geol. Pén. Bal., Belgrade, 159-173.
-- & SLADIC-TRIFUNOVIC, M. (1984): Mesozoic floral Provinces of Tethys Ocean and its Margins with Respect to Plate tectonics.- 27th Internat. Geol. Congr., Abstracts, I.
-- & STEFANOVIĆ, D. (1984): Complex interaction of cosmic and geological events that affect the variation of Earth Climate through the geologic history.- in: BERGER, L. et al. (eds.): Milankovitch and Climate, Part I, 251-164, 1984, Reidel Publ. Co., Dordrecht.
RAMPING, R. et al. (1979): Can Rapid Climatic Change cause Volcanic Eruption?.- Science 206, 826-828.
ROZANOV, I.A. (1985): Evolucija zemnoi kori (Russ.).- Nauka, Moskva, 1-143.
STEININGER, F. et al. (1976): Current Oligocene - Miocene biostratigraphy of the Central Paratethys (Middle Europe).- Newsl. Stratigr. 4, 174-202.
TRAVERSE, A. (1982): Response of world vegetation to Neogene Tectonic and climatic events.- Alcheringa 6, 197-209.
VAKHRAMEEV, I. et al. (1978): Paläozoische und mesozoische Floren Eurasiens und die Phytogeographie dieser Zeit.- Fischer, Jena.
WEERD,A. (1983): Palynology of some Upper Miocene and Pliocene Formations in Greece.- Geol. Jb. B 48, 3-63.
WOLFE, J. (1978): A Paleobotanical Interpretation of Tertiary Climates in the Northern Hemisphere.- Amer. Sc. 66, 694-703.
-- (1980): Tertiary Climates and floristic relationships at high latitudes in the northern Hemisphere.- Palaeogeogr., Palaeoclim., Palaeoecol. 30, 313-323.

REFLECTIONS UPON THE CHANGES OF LOCAL
TERTIARY HERPETOFAUNAS TO GLOBAL EVENTS

SCHLEICH, H. Hermann *)

A contribution
to Project
GLOBAL
BIO-
EVENTS

Abstract: For the Tertiary as well for Europe as for southern Germany
herptile distributions are shown. Their distributional patterns are discussed in context with geological events, here mainly the Ries Event.
Reflections supporting the latter are made with palaeobotany, palaeozoo-
logy and palaeoclimatology. Faunal changes of the European Tertiary based
on palaeoherpetological analyses are mentioned particularly for the Eocene-Oligocene boundary and for the mid-Miocene Ries Event.

Introduction to Tertiary herpetofaunas of Europe

During the last decade a surprising amount of new information on the
palaeoherpetology especially for the European Tertiary, has been gained
from the most worthwhile issues of the different systematic groups in
the Handbook of Paleoherpetology. Here particularly mentionable are
the Crocodylia (Steel 1973), Testudines (Mlynarski 1976), Sauria terrestria incl. Amphisbaenia (Estes 1983), Serpentes (Rage 1984) and the
Gymnophiona, Caudata (Estes 1981). With these contributions new light
was spread on the systematical and stratigraphical distribution of amphibians and reptiles. Such contributions as provided by the respective
handbooks supply additional new basis for further systematic investigations of these long neglected fields. Still more detailed informations
may be expected in the near future with the handbook of Anura.

Figues 1 to 3 represent a comparative scope of the knowledge of
systematics, stratigraphy and distribution of Tertiary European amphibians and reptiles. These illustrations are compiled generally after
Schleich (1985) and including new compilations for the Testudines after
Mlynarski (op.cit.) and Broin (1977). Own studies based mainly on Tertiary herptiles**) of central Europe yielded initial information as regards
to the palaeoherpetological records under different, but only systematical or local, views.

Results which were published (Schleich 1984, 1985) contained information of a not only descriptive but also quantitative and stratigraphical nature and were discussed as the Ries Event. Registrations like in
these studies (e.g. Schleich 1984b) provide evidence of subsequent local
faunal shiftings (Schleich 1985). Therefore with the use of different
methods and views on systematics one could soon obtain an extraregional
scope.

**) term here generally used for amphibians and reptiles together

*) Institut für Paläontologie & historische Geologie, D-8000 München 2,
    F.R.G.

Lecture Notes in Earth Sciences, Vol. 8
Global Bio-Events. Edited by O. Walliser
© Springer-Verlag Berlin Heidelberg 1986

Ordnung SAURIA

	PLEISTOZAN	PLIOZAN	MIOZAN	OLIGOZAN	EOZAN	PALEOZAN
Fam. IGUANIDAE					Geiseleilatella	
Fam. AGAMIDAE / CHAMAELEONIDAE			Chamaeleo Uromastyx	Agama	Tinosaurus	
Fam. GEKKONIDAE		GEKKONIDAE	Gerandogekko ?Phyllodactylus Cadurcogekko	Cadurcogekko	Rhodanogekko	
Fam. LACERTIDAE	Lacerta	Lacerta	Lacerta Eremias	Dracaenosaurus Lacerta Plesiolacerta Pseudeumeces	Eolacerta Plesiolacerta	?Plesiolacerta ?Pseudeumeces
Fam. SCINCIDAE			?SCINCIDAE			
Fam. CORDYLIDAE				Pseudolacerta	Pseudolacerta	
Fam. ANGUIDAE	Anguis Ophisaurus	?Anguis Ophisaurus	Ophisaurus	Ophisaurus	Ophisaurus Ophisauriscus Ophispseudopus Placosaurus Xestops ?Melanosaurus	GLYPTOSAURINAE
Fam. NECROSAURIDAE					Necrosaurus Eosaniwa	Necrosaurus
Fam. HELODERMATIDAE					Eurheloderma	
Fam. VARANIDAE	?Varanus	Varanus	Varanus Iberovaranus	VARANITDAE	Saniwa	

Ordnung AMPHISBAENIA

	PLEISTOZAN	PLIOZAN	MIOZAN	OLIGOZAN	EOZAN	PALEOZAN
AMPHISBAENIA	AMPHISBAENIA	AMPHISBAENIA	Blanus AMPHISBAENIA Omoiotyphlops	AMPHISBAENIA Omoiotyphlops	AMPHISBAENIA	

Ordnung SERPENTES

	EOZAN	OLIGOZAN	MIOZAN	PLIOZAN	PLEISTOZAN
Unterordnung SCOLECOPHIDIA — Familie TYPHLOPIDAE			?Typhlops		
Unterordnung ALETHINOPHIDIA — Familie ANILIIDAE	Coniophis Eoanilius				
U.-Fam. BOIDEA — Familie BOIDAE	Cadurceryx Calamagras Cadurcoboa Dunnophis Paleryx Palaeopython	Bransateryx Platyspondylia Plesiotortrix	?Python Albaneryx Ogmophis		
Familie PALAEOPHEIDAE	Archaeophis Palaeophis				
U.-Fam. ACROCHORDOIDEA — Familie NIGEROPHEIDAE	Woutersophis				
U.-Fam. COLUBROIDEA — Familie ANOMALOPHEIDAE	Archaeophis				
Familie RUSSELLOPHEIDAE	Russellophis				
Familie COLUBRIDAE		Coluber	Coluber Dolniceophis Elaphe Palaeonatrix Texcophis Protropidonotus	Coluber Elaphe Malpolon Natrix	Coluber Elaphe Natrix
Familie ELAPIDAE			Palaeonaja	Palaeonaja	
Familie VIPERIDAE			VIPERIDAE		Vipera

Figure 1. Systematical and stratigraphical distribution of reptiles in the Tertiary of Europe. left: Sauria and Amphisbaenia, right: Serpentes; after Schleich (1985) from Estes (1983) and Rage (1984).

Mlynarski (1976):

Taxon	PLEISTOCENE	PLIOCENE	MIOCENE	OLIGOCENE	EOCENE
Ordo TESTUDINES					
Subordo CRYPTODIRA					
Suprafamilia CHELYDROIDEA					
Familia DERMATEMYDIDAE			*Trachyaspis*	*Trachyaspis*	*Trachyaspis*
Familia CHELYDRIDAE			*Chelydropsis*	*Chelydropsis*	
Suprafamilia TRIONYCHOIDEA					
Familia CARETTOCHELYIDAE			*Anosteira*	*Anosteira*	*Allaeochelys* *Allaeochelys* *Anosteira*
Familia TRIONYCHIDAE			*Trionyx*	*Trionyx*	*Trionyx*
Suprafamilia TESTUDINOIDEA					
EMYDIDAE inc. sedis	*Emys*	*Emys*			
Familia EMYDIDAE	*Emys* *Sakya*	*Clemmydopsis* *Geoemyda* *Mauremys* *Sakya*	*Broilia* *Chinemys* *Chrysemys* *Clemmydopsis* *Cuora* *Geoemyda* *Mauremys* *Ocadia* *?Palaeochelys* *Psychogaster*	*Broilia* *Chinemys* *Chrysemys* *Geoemyda* *Ocadia* *?Palaeochelys* *Psychogaster*	*Chrysemys* *Geoemyda* *Ocadia* *Psychogaster*
Familia TESTUDINIDAE	"*Geochelone*" *Testudo*	"*Geochelone*" *Testudo*	*Cheirogaster* "*Geochelone*" *Stylemys* *Testudo*	*Cheirogaster* "*Geochelone*" *Testudo*	*Cheirogaster* "*Geochelone*"
TESTUDINIDAE inc. sedis		"*Palaeochelys*"	"*Palaeochelys*"	"*Palaeochelys*" *Podocnemis* *Polysternon*	*Neochelys* *Palaeaspis* *Polysternon*
Subordo PLEURODIRA					
Familia PELOMEDUSIDAE					

Broin (1977):

Taxon	PLEISTOCENE	PLIOCENE	MIOCENE	OLIGOCENE	EOCENE	PALEOCENE
Familia DERMATEMYDIDAE						
Familia CHELYDRIDAE		*Chelydropsis*	*Chelydropsis*	*Chelydropsis*		
Familia CARETTOCHELYIDAE					*Allaeochelys*	
Familia TRIONYCHIDAE		*Trionyx*	*Trionyx*	*Trionyx*	*Trionyx* "*Palaeotrionyx*" *Eurycephalochelys*	*Trionyx* "*Palaeotrionyx*"
Familia EMYDIDAE	*Mauremys*	*Clemmydopsis* *Mauremys* *Psychogaster*	*Clemmydopsis* *Mauremys* *Palaeochelys* *Psychogaster*	*Mauremys* *Palaeochelys* *Psychogaster*	*Geiselemys*	
Familia TESTUDINIDAE		*Geochelone* *Testudo*	*Cheirogaster* *Geochelone* *Testudo*	*Cheirogaster* *Ergilemys*	*Cheirogaster* *Hadrianus* *Dithyrosternon*	
TESTUDINIDAE inc. sedis	*Testudo*			*?Testudo*		
Familia PELOMEDUSIDAE				*Neochelys*	*Eurysternum* *Neochelys* *Palaeaspis* ... *Taphrosphys*	

Figure 2. Systematical and stratigraphical distribution of Testudines in the Tertiary of Europe. left: compiled after Mlynarski (1976) and right: compiled after Broin (1977).

	PLIOZÄN	MIOZÄN	OLIGOZÄN	EOZÄN	PALEOZÄN
Ordnung **C R O C O D I L I A**					
Unterordnung *SEBECOSUCHIA*					
Familie BAURUSUCHIDAE				Bergisuchus	
Unterordnung *EUSUCHIA*					
Familie CROCODYLIDAE					
U.-Fam. CROCODYLINAE		Crocodilus	Crocodilus	Asiatosuchus Crocodilus Kentisuchus Megadontosuchus	Crocodilus
U.-Fam. ALLIGATORINAE	Diplocynodon	Diplocynodon	Diplocynodon Hispanochasma	Allognathosuchus Arambourgia ?Caimanosuchus Diplocynodon ?Eocenosuchus Menatalligator	
U.-Fam. PRISTICHAMPSINAE				Pristichampsus	
U.-Fam. THORACOSAURINAE		Gavialosuchus Tomistoma		Dollosuchus Eosuchus Tomistoma	Thoracosaurus

	PLEISTOZÄN	PLIOZÄN	MIOZÄN	OLIGOZÄN	EOZÄN	PALEOZÄN
Ordnung **C A U D A T A**						
CRYPTOBRANCHIDAE		Andrias	Andrias	Andrias		
PROSIRENIDAE			Albanerpeton			
PROTEIDAE	Proteus		Orthophyia Mioproteus			
BATRACHOSAUROIDIDAE					Palaeoproteus	Palaeoproteus
DICAMPTODONTIDAE			Bargmannia			Geyeriella Wolterstorffiella
SALAMANDRIDAE	Salamandra Euproctus Triturus	Mertensiella Salamandra Pleurodeles Triturus	Archaeotriton Brachycormus Chelotriton Chioglossa Oligosemia Salamandra Salamandrina Triturus	Archaeotriton Chelotriton Chioglossa Megalotriton Palaeopleurodeles Salamandra Triturus	Chelotriton Koalliella Megalotriton Salamandra cf. Triturus Tylototriton	

Figure 3. Systematical and stratigraphical distribution of Crocodylia and Caudata in the Tertiary of Europe.
left: Crocodylia after Schleich (1985) from Steel (1973) and right: Caudata modified after Schleich (1985) from Estes (1981).

Changes in Tertiary herpetofaunas

Tihen (1964), still found it necessary to mention about the Tertiary herpetofaunas of North America that the "knowledge of Tertiary history of North American reptiles and amphibians is sadly deficient". He demonstrated by 4 figures, showing the distribution of lizards, snakes, anurans and salamanders, the general composition of the herpetological faunas of the area, at various stages in the Tertiary.

But in spite of his restrictions he could report (op.cit.: 267) about
"the abruptness and magnitude of the change that centers around the
early part of the Miocene epoch" and that (p. 270) "the Miocene faunas
present a very modern appearance, with relatively few archaic forms
represented, and this primarily in the lower and middle Miocene." For
earlier times he talks about another abruptness, of the change from
"the archaic to the basically modern faunas, probably beginning in the
Oligocene".

As causative reasons Tihen sees for the abruptness of the change that
"... it seems likely that major faunal movements are the basic factor
involved in the transition, not rapid evolution in situ."!!

Again for North America Tihen (op.cit., p. 273) mentions the sharp
differences between lizards' and snakes' pre-Miocene diversity and that
of the anurans where only few of any kind were known for Eocene or Oligocene. He summarizes for North America that "major herpetofaunal changes
must have taken place in North America between the Middle Oligocene and
the close of the Miocene." but without "evidence of any rapid burst of
evolution at this time" and concluding finally "... there is nonetheless
a very strong general correspondence between the major faunal movements
that have been postulated and the varying extents of the major floras
of the Tertiary". Similarities in these aspects are also present for the
general distributional patterns for Europe (see Fig. 1 - 3, 4). As an
extract of the previously shown distributions of Tertiary herptiles in
Europe the following graphs (Fig. 4) shall initiate further discussions.
The differences in taxa diversities of the different categories at
different stratigraphical levels are evident.

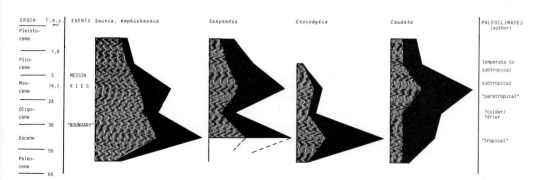

Figure 4.   Approximate taxa diversity of herptiles in the Tertiary of
Europe. Inspite that for real events one expects drastic incisions, a
type of graph is chosen that merely links conservatively the average
values in time and diversity. But I do consider these as the easier and
more generally recognizable demonstration as also events may cause in a
more longterm change to fauna or flora.

Detailed investigations on phylogeny, palaeobiogeography, and systematics must still be matters for further future investigations.
Of course, there remains the problem of just how reliable these distributional patterns are. But as there seems to be an apparently good correlation of the different categories one might allow their use for the initial basis of further reflections.

Local or global characteristics

As first shown in Schleich (1984), by subsequent registrations of detailed quantitative and qualitative studies local faunal shiftings were proved (Schleich 1985).
Thus, through different methods and views on systematics one could get soon an overregional frame.
Its relevance to the Ries Event and the effects upon flora and fauna of the area, was favoured by the extensive systematical studies which concentrated upon adjacent areas.
The studies are linked to recent research programmes such as the "Molasse Basin" or the "Tertiary Fissure Fillings". Thus by these primary qualitative and following quantitative analysis of huge amounts of fossil material and also bibliographical records, the hypothesis for the Ries Event was recorded by fossil reptiles. Particularly the drastic reduction or complete extinction of species or finally the changes on species level were discussed by Schleich in 1984, 1985. Meanwhile these effects led to further studies, and the following discussed aspects are given to fundament the theory of "(herpeto)faunistic changes" after this event. Not only these aspects indicate a much wider significance of the Ries Event and may even rank it to a more global scope.
Other characteristics were also shown (Schleich 1985, Fig. 20), for this regional scale, and might, besides their local influencing factors, become discussed as a response on another level.
Here we might understand or interpret those fluctuating curves (Fig. 5) as reflections of palaeoclimate and/or (Schleich, op.cit.) as response to changing palaeogeography. That the last parameter not only has local importance is evident and their causes may be found on a higher scale as shown in the previous figure. Of course, this again may only indicate tendencies or trends and not yet give information on a scale what can be expected for further comprehensive local studies. Finally, the local characteristics with their much more detailed records in time, localities, and systematic diversification, show, within the (partially) global features, a more detailed raster for the study areas.

435

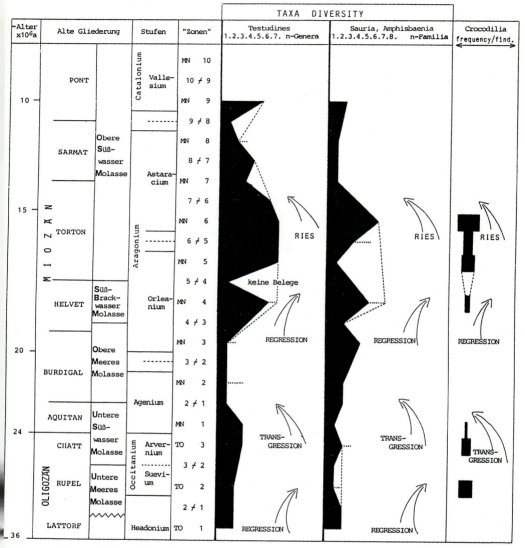

Figure 5. Distribution of reptilian taxa for the Tertiary of southern Germany and their hypothetical correlation to trans- or regressions of the Paratethys and the Ries Event (modified after Schleich 1985, Fig. 20).

The Ries impact - hypothesis for an event or only a basis for necessary investigations

After Gretener's (1984) definition no discussion is necessary for accepting the Ries impact as being a geological event. Especially considering his definitions, e.g. (p. 83) "... the duration of an event occupies no more than 1/100 of the total time span being considered", and still more

sophisticated, after Gretener (op.cit. p. 87) it could be termed as a
"rare event", "defined as a spasm, an episode, or a punctuation with
such a low rate of occurrence that it has taken place, at most, a few
times through all of earth history....".
Till now there are few studies which are dealing with, or might, at
least, be interpreted as important possible causes of the Ries impact
as a biological event.

Spitzlberger (1984) augmented in his recent study, "Die Rieskatastrophe in ihrer Auswirkung auf die Florengeschichte Mitteleuropas", that
by the catastrophe wide destructions of areas in middle Europe might
have been caused (p. 175).
In this contribution Spitzlberger (p. 174, 175), mentions also the studies of Knobloch (1969), Jung & Mayr (1980), and Mai (1981), as being
representative for drastic changes caused most probably by the Ries
Event.

Schleich (1984, 1985) discussed faunal extinctions of shiftings
after the Ries impact for the first time as an event and demonstrated
this by quantitative and qualitative analyses of southern German herptile distributions. A feature that obviously fits into the afore mentioned documentations. Geological aspects on this topic were extensively
discussed by various authors (see lit. in Schleich 1984, Spitzlberger
1984, Lemcke 1981, 1984), and also issued in a comprehensive volume in
Geologica Bavarica 1977 (Vol. 76).
Inspite of the fact that following chapters merely consider evidence or
hints plausible to the Ries Event, other authentic Tertiary events, like
e.g. the "Messinian"- or "Eocene/Oligocene-Boundary-Event", should be
subject of further presentations.

Jung & Mayr (1980) talk about climaxes of evolution in the zonations
MN5-6 and MN7-9, separated by a period of species scarcity. Facts about
possible evidence for regarding the Ries impact generally as a biological event are discussed in the following chapter.

Herptiles amongst others

Differentiation on lower vertebrates such as amphibians and reptiles
in "pre-riesian" and "post-riesian" faunas was possible using diversity,
composition or species levels. The most drastic documentation is
supplied by the (? almost) complete lack of larger reptiles such as
crocodiles or giant tortoises after the Ries Event. Other faunal differentiation using diversity was shown in Schleich (1984b, 1985) and on
taxonomic levels in Schleich (1984a).

Palaeofaunistical records

Only a few, and then hardly representative studies on the distributional composition of mammals throughout particularly the middle Miocene period of southern Germany have been made. But the ones mentioned hereafter may be indicative for documenting the Ries Event:

Jung & Mayr (op. cit.) do not only show palaeobotanical differentiations in a time equivalent of the Molasse Basin that might be interpreted as a "pre-riesian" and a "post-riesian" range, but they also demonstrate the changes in micromammalian faunal distributions for those times.

Dehm (1955, 1960) showed in his subdivisions of the Molasse Basin by the distribution of larger mammals three units denoted clearly by different faunistic spectra (he also mentions amphibians and reptiles (1960: 38), due to chronologically and palaeoclimatologically different conditions.

Perhaps this might be of some importance, and should be reexamined for its capability to document the mid-Miocene incision by the Ries impact, too. At least one clearly expressed evidence for probable faunal separation is stated by van Couvering (1977): "In Eurasia the beginnings of a savanna-adapted Chronofauna is first known in the Middle Miocene..."

By micromammalian recording at least one author (Mayr 1979: 356 ff) documents faunistic and floristic differentiations in the time ranges before and after the Ries Event.

Palaeobotanical records

Most of the fossil records that indicate any characteristic change which may possibly link them with, or be interpreted for the Ries Event are palaeobotanical. Not surprising, because till now most climatological interpretations have also been made from the palaeobotanical side. Also some authors mention potential hints of floristic changes during the Miocene:

Knobloch (1969) sees floristic changes for all of middle Europe in the Upper Tortonian,

Jung & Mayr (1980) show that there are two stages of forestation separated by a chronological gap between the stages MN5/6 and MN7/9, of more open vegetation.

Mai (1981) talks about dramatic changes in the Lower Sarmatium and about the evolutionary and climatical differentiation of the deciduous forest floras of middle Europe during the Tertiary. To summarize after this author for a Ries Event interpretation might be useful: (p. 555) "changes in the Upper Miocene European floras and appearances of modern species" and "real drastic changes are recognizable in the Lower Miocene

(Lower Sarmatium) with an apparent xerophytisation in larger areas of southern and southeastern Europe together with a reduction of the humid/subtropical elements in Middle and West Europe."

Events and their relations to palaeoclimates

As herptiles are already known for being utmost sensitive climato-ecological indicators an attempt is made to list here under this topic other relevant subjects, too.

I'll try to favour here other subjects more likely to demonstrate possible changes capable of being combined with the Ries Event. Many more records are yielded from palaeoclimatological interpretations than from any other field.

The following authors mentioned here represent only a small selection of scientists as having recorded changes around mid-Miocene times:

Bizon & Müller (1977: 383): "The observations indicate an important cooling at about 13 to 11 m.y. (NN7 to NN9, respectively N13 to N15) with the maximum at about 12 to 11 m.y." and state further "this conclusion is supported by the results of palynological investigations by Benda (1977), who reported: "The most remarkable climatic changes to a drier and cooler climate must have taken place before the Tortonian or in the transitional interval between the Serravalian and Tortonian" and continued "a very distinct cooling at this time is also described from California for the Molmian formation at 12 m.y.".

Berggren & Couvering, v. (1974) mention the (p. 12, Fig. 1) "climatic deterioriation began to accelerate during the late Neogene (about 10 m.y. ago) and show also an evidently clear temperature drop in the D-O 18 Paleotemperature curve for New Zealand (after Devereux 1967)".

Beu (1966) documents a drastic climatic deterioriation for New Zealand sea temperatures after a tropical climax (p. 185, Fig. 1) in the Otaian/Awamoan Stage (corresponds to European Aquitan/Burdigal) to cool temperate Pleistocene conditions through the range of the Middle and Upper Miocene by the use of molluscs.

Buchardt (1978) demonstrates in his widely discussed temperature curve for the Tertiary North Sea basin a drop in temperature rather exact at the 14,7 (Ries-Event!) mio years level with a following deterioriation for the end of the Middle Miocene.

Cracraft (1972): Return to warmer conditions in the Lower Miocene with climatic deterioriations in the later Miocene.

Even in more general botanic literature (Göttlich, K., Hrsg. 1976) besides a continuous cooling for the late Tertiary particularly a climatic deterioriation at the end of the Middle Miocene is mentioned.

Jung & Mayr (1980: 170) again mention an obvious break in fauna and flora but regard this as being due to differences in precipitation not temperatures. They also express the existence of two climaxes in the lower to middle Miocene units MN4/5 and in the upper MN8/9).

Kemp (1978: 195): "The oxygen isotope curve suggest that growth of the major ice-sheet of East Antarctica probably commenced or intensified in the late early Miocene...".

Königswald (1930: 11): "Durch das Jungtertiär läuft aber ein ganz markanter Schnitt, der sich mit der oben angeführten Ansicht einer allmählichen und allgemeinen Klimaverschlechterung nicht erklären läßt." and p. 12: "Unser Miozänklima darf, bei dem Vorkommen von Zimmetbäumen und Palmen, als tropisch bis subtropisch angesprochen werden."

Shackleton & Kennett (1975) see in accordance to D O16/18 studies only after (!) the late Middle Miocene an Antarctic glaciation.

Schwarzbach (1968a, Fig. 143) shows in a climate curve of the middle European Tertiary a drastic decline after the Middle Miocene.

Tanai (1967) showed even for East Asia also floral differentiations between middle and late Miocene palaeobotanical records combined with a climatic and sea level regression at the same time.

Wolfe (1971) discusses an average temperature decline since the Middle Miocene of at least 6 centigrades.

Wolfe & Hopkins (1967) (from Axelrod & Bailey) demonstrate an apparent drop in temperature conditions at the 15 m.y. time range as indicated by palaeobotanical evidence which again is also expressed in Dorf's (1955) illustration (from Axelrod & Bailey) for Tertiary floras in the United States.

Discussing the palaeobiologically supposed Ries Event for example with its possible influences on climatology one has to document fully another field of research which due to lack of space is not possible here.

However, some definite points of discussion to this topic may be the influence or changes of the troposphere and its climatological significant correlations, as well as recent observations such as mere volcanic eruptions (Lamb 1971) with their known influences on climatic deterioriations, e.g. the historical eruptions of Krakatau and Mt. Tambora (Schneider & Londer 1984). How such phenomena of, for example tropospheric influences may result in climatic deterioriation and consequently in the smallest initial global shiftings like to maximal causes in longer term equivalents such as ice raftings or glaciations must be the subject of other studies.

Conclusions

As shown in the afore mentioned chapters events may become expressed
in both qualitative and quantitative analytical studies. Representative
analyses are not obtained by only systematical studies or mere descriptions of some few faunulae. Problems may be found in different aspects,
firstly in the scale of studies or observations and secondly the either
local or global influences. While it might be demonstrated, for example
by the distribution of the higher categories of reptile taxa for Europe
that some shifting trends are observed, the local analyses as shown
for the herptile distributions of southern Germany may reflect more
correctly the influences on their distributional patterns. But the main
problems here are surely superimposed by local aspects. E.g. Rögl &
Steininger (1983) showed for the changing palaeogeographical conditions
of the Paratethys to Mediterranean that severe changes caused primarily
by trans- or regressions, and/or other geological effects may influence
the local faunas more prominent than global effects could do. On the
other hand the question must arise as to how these more local occurrence
are due to, or at least linkable with the global events. However, if
real events had occurred elsewhere the still unstudied problems remain.
How long will it last until, or whether at all, they have an effect on
local recognizable biological (palaeontological) events? For the above
shown palaeoherpetofaunal shiftings of the southern German Tertiary both
effects might be hidden in the results. Primarily there are the influen
ces of the parathetical changes and their regional palaeogeographical
characteristics to combine with the distributional curves, and secondarily they might be due to overranking global events. If not applicable
to the latter, at least a local effect caused most probably, and discussed here, by the Ries impact is another result of local occurrences
that might again be used for extrapolation to a global scope. E.g. the
different authentic "Tertiary glaciations" might be a good starting
point for discussion on this topic.
I am fully aware that various colleagues regard my interpretation of th
Ries Event as being speculative or even as a wrong hypothesis (e.g.
Heissig 1985, 1986 verbal comm. and lit. in press).
    It is however not my aim here, to support a dinosaur extinction
theory by this more actual or evident event theory. Arguments supplied
by, for example, a single stratigraphical unsuitable locality do not
make much sense and even detract from these meanwhile evident recorded
general trends of shifting after the Ries Event.

REFERENCES

BERGGREN, W.A. & COUVERING, J.A. v. (1974): The Late Neogene: Biostrati-
    graphy, geochronology and paleoclimatology of the last 15 million
    years in marine and continental sequences.- Palaeogeogr., Palaeoclim.,
    Palaeoecol. 16, 1-216.
BEU,   (1966): Sea Temperatures in New Zealand During the Cenozoic Era,
    as indicated by Molluscs.- Trans. roy. Soc. N.Z. Geol. 4, 177-187.
BIZON, G. & MÜLLER, C. (1977): Remarks on some Biostratigraphic Problems
    in the Mediterranean Neogene.- Internat. Symp. Structural Hist. Medit.
    Basins, 381-390, Edit. Technip, Paris.
BROIN, F. de (1977): Contribution à l'étude des Chéloniens.- Mem. Mus.
    Nat. Hist. Nat. Ser. C. 38, 366 p.
BUCHARDT, B. (1978): Oxygen isotope palaeotemperatures from the Tertiary
    period in the North Sea area.- Nature 275, 121-123.
COUVERING, J.A.H. van (1977): The Collapse of the Continental Savanna
    Mosaic Chronofauna in Holarctica: The End-Miocene Terrestrial Event.-
    Messinian Seminar No. 3, Abstracts of the Papers, 2 p. Malaga
    (IGCP Proj. 96, Messinian Correlation).
CRACRAFT, J. (1972): Vertebrate Evolution and Biogeography in the Old
    World Tropics: Implications of Continental Drift and Palaeoclimato-
    logy.- Implications of Continental Drift to the Earth Sciences 1,
    373-393, Acad. Press, London.
DEHM, R. (1955): Die Säugetier-Faunen in der Oberen Süßwasser-Molasse
    und ihre Bedeutung für die Gliederung.- Erl. Geol. Übersichtskarte
    Süddeutsche Molasse 1 : 300.000, 81-88.
-- (1960): Zur Frage der Gleichaltrigkeit bei fossilen Säugerfaunen.-
    Geol. Rdsch. 49, 36-40.
DORF, E. (1955): Plants and the geological time scale.- Geol. Soc. Amer.,
    Spec. Pap. 62, 575-592.
ESTES, R. (1981): Gymnophiona, Caudata.- Handbuch der Paläoherpetologie.
    2, 115 p., Stuttgart, New York, Fischer Verl.
-- (1983): Sauria terrestria, Amphisbaenia.- Handbuch der Paläoherpeto-
    logie. 10A, 249 p., Stuttgart, New York, Fischer Verl.
GÖTTLICH, K. (1976): (Hrsg.) Moor- und Torfkunde.- 269 p., Stuttgart,
    Schweizerbart Verl.
GRETENER, P.E. (1984): Reflections on the "Rare Event" and Related Con-
    cepts in Geology.- in: BERGGREN, W.A. & COUVERING, J.A. van (eds.):
    Catastrophes and Earth History. 77-90, Princeton Univ. Press.
JUNG, W. & MAYR, H. (1980): Neuere Befunde zur Biostratigraphie der
    Oberen Süßwassermolasse Süddeutschlands und ihre palökologische Deu-
    tung.- Mitt. Bayer. Staatsslg. Paläont. hist. Geol. 20, 159-173.
KEMP, E. (1978): Tertiary Climatic Evolution and Vegetation History in
    the Southeast Indian Ocean Region.- Palaeogeogr., Palaeoclim., Palaeo-
    ecol 24, 169-208.
KNOBLOCH, E. (1969): Tertiäre Floren von Mähren.- Brno.
KÖNIGSWALD, R. v. (1930): Die Klimaänderung im Jungtertiär Mitteleuropas
    und ihre Ursachen.- Z. Geschiebeforsch. 6, 11-21.
LAMB, H.H. (1971): Volcanic Activity and Climate.- Palaeogeogr., Palaeo-
    clim., Palaeoecol. 10, 203-230.
LEMCKE, K. (1981): Unübliche Gedanken zum Einschlag des Ries-Meteoriten.-
    Bull. Ver. schweiz. Petrol. Geol. u. Ing. 46, 17.
-- (1984): Geologische Vorgänge in den Alpen ab Obereozän im Spiegel vor
    allem der deutschen Molasse.- Geol. Rdsch. 73, 371-397.
MAI, D.H. (1981): Entwicklung und klimatische Differenzierung der Laub-
    waldflora Mitteleuropas im Tertiär.- Flora 171, 525-582.
MAYR, H. (1979): Gebißmorphologische Untersuchungen an miozänen Gliriden
    (Mammalia, Rodentia) Süddeutschlands.- 360 p., München, Diss. Druck.
MLYNARSKI. M. (1976): Testudines.- Handbuch der Paläoherpetologie.- 7,
    130 p., Stuttgart, New York, Fischer Verl.
RAGE, J.Cl. (1984): Serpentes.-Handbuch der Paläoherpetologie 11, 80 p.
    Stuttgart, New York, Fischer Verl.

RÖGL, F. & STEININGER, F.F. (1983): Vom Zerfall der Tethys zu Mediterran und Paratethys.- Ann. Naturhist. Mus. Wien 85/A, 135-163.

SCHLEICH, H.H. (1984a): Neogene Testudines of Germany. Their stratigraphical and ecological evaluation.- Stud. Geol. Salmanticensia, Vol. espec. Stud. Palaeocheloniologica 1, 249-267.

-- (1984b): Neue Reptilienfunde aus dem Tertiär Deutschlands. 1. Schildkröten aus dem Jungtertiär Süddeutschlands.- Nat. wiss. Z. Niederbayern 30, 63-93.

-- (1985): Zur Verbreitung tertiärer und quartärer Reptilien und Amphibien. I. Süddeutschland.- Münchner Geowiss. Abh. (A) 4, 67-149.

SCHNEIDER, St. H. & LONDER, R. (1984): The Coevolution of Climate and Life.- 563 p., San Franzisco, Sierra Club.

SCHWARZBACH, M. (1968): Das Klima des rheinischen Tertiärs.- Z. dt. geol Ges. 118, 33-68.

SHACKLETON, & KENNETT, (1975): Paleotemperature history of the Cenozoic and the initiation of Antarctic glaciation: oxygen and carbon isotope analyses in DSDP sites 277, 279 and 281.- Init. Rep. Deep Sea Drill. Proj. 29, 743-755.

SPITZLBERGER, G. (1984): Die Rieskatastrophe in ihrer Auswirkung auf die Florengeschichte Mitteleuropas.- Nat. wiss. Z. Niederbayern 30, 173-174.

STEEL, R. (1973): Crocodylia.- Handbuch der Paläoherpetologie 16, 116 p. Stuttgart, Portland, Fischer Verl.

TANAI, T. (1967): Miocene floras and climate in East Asia.- Abh. zentr. geol.Inst. 10, 195-205.

TIHEN, J.A. (1964): Tertiary changes in the herpetofaunas of temperate North America.- Senck. biol. 45, 265-279.

WOLFE, J.A. (1971): Tertiary climatic fluctuations and Methods of Analysis of Tertiary Floras.- 9, 27-57.

-- & HOPKINS, D.M. (1967): Climatic changes recorded by Tertiary land floras in northwestern North America.- Tertiary Correlations and Climatic Changes in the Pacific. Pacific Sci. Congr. 11, 67-76.